Advances in Industrial Control

AF794020

For further volumes:
http://www.springer.com/series/1412

Mi-Ching Tsai • Da-Wei Gu

Robust and Optimal Control

A Two-port Framework Approach

Springer

Mi-Ching Tsai
Department of Mechanical Engineering
National Cheng Kung University
Tainan, Taiwan

Da-Wei Gu
Department of Engineering
University of Leicester
Leicester, UK

ISBN 978-1-4471-7236-9 ISBN 978-1-4471-6257-5 (eBook)
DOI 10.1007/978-1-4471-6257-5
Springer London Heidelberg New York Dordrecht

© Springer-Verlag London 2014
Softcover reprint of the hardcover 1st edition 2014
This work is subject to copyright. All rights are reserved by the Publisher, whether the whole or part of the material is concerned, specifically the rights of translation, reprinting, reuse of illustrations, recitation, broadcasting, reproduction on microfilms or in any other physical way, and transmission or information storage and retrieval, electronic adaptation, computer software, or by similar or dissimilar methodology now known or hereafter developed. Exempted from this legal reservation are brief excerpts in connection with reviews or scholarly analysis or material supplied specifically for the purpose of being entered and executed on a computer system, for exclusive use by the purchaser of the work. Duplication of this publication or parts thereof is permitted only under the provisions of the Copyright Law of the Publisher's location, in its current version, and permission for use must always be obtained from Springer. Permissions for use may be obtained through RightsLink at the Copyright Clearance Center. Violations are liable to prosecution under the respective Copyright Law.
The use of general descriptive names, registered names, trademarks, service marks, etc. in this publication does not imply, even in the absence of a specific statement, that such names are exempt from the relevant protective laws and regulations and therefore free for general use.
While the advice and information in this book are believed to be true and accurate at the date of publication, neither the authors nor the editors nor the publisher can accept any legal responsibility for any errors or omissions that may be made. The publisher makes no warranty, express or implied, with respect to the material contained herein.

Printed on acid-free paper

Springer is part of Springer Science+Business Media (www.springer.com)

Series Editors' Foreword

The series *Advances in Industrial Control* aims to report and encourage technology transfer in control engineering. The rapid development of control technology has an impact on all areas of the control discipline: new theory, new controllers, actuators, sensors, new industrial processes, computer methods, new applications, new philosophies..., new challenges. Much of this development work resides in industrial reports, feasibility study papers, and the reports of advanced collaborative projects. The series offers an opportunity for researchers to present an extended exposition of such new work in all aspects of industrial control for wider and rapid dissemination.

The *Advances in Industrial Control* monograph series started in 1992, and in many ways the sequence of volumes in the series provides an insight into what industries and what control system techniques were the focus of attention over the years. A look at the series titles on *robust* control yields the following list:

- *Robust Multivariable Flight Control* by Richard J. Adams, James M. Buffington, Andrew G. Sparks, and Siva S. Banda (ISBN 978-3-540-19906-9, 1994)
- H_∞ *Aerospace Control Design* by Richard A. Hyde (ISBN 978-3-540-19960-1, 1995)
- *Robust Estimation and Failure Detection* by Rami S. Mangoubi (ISBN 978-3-540-76251-5, 1998)
- *Robust Aeroservoelastic Stability Analysis* by Rick Lind and Marty Brenner (ISBN 978-1-85233-096-5, 1999)
- *Robust Control of Diesel Ship Propulsion* by Nikolaos Xiros (ISBN 978-1-85233-543-4, 2002)
- *Robust Autonomous Guidance* by Alberto Isidori, Lorenzo Marconi, and Andrea Serrani (ISBN 978-1-85233-695-0, 2003)
- *Nonlinear* H_2/H_∞ *Constrained Feedback Control* by Murad Abu-Khalaf, Jie Huang, and Frank L. Lewis (ISBN 978-1-84628-349-9, 2006)
- *Structured Controllers for Uncertain Systems* by Rosario Toscano (ISBN 978-1-4471-5187-6, 2013)

And from the sister series, *Advanced Textbooks in Control and Signal Processing* come:

- *Robust Control Design with MATLAB®* by Da-Wei Gu, Petko Hr. Petrov, and Mihail M. Konstantinov (2nd edition ISBN 978-1-4471-4681-0, 2013)
- *Robust and Adaptive Control* by Eugene Lavretsky and Kevin Wise (ISBN 978-1-4471-4395-6, 2013)

Clearly, robust control has seen a steady stream of monographs and books in both series. There is no doubt that the work of George Zames, Bruce Francis, John Doyle, Keith Glover, and many others created a paradigm change in control systems theory. Also note the number of aerospace-industry applications in the above list of texts. This emphasis can be ascribed to the availability of accurate high-dimensional multivariable models of aerospace systems within the industry, to the wide range of operating envelopes and therefore models, that aerospace vehicles traverse during a flight and the facility of optimization-based robust-control techniques in dealing with multivariable systems and their operational constraints.

From time to time, the *Advances in Industrial Control* series publishes a monograph that is theoretical and tutorial in content. This contrasts with most entries to the series that contain a mix of the theoretical, the practical, and the industrial. This monograph *Robust and Optimal Control* by Mi-Ching Tsai and Da-Wei Gu is one of those exceptions. The authors themselves actually raise the question "Why another book on the topic of Robust Control?" and their answer is that they have devised a new route to understanding the derivations and computation of robust and optimal controllers that they believe is a valuable addition to the literature of the subject. Their two-port approach is claimed to be more accessible to an engineering readership and to resonate in particular with an electrical- and electronic-engineering readership. The theoretical developments reported in the monograph are fully supported by detailed chapters covering all the background material and MATLAB® code and illustrated in a simple but persuasive servo-motor-control problem in the final chapter of the monograph. The list of monographs and textbooks on robust control shows that there is a continuing industrial interest in this field and for this reason this monograph is a valuable entry to the *Advances in Industrial Control* monograph series.

Industrial Control Centre
Glasgow, Scotland, UK

M.J. Grimble
M.A. Johnson

Preface

This is a book on robust and optimal control of linear, time-invariant systems.

The human being always seeks better results in all activities, and this desire pushes the advance of science and technology. This happens in the control systems and control engineering area as well. It is desirable to develop a control strategy, a controller, for a dynamic system under consideration, to satisfy all possible constraints and to optimize a certain cost function which reflects the design objectives. This is so-called an optimal control problem. Such problems can be traced back to as early as the seventeenth century, in the Brachistochrone curve problem raised by Johann Bernoulli. Solution approaches towards such problems include the classical Calculus of Variations, Pontryagin's Maximum Principle, Dynamic Programming, (Differential) Game Theory, and Nonsmooth Optimization. These procedures are complicated, and solutions do not always exist. Fortunately, for linear time-invariant (LTI) systems, many cases would have satisfactory results, and this book presents a powerful approach which is also easy to understand, for electrical engineers in particular, we hope.

On the other hand, robustness is another vitally important issue in control systems design. A successfully designed automatic control system should be always able to maintain stability and an acceptable performance level despite uncertainties in system dynamics and/or in the operation environment to a certain degree, while such uncertainties inevitably exist in any real-world control system. In the late 1970s and early 1980s with the pioneering work by Zames [8] and Zames and Francis [9], a theoretic development, now known as the H_∞ optimal control theory, was taking shape. Robust controllers for LTI systems can be found by solving corresponding optimization problems. Robustness is thus achieved by designing a controller which attains certain optimality of the closed-loop system. The H_∞ and related optimization approaches are well developed and elegant. They provide systematic design procedures, in particular, for multi-input, multi-output linear systems.

There have been a number of books on this subject. Some books are on the underlying theories and the derivation of solution formulae [1, 3, 5, 7, 10]. Others are more on design methodologies, application of such theories, and implementation software [2, 6]. Naturally, a question arises, "Do we need another book on this subject?"

It seems satisfactory that practicing control engineers can use available solution formulae and software routines to work out robust and optimal controllers for given design problems, when they know well the underlying control systems and design specifications. However, are we happy with such designed controllers without knowing exactly how the formulae are derived and on what grounds the solution procedures are based? As control engineers, are we confident enough to implement such designed controllers? Answers to above queries might be obvious, and there are sources for us to know the theories of design approaches, as pointed out earlier. The problem is that the theory behind the state-space approaches presented in [10] and other books is very mathematically oriented and difficult for engineers, and students as well, to understand. Hence, is it possible to present the robust and optimal control theory for LTI systems in a way such that engineers and students can follow and grasp the essence of the solution approach? This motivated the research and writing of the present book.

This book presents an alternative approach to find a robust controller via optimization. This approach is based on the chain scattering decomposition (CSD), initiated by Professor Hidenori Kimura [4] and references therein who also named this as chain scattering description. CSD uses the configuration of two-port circuits which is a fundamental ingredient of electrical engineering and is familiar to all electrical engineers and students with basic electrical engineering knowledge. It is shown in the book that (sub)optimal H_∞, H_2 as well as stabilizing controllers can be synthesized following the CSD approach. The book starts from the well-know linear fractional transformation (LFT), in which a control design problem can easily be formulated, and then converts LFT into CSD format. From the CSD formulation, the desired controller can be directly derived by using the framework proposed in the book in an intuitive and convenient way. The results are complete and valid for general system settings. The derivation of solution formulae is straightforward and uses no mathematics beyond linear algebra. It is hoped that readers may obtain insight from this robust and optimal controller synthesis approach, rather being bewildered in a mathematics maze.

The prerequisites for reading this book are classical control and state variable control courses at undergraduate level as well as elementary knowledge of linear algebra and electrical circuits. This book is intended to be used as a textbook for an introductory graduate course or for senior undergraduate course. It is also our intention to prepare this book for control engineers training courses on robust and optimal control systems design for linear time-invariant systems. With the above consideration in mind, we use plenty of simple yet illustrative worked examples throughout the book to help readers to understand the concepts and to see how the theory develops. Where appropriate, *MATLAB* codes for the examples are also included for readers to verify the results and to try on their own problems. Most chapters are followed with exercises for readers to digest the contents covered in the chapter. To further demonstrate the proposed approaches, in the last chapter, an application case study is presented which shows wider usage of the framework.

References

1. Green M, Limebeer DJN (1995) Linear robust control. Prentice Hall, Englewood Cliffs
2. Gu DW, Petkow PH, Konstantinov MM (2005) Robust control design with Matlab. Springer, London
3. Helton JW, Merino O (1998) Classical control using H_∞ methods-theory, optimization, and design. Society for industrial and Applied Mathematics, Baltimore
4. Kimura H (1997) Chain-scattering approach to H_∞ control. Birkhäuser, Boston
5. Maciejowski JM (1989) Multivariable feedback design. Addison-Wesley, Wokingham/Berkshire
6. Skogestad S, Postlethwaite I (2005) Multivariable feedback control: analysis and design. Wiley, New York
7. Stoorvogel AA (1992) The H_∞ control problem: a state space approach. Prentice Hall, Englewood Cliffs
8. Zames G (1981) Feedback and optimal sensitivity: model reference transformation, multiplicative seminorms, and approximated inverse. IEEE Trans Autom Control 26:301–320
9. Zames G, Francis BA (1983) Feedback, minimax sensitivity, and optimal robustness. IEEE Trans Autom Control 28:585–601
10. Zhou K, Doyle JC, Glover K (1996) Robust and optimal control. Prentice Hall, Upper Saddle River

National Cheng Kung University, Tainan, Taiwan — Mi-Ching Tsai
University of Leicester, Leicester, UK — Da-Wei Gu

Acknowledgements

The authors started research on robust and optimal control using the CSD approach about two decades ago. Many interesting results were however discovered while the first author was taking his sabbatical leave (2003–2004) in the University of Cambridge. The main parts of the book were written during the second author's recent visits at the National Cheng Kung University (NCKU), Taiwan. The authors thank NCKU and the University of Leicester for the support they received. The authors are greatly indebted to many individual people. The book would not be completed without their help.

All the contributors in the development of the CSD approach and related robust and optimal control theory, Professor Keith Glover, Professor Hidenori Kimura, Professor Ian Postlethwaite, Professor Malcolm Smith, Professor Jan Maciejowski, Professor Fang Bo Yeh, and Professor Shinji Hara, to name just a few, are very gratefully thanked. The authors are also indebted to their colleagues and students, past and present, for finding time in their busy schedules to help in editing, reviewing, and proofreading of the book. The long list includes Dr. Chin-Shiong Tsai, Dr. Bin-Hong Sheng, Dr. Jia-Sen Hu, Dr. Fu-Yen Yang, Dr. Wu-Sung Yao, Chun-Lin Chen, Ting-Jun Chen, Chia-Ling Chen, and many others.

Finally and most importantly, the authors owe their deepest gratitude and love to their families for their understanding, patience, and encouragement throughout the writing of this book.

Contents

Chapter 1
Introduction

This book presents a fresh approach to optimal controller synthesis for linear time-invariant (LTI) control systems. The readers are assumed to have taken taught modules on automatic control systems, including classical control in the frequency domain and state variable control, in a first-degree course (BEng or BSc). Knowledge of electrical and electronic engineering will be beneficial to understanding of the approach.

Consider the negative unity feedback control system configuration in Fig. 1.1 with the plant G and controller K. The performance specification of tracking control requires the output y to follow the reference input r closely. This closeness can be found in the error signal e. It is obvious that

$$E = (I + GK)^{-1}R \tag{1.1}$$

where $E(s)$ and $R(s)$ are the Laplace transforms of the time signals $e(t)$ and $r(t)$. Therefore, it is required that the error signal e to be "small." An adequate measurement to judge the "size" of a signal is the 2-norm, i.e., the square root of the "energy" of the signal. For a given, fixed input signal r, good tracking asks for that the 2-norm of e should be as small as possible. That is, one wants to find a stabilizing controller K which makes $\|(I + GK)^{-1}R\|_2$ small. If one wants "best" tracking, the controller K should be sought via solving the following optimization problem:

$$\min_{\text{stabilizing } K} \left\|(I + GK)^{-1}R\right\|_2. \tag{1.2}$$

In the case that the reference r is not explicitly known, rather it belongs to an *energy-bounded* set, the objective of good tracking then requires the infinity norm of the transfer function (matrix) from r to e to be small. That is, a stabilizing controller K is sought to make $\|(I + GK)^{-1}\|_\infty$ small, or to solve

M.-C. Tsai and D.-W. Gu, *Robust and Optimal Control: A Two-port Framework Approach*, Advances in Industrial Control, DOI 10.1007/978-1-4471-6257-5_1,
© Springer-Verlag London 2014

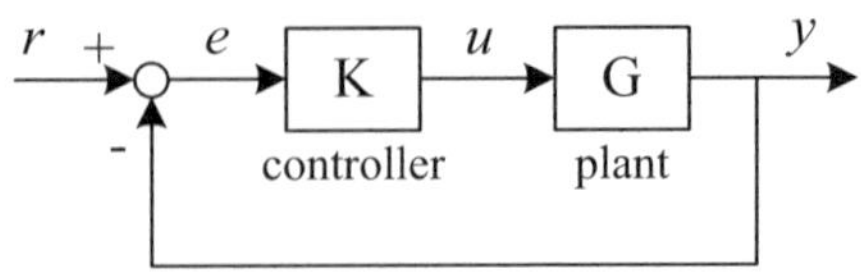

Fig. 1.1 The unity feedback control system

$$\min_{\text{stabilizing } K} \left\| (I + GK)^{-1} \right\|_{\infty}. \tag{1.3}$$

How to find such controllers K is the theme of this book. (Definitions and details of the solution deduction can be found in Chaps. 2, 8, and 9.)

In addition to the performance requirement, robustness is another central issue in control systems design. Most effective and efficient control design methods are model based. That is, the controller is designed based on a model which represents the dynamics of the system (the plant) to be controlled. It should be noted that in almost all cases in reality, a model will never completely and truly describe the dynamic behavior of the actual system because of unintentionally or intentionally excluded dynamics (so-called unmodeled plant dynamics) and/or because of wear-and-tear effects upon the plant (plant parameter variation). On the other hand, industrial control systems operating in real world are vulnerable to external disturbances and measurement noises. Hence, a designed controller should be able to keep the control system stable as well as maintain certain performance level even in the presence of unmodeled dynamics, parameter variations, disturbances, and noises. That is the commonly acceptable definition for the robustness of a control system.

Consideration of robustness in control systems design is not a new topic. Up to the middle of the last century, the 1950s, control systems analysis and design had been dominated by frequency domain methods. Single-input-single-output (SISO) cases were the main subject. Certain stability robustness can be achieved by ensuring good gain and phase margins in the design. Good stability margins usually lead to good, well-damped time responses of SISO systems, i.e., good performance. Rapid developments and needs in aerospace engineering in the 1960s have greatly propelled the development of control system theory and design methodology, particularly in the state-space approach which is powerful for multi-input-multi-output (MIMO) systems. Because the system models of aerospace systems are relatively accurate, the multivariable techniques developed at that time placed emphasis on achieving good system performance, rather on how to deal with system dynamics perturbations. These techniques are based on linear quadratic performance criteria with Gaussian disturbance (noise) considered adequate in such systems and proved to be successful in many aerospace applications. However, applications of these techniques, commonly referred to as linear-quadratic-Gaussian (LQG) methods, to other industrial problems make apparent the poor robustness properties exhibited by LQG controllers. This led to a substantial research effort to develop theory and methodology that could explicitly address the robustness issue in general feedback systems design. The pioneering work in the development now known as the H_∞ optimal control theory was published in the early 1980s by Zames [6] and Zames and Francis [7].

In the H_∞ framework, the designer from the outset specifies a model of system uncertainty, such as additive perturbation together with output disturbance, which is most suitable to the problem at hand. A constrained optimization is then formulated to maximize the robust stability of the closed-loop system to the type of uncertainty chosen, the constraint being the internal stability of the feedback system. In most cases, it would be sufficient to seek a feasible (suboptimal) controller such that the closed-loop system achieves certain robust stability. Performance objectives can also be included in the cost function of the optimization problem. Mature theory and elegant solution formulae in the state-space approach [1, 2, 8] have been developed, which are based on the solutions of certain algebraic Riccati equations, and are available in software packages such as *MATLAB* and *Slicot* [5].

However, many people, including students and practicing engineers, have experienced difficulties in understanding the underlying theory and solution procedures due to the complex mathematics involved in the advanced approaches that very much hinders the wide application of this advanced methodology in real world and industrial control systems design. In this book an alternative approach is presented for the synthesis of H_∞ and H_2 (sub)optimal controllers. This approach uses the so-called chain scattering decomposition or chain scattering description (CSD) and provides a unified synthesis framework. CSD is based on the two-port circuit formulation which is familiar to electrical and electronic engineering students and engineers. With this engineering background, readers are guided through a general yet rigorous synthesis procedure. Complemented by illustrative examples and exercise problems, readers can not only learn detailed steps in synthesizing a robust controller but also see and obtain an insight to the methodology.

The chain scattering description approach was first proposed by Kimura [4] in an attempt to provide a unified, systematic, and self-contained exposition of H_∞ control. In Kimura's approach, to deal with general systems, signal augmentation is required to formulate a single CSD framework which leads to extra calculations in the synthesis of controllers. This book defines right and left CSD matrices instead and proposes to use coupled CSD and coprime factorizations with specific requirements in controller synthesis. This approach is capable of dealing with all general systems and avoids unnecessary computation load and is of a transparent and intuitive nature. Such a framework covers the synthesis of all stabilizing controllers, H_2 optimal as well as H_∞ (sub)optimal control synthesis problems.

The CSD approach is straightforward in the derivation of the resultant controller synthesis procedures. It starts with a standard configuration (SCC, standard control configuration) of robust control scheme, converts it into a chain scattering decomposition format, and then uses coprime factorization and spectral factorization to characterize the required stabilizing controllers. The mathematical tools required are limited within linear algebra (matrix and vector manipulations) and algebraic Riccati equations.

A brief outline of the content of this book is as follows.

Following the Introduction is a chapter on prerequisite material readers should be familiar with. This includes linear algebra basics (vector and matrix manipulations, eigenvalues and eigenvectors, linear spaces, vector and matrix norms), function spaces (signals and operators, norms), linear control system basics (realization of LTI state-space model, stability, controllability, observability, system manipulations), and a brief description of linear fractional transformation (LFT) and chain scattering matrices. Most books on robust control would start with an introduction of uncertainties considered in the design and illustrate how such uncertainties can be incorporated in the so-called standard control (or compensation) configuration (SCC) which is in the linear fractional transformation (LFT) form. This book will however exclude these topics; rather, it is assumed that readers are already familiar with how to formulate an appropriate SCC. Starting with a standard control configuration, this book explains how to equivalently convert it into a chain scattering description and how to find solutions to stabilization problems, H_2 and H_∞ (sub)optimization problems, and other synthesis problems in the CSD form.

Chapter 3 introduces one-port and two-port networks which are widely used in electrical circuits. With basic electrical signals (voltage, current, impedance, and admittance), the network structures, input/output variables, and system parameters are explained. It is interesting to see how abstract concepts, such as LFT, can be illustrated with simple electrical circuit examples, while in most books these concepts are introduced without any practical explanations, and readers would most likely fail to catch the essence of the concepts. This in turn helps readers understand the theories and methodology developed later in this book.

Chapter 4 focuses further on linear fractional transformations (LFTs), its structure, usage in control systems analysis and synthesis, and its properties when the interconnected system is inner (co-inner). A direct method is introduced to formulate LFT configurations from a block diagram description of feedback systems. The chapter also describes the relation between LFT and Mason's Rule, which helps readers to link classical control techniques with the modern, state-space tools in control systems analysis and synthesis.

As mentioned earlier, the alternative synthesis approach adopted to find robust (sub)optimal controllers employs chain scattering descriptions (CSDs) of control systems. Chapter 5 discusses CSD in detail, its definitions, manipulations, state-space representations, its relationship with LFT, etc. Using CSD formulation to represent a LFT is not a new idea [3, 4]. However, previously it either requires certain assumptions on the LFT system or has to include "artificial" input/output in conversion. One novelty of this book is to present an equivalence conversion between LFT and CSD generally.

Chapters 6 and 7 are devoted to coprime factorization, spectral factorization, and algebraic Riccati equations. General solution formulae as well as special solution formulae particularly used in the CSD procedure, all in state-space form, are derived. In the derivation, block diagram manipulations and illustrative examples are extensively used to help readers obtain insight of the development of theoretic results and methodology.

With all the ingredients prepared in previous chapters, Chaps. 8 and 9 readily present solution procedures and formulae to compute all the stabilizing controllers, the H_2 optimal controller and H_∞ suboptimal controllers. Synthesis of all these controllers follows a unified framework. Again, all the derivations are accompanied with graphical illustrations and simple numerical examples.

Following the introduction of the controller synthesis problems in previous chapters, the last chapter focuses on presenting a more practical case study. It shows how the CSD approach can be applied in many industrial applications. Again, the solution procedure is transparent, logical, and explicit.

For readers' convenience, *MATLAB* codes for most worked examples are included in this book.

The material in this book can be used for postgraduate as well as senior undergraduate teaching.

References

1. Doyle JC, Glover K, Khargonekar PP, Francis BA (1989) State-space solutions to standard H_2 and H_∞ control problems. IEEE Trans Autom Control 34:831–847
2. Francis BA (1987) A course in H_∞ control theory. Springer, Berlin
3. Green M (1992) H_∞ controller synthesis by J-lossless coprime factorization. SIAM J Control Optim 30:522–547
4. Kimura H (1996) Chain-scattering approach to H_∞ control. Birkhäuser, Boston
5. Niconet e. V (2012) SLICOT – Subroutine Library in Systems and Control Theory. http://slicot.org/
6. Zames G (1981) Feedback and optimal sensitivity: model reference transformation, multiplicative seminorms, and approximated inverse. IEEE Trans Autom Control 26:301–320
7. Zames G, Francis BA (1983) Feedback, minimax sensitivity, and optimal robustness. IEEE Trans Autom Control 28:585–601
8. Zhou K, Doyle JC, Glover K (1995) Robust and optimal control. Prentice Hall, Upper Saddle River

Chapter 2
Preliminaries

Classical control design and analysis utilizes the frequency domain tools to specify the system performance. The background of operator theories and single-input and single-output, linear systems is required. In modern control, the time domain approach can be used to deal with multi-input and multi-out cases. Moreover, concepts of linear algebra and matrix-vector operations are used in system analysis and synthesis. Some useful fundamentals will be therefore reviewed in this chapter.

2.1 Linear Algebra and Matrix Theory

This section presents useful and well-known fundamentals of linear algebra and matrix theory, which facilitate the understanding of the subsequent control system concepts and methodology introduced. The stated results can be considered to be purely preliminary in nature, and hence, their proofs are omitted.

2.1.1 Vectors and Matrices

Control systems are, in general, multivariable. That means one deals with more than one variable in input, output, and state. Hence, vectors and matrices are frequently used to represent systems and system interconnections. In engineering and science, one usually has a situation where more than one quantity is closely linked to another. For instance, in specifying the location of a robot on a flat floor, one may use the numbers 2 and 3 to indicate the robot is at 2 units east and 3 units north from where one stands, and following the same logic, one may use -1 and -2 to indicate that the robot is at 1 unit west and 2 units south. Here, (2, 3) and $(-1, -2)$ represent two different locations, and the numbers 2 and 3 are in a fixed order to show that

M.-C. Tsai and D.-W. Gu, *Robust and Optimal Control: A Two-port Framework Approach*, Advances in Industrial Control, DOI 10.1007/978-1-4471-6257-5_2,
© Springer-Verlag London 2014

particular location, while (3, 2) would represent the position 3 units east and 2 units north. Such a group of numbers in a certain order forms a vector, and the dimensions of a vector correspond to how many numbers there are in the vector. Hence, (2, 3) is a 2-dimensional vector. Conventionally, a vector is defined as a column vector. In the above example, the position vector is thus written as $\begin{bmatrix} 2 \\ 3 \end{bmatrix}$ or $\begin{bmatrix} -1 \\ -2 \end{bmatrix}$. For any positive integer n, an n-dimensional (usually shortened as n-dim or n-D) vector x is denoted by $x = \begin{bmatrix} x_1 \\ \vdots \\ x_n \end{bmatrix}$.

The transpose of a vector x is denoted as x^{T} and is defined by $x^{\mathrm{T}} = \begin{bmatrix} x_1 \; x_2 \cdots x_n \end{bmatrix}$, a row vector. A group of vectors of the same dimension in a certain order forms a matrix. For example, for $M_i = \begin{bmatrix} m_{i1} \\ \vdots \\ m_{in} \end{bmatrix}$, $1 \le i \le p$, M_i is an n-dim vector and $M = \begin{bmatrix} M_1 \; M_2 \cdots M_P \end{bmatrix}$ is an $n \times p$ matrix. Obviously, a vector is a special case of matrices. $x^{\mathrm{T}} = \begin{bmatrix} x_1 \; x_2 \cdots x_n \end{bmatrix}$ is simply a matrix of $1 \times n$ dimensions. The elements or entries in a matrix can be real numbers or complex numbers. One uses $M \in \mathbb{R}^{n \times p}$ to show the matrix M is of $n \times p$ dimensions and all the elements of M are real numbers; $M \in \mathbb{C}^{n \times p}$ shows a $n \times p$ dimensional matrix M with complex numbers. It is clear that $\mathbb{R}^{n \times p} \subset \mathbb{C}^{n \times p}$. It is also a convention to use capital English letters to show a matrix such as M, whereas lower case letters (sometimes in bold face) are employed to show a vector such as x and lower case letters to show a scalar number. A matrix $M = \begin{bmatrix} m_{11} & \cdots & m_{1p} \\ \vdots & \cdots & \vdots \\ m_{n1} & \cdots & m_{np} \end{bmatrix}$ of the $n \times p$ dimension can be abbreviated as $M = \{m_{ij}\}_{n \times p}$. Similar to vectors, the transpose of a matrix M is $M^{\mathrm{T}} = \{m_{ji}\}_{p \times n}$. When M is in $\mathbb{C}^{n \times p}$, the complex conjugate transpose of M is defined by $M^* = \{\overline{m}_{ji}\}_{p \times n}$ when $\overline{m}_{ji}$ is the complex conjugate of m_{ji}.

A few manipulations can be defined for vectors and matrices. Two matrices of the same dimensions, e.g., $M = \begin{bmatrix} m_{11} & \cdots & m_{1p} \\ \vdots & \cdots & \vdots \\ m_{n1} & \cdots & m_{np} \end{bmatrix}$ and $N = \begin{bmatrix} n_{11} & \cdots & n_{1p} \\ \vdots & \cdots & \vdots \\ n_{n1} & \cdots & n_{np} \end{bmatrix}$, can be added together, i.e., $P = M + N$ where $P = \begin{bmatrix} p_{11} & \cdots & p_{1p} \\ \vdots & \cdots & \vdots \\ p_{n1} & \cdots & p_{np} \end{bmatrix} = \begin{bmatrix} m_{11} + n_{11} & \cdots & m_{1p} + n_{1p} \\ \vdots & \cdots & \vdots \\ m_{n1} + n_{n1} & \cdots & m_{np} + n_{np} \end{bmatrix}$. A multiplication is defined for two matrices only

Table 2.1 Classification of normal matrices

Matrix (*A*)	Definition	Eigenvalues	Diagonal elements	Determinant
Hermitian	$M^* = M$	$\lambda \in \mathbb{R}$	$a_{ii} \in \mathbb{R}$	$\det(M) \in \mathbb{R}$
Positive definite	$x^*Mx > 0, \forall\, x \neq 0$	$\lambda > 0$	$a_{ii} > 0$	$\det(M) > 0$
Positive semi-definite	$x^*Mx \geq 0, \forall\, x$	$\lambda \geq 0$	$a_{ii} \geq 0$	$\det(M) \geq 0$
Unitary $M \in C^{n\times n}$	$M^*M = I$	$\lvert\lambda\rvert = 1$	NA	$\lvert\det(M)\rvert = 1$
Orthogonal $M \in R^{n\times n}$	$M^{\mathrm{T}}M = I$	$\lvert\lambda\rvert = 1$	NA	$\lvert\det(M)\rvert = 1$

when their dimensions are compatible. That is, for $M = \{m_{ij}\}_{n\times p}$, $N = \{n_{kl}\}_{k\times l}$, only when $p = k$, one may have the product $P = MN$, where $P = \{p_{ij}\}_{n\times l}$, with $p_{ij} = \sum_{r=1}^{p(=k)} m_{ir} n_{rj}$. The following paragraph summarizes a few more aspects of vector/matrix manipulations [4].

1. A square matrix M is called nonsingular if a matrix B exists, such that $MB = BM = I$. Define $B = M^{-1}$. The inverse matrix M^{-1} exists if $\det(M) \neq 0$, where $\det(M)$ is the determinant of M. If M^{-1} does not exist, M is said to be singular. If the inverse of M, B, and MB all exist, then $(MB)^{-1} = B^{-1}M^{-1}$.
2. A complex square matrix is called unitary if its inverse is equal to its complex conjugate transpose $M^*M = MM^* = I$, where I denotes the identity matrix of the appropriate dimensions. A square matrix M is called orthogonal if it is real and satisfies $M^{\mathrm{T}}M = MM^{\mathrm{T}} = I$. For an orthogonal matrix, the inverse is its transpose.
3. An $n \times p$ matrix M is of rank m if the maximum number of linearly independent rows (or columns) is m. This equals to the dimension of $\mathrm{img}(M) := \{Mx | x \in \mathbb{R}^p\}$.
4. An $n \times p$ matrix M is said to have full row rank if $n \leq p$ and $\mathrm{rank}(M) = n$. It has a full column rank if $n \geq p$ and $\mathrm{rank}(M) = p$.
5. A symmetric matrix M of $n \times n$ dimension is positive definite if $x^{\mathrm{T}}Mx \geq 0$, where x is any n-dimensional (real) vector, and $x^{\mathrm{T}}Mx = 0$, only if $x = 0$. If for any n-dimensional vector x, $x^{\mathrm{T}}Mx \geq 0$ always holds, then M is positive semi-definite. A positive (semi-)definite matrix M may be denoted as $M > 0 (M \geq 0)$. Similarly, negative definite and negative semi-definite matrices may be defined.
6. For a positive definite matrix M, its inverse M^{-1} exists and is also positive definite.
7. All eigenvalues of a positive definite matrix are positive.
8. For two positive definite matrices M_1 and M_2, one has $\alpha M_1 + \beta M_2 > 0$ when α, β are nonnegative and both not zero.
9. A square matrix M is called normal if $MM^* = M^*M$. A normal matrix has the decomposition of $M = U\Lambda U^*$, where $UU^* = I$ and Λ is a diagonal matrix. The following Table 2.1 summarizes the classification of normal matrices.

2.1.2 Linear Spaces

Let $\mathbb{R}$ and $\mathbb{C}$ be real and complex scalar fields, respectively. A linear space V over a field F consists of a set on which two operations are defined. The first one is denoted by "addition (+)"; for each pair of elements x and y in V, there exists a unique element $x+y$ in V. And the second one is a scalar "multiplication (·)"; for each element α in F and each element x in V, there is a unique element αx in V. The following conditions hold with respect to the above two operations.

1. For each element x in V, $1 \cdot x = x$.
2. For all x, y, z in V, $(x+y)+z = x+(y+z)$.
3. For all x, y in V, $x+y = y+x$.
4. For each element x in V, there exists an element y in V, such that $x+y=0$.
5. There exists an element in V denoted by 0, such that $x+0=x$ for each x in V.
6. For each element α in F and each pair of elements x and y in V, $\alpha(x+y) = \alpha x + \alpha y$.
7. For each α, β in F and each element x in V, $(\alpha\beta)x = \alpha(\beta x)$.
8. For each α, β in F and each element x in V, $(\alpha+\beta)x = \alpha x + \beta x$.

Note that one uses the same symbol "0" to denote the element zero and scalar number zero in V and F, respectively. In the following, some basic concepts are reviewed first. These definitions can be easily found in standard linear algebra textbooks, for example see [8].

1. As mentioned in the earlier paragraph, the elements $x+y$ and αx are called the sum of x and y and the product of α and x, respectively, where $x, y \in V, \alpha \in F$.
2. A subset W of a vector space V over a field F is called a subspace of V if W itself is a vector space over F under the operations of addition and scalar multiplication defined on V.
3. Let $x_1, x_2, \ldots, x_k$ be vectors in V, then an element of the form $\alpha_1 x_1 + \alpha_2 x_2 + \cdots + \alpha_k x_k$ with $\alpha_i \in F$ is a linear combination over F of $x_1, x_2, \ldots, x_k$.
4. The set of all linear combinations of $x_1, x_2, \ldots, x_k \in V$ is a subspace called the span of $x_1, x_2, \ldots, x_k$, denoted by

$$\operatorname{span}\{x_1, x_2, \ldots, x_k\} = \left\{x \,\middle|\, x = \alpha_1 x_1 + \alpha_2 x_2 + \cdots + \alpha_k x_k; \alpha_i \in F\right\}. \tag{2.1}$$

5. Vectors $x_1, x_2, \ldots, x_k$ are said to be linearly dependent if there is at least one x_i that can be expressed as a linear combination of $\{x_j, j=1,2,\ldots,k, j \neq i\}$ or there exist constants $c_1, c_2, \ldots, c_k$ which are not all zero, such that $c_1 x_1 + c_2 x_2 + \cdots + c_k x_k = 0$. The vectors $x_1, x_2, \ldots, x_k$ are linearly independent if $c_1 x_1 + c_2 x_2 + \cdots + c_k x_k = 0$ indicates that all $c_1, c_2, \ldots, c_k$ are zero.
6. The vectors $x_1, x_2, \ldots, x_k$ are orthonormal if $x_i{}^* x_j = \delta_{ij} = \begin{cases} 1 & \text{when } i=j \\ 0 & \text{otherwise} \end{cases}$, where δ_{ij} is usually called the Kronecker delta.

7. Let W be a subspace of a vector space V, then a set of vectors $\{x_1, x_2, \ldots, x_k\} \in W$ is said to be a basis of W if $x_1, x_2, \ldots, x_k$ are linearly independent and $W = \text{span}\{x_1, x_2, \ldots, x_k\}$. The dimension of a vector subspace W equals to the number of basis vectors.
8. Let W be a subspace of V. The set of all vectors in V that are orthogonal to every vector in W is the orthogonal complement of W and is denoted by $W^{\perp}$. Hence,

$$W^{\perp} = \{y \in V : y^{*}x = 0, \forall x \in W\}. \tag{2.2}$$

Each vector x in V can be expressed uniquely in the form $x = x_W + x_{W^{\perp}}$ for $x_W \in W$ and $x_{W^{\perp}} \in W^{\perp}$.
9. A set of vectors $\{u_1, u_2, \ldots, u_k\}$ is said to be an orthonormal basis for a k-dimensional subspace W if the vectors form a basis and are orthonormal. Suppose that the dimension of V is n, it is then possible to find a set of orthonormal basis $\{u_{k+1}, \ldots, u_n\}$ such that

$$W^{\perp} = \text{span}\{u_{k+1}, \ldots, u_n\}. \tag{2.3}$$

10. A collection of subspaces $W_1, W_2, \ldots, W_k$ of V is mutually orthogonal if $x^{*}y = 0$ whenever $x \in W_i$ and $y \in W_j$ for $i \neq j$.
11. The kernel (or null) space of a matrix $M \in \mathbb{R}^{n \times p}$, which can be viewed as a linear transformation from $\mathbb{R}^p$ to $\mathbb{R}^n$, is defined as

$$\ker M = N(M) = \left\{x \middle| x \in \mathbb{R}^p : Mx = 0\right\}. \tag{2.4}$$

12. The image (or range) of M is

$$\text{img}(M) = \left\{y \middle| y \in \mathbb{R}^n : y = Mx, \forall x \in \mathbb{R}^p\right\}. \tag{2.5}$$

13. Let M be an $n \times p$ real, full rank matrix with $n > p$, the orthogonal complement of M is a matrix $M^{\perp}$ of dimension $n \times (n-p)$, such that $\begin{bmatrix} M & M^{\perp} \end{bmatrix}$ is a square, nonsingular matrix with the following property: $M^{\mathrm{T}} M^{\perp} = 0$.
14. The following properties hold:

$$(\ker M)^{\perp} = [\text{img}(M)]^{\mathrm{T}} \quad \text{and} \quad [\text{img}(M)]^{\perp} = \ker M^{\mathrm{T}}. \tag{2.6}$$

2.1.3 Eigenvalues and Eigenvectors

A matrix can be interpreted as a mapping between two linear spaces. For example, a 2×2 matrix $M = \{m_{ij}\}_{2 \times 2}$, $y = Ax$, where $x = \begin{bmatrix} x_1 \\ x_2 \end{bmatrix}$ and $y = \begin{bmatrix} y_1 \\ y_2 \end{bmatrix}$ are both in $\mathbb{R}^{2\times 1}$ (the two spaces in this case are the same). For most x, the image y would show a rotation of x plus an expansion or reduction in length, which is decided by the matrix M. However, there are some vectors in the space of which the images generated by the mapping M will remain at the same direction as the original vectors. These vectors are the eigenvectors of M, showing somehow the essence (eigen) of the mapping M. The factors of the length change are the eigenvalues of M. Rigorous definitions are given below.

For an $n \times n$ square matrix M, the determinant $\det(\lambda I - M)$ is called the *characteristic polynomial* of M. The *characteristic equation* is given by

$$\det(\lambda\, I - M) = 0. \tag{2.7}$$

The n roots of the characteristic equation are the *eigenvalues* of M. For an eigenvalue λ of matrix M, there is a nonzero vector ξ such that

$$M\xi = \lambda\xi \tag{2.8}$$

where ξ is called the eigenvector of M corresponding to the eigenvalue λ.

Definition 2.1 The spectral radius of matrix M is defined as

$$\rho(M) = \max_i |\lambda_i(M)| \tag{2.9}$$

where $\{\lambda_i\}$ is the eigenvalue set of M and $|\bullet|$ is the modulus of $\bullet$.

It is easy to show that if M is a Hermitian matrix, i.e., $M = M^*$, then all eigenvalues of M are real. The spectral radius indicates the size of the set which contains all the eigenvalues of M.

Definition 2.2 If M is Hermitian, then there exists a unitary matrix U (i.e., $U^*U = UU^* = I$) and a real diagonal matrix Λ, such that

$$M = U\Lambda U^*. \tag{2.10}$$

In this case, U is the right eigenvector matrix of M.

2.1.4 Matrix Inversion and Pseudoinverse

Matrix inversion is unavoidable and essential in control system manipulation. In this section, the useful formulae of the matrix inversion can be found [4].

Let M be a square $n \times n$ matrix partitioned as

$$M = \begin{bmatrix} M_{11} & M_{12} \\ M_{21} & M_{22} \end{bmatrix} \tag{2.11}$$

where $M_{11} : n_1 \times n_1, M_{12} : n_1 \times n_2, M_{21} : n_2 \times n_1, M_{22} : n_2 \times n_2$, and $n_1 + n_2 = n$.

Suppose that M_{11} is nonsingular, then M can be decomposed (block diagonalized) as

$$M = \begin{bmatrix} I & 0 \\ M_{21}M_{11}^{-1} & I \end{bmatrix} \begin{bmatrix} M_{11} & 0 \\ 0 & S \end{bmatrix} \begin{bmatrix} I & M_{11}^{-1}M_{12} \\ 0 & I \end{bmatrix} \tag{2.12}$$

where $S = M_{22} - M_{21}M_{11}^{-1}M_{12}$ is the Schur complement of M_{11} in M. Then, if M is nonsingular, it can be derived that

$$M^{-1} = \begin{bmatrix} M_{11}^{-1} + M_{11}^{-1}M_{12}S^{-1}M_{21}M_{11}^{-1} & -M_{11}^{-1}M_{12}S^{-1} \\ -S^{-1}M_{21}M_{11}^{-1} & S^{-1} \end{bmatrix}. \tag{2.13}$$

Dually, if M_{22} and M are nonsingular, then

$$M = \begin{bmatrix} I & M_{12}M_{22}^{-1} \\ 0 & I \end{bmatrix} \begin{bmatrix} \widehat{S} & 0 \\ 0 & M_{22} \end{bmatrix} \begin{bmatrix} I & 0 \\ M_{22}^{-1}M_{21} & I \end{bmatrix} \tag{2.14}$$

and

$$M^{-1} = \begin{bmatrix} \widehat{S}^{-1} & -\widehat{S}^{-1}M_{12}M_{22}^{-1} \\ -M_{22}^{-1}M_{21}\widehat{S}^{-1} & M_{22}^{-1} + M_{22}^{-1}M_{21}\widehat{S}^{-1}M_{12}M_{22}^{-1} \end{bmatrix} \tag{2.15}$$

where $\widehat{S} = M_{11} - M_{12}M_{22}^{-1}M_{21}$ is called the Schur complement of M_{22} in M. The matrix inversion formulae can be further simplified if M is block triangular as

$$\begin{bmatrix} M_{11} & 0 \\ M_{21} & M_{22} \end{bmatrix}^{-1} = \begin{bmatrix} M_{11}^{-1} & 0 \\ -M_{22}^{-1}M_{21}M_{11}^{-1} & M_{22}^{-1} \end{bmatrix}, \tag{2.16}$$

$$\begin{bmatrix} M_{11} & M_{12} \\ 0 & M_{22} \end{bmatrix}^{-1} = \begin{bmatrix} M_{11}^{-1} & -M_{11}^{-1}M_{12}M_{22}^{-1} \\ 0 & M_{22}^{-1} \end{bmatrix}. \tag{2.17}$$

If both M_{11} and M_{22} are nonsingular, then $\widehat{S}^{-1}$ can be represented by

$$\begin{aligned} \widehat{S}^{-1} &= \left(M_{11} - M_{12}M_{22}^{-1}M_{21}\right)^{-1} \\ &= M_{11}^{-1} + M_{11}^{-1}M_{12}\left(M_{22} - M_{21}M_{11}^{-1}M_{12}\right)^{-1}M_{21}M_{11}^{-1}. \end{aligned} \tag{2.18}$$

The pseudoinverse (also called Moore-Penrose inverse) of a matrix M is denoted as M^+ which satisfies the following conditions:

$$MM^+M = M, \tag{2.19}$$

$$M^+MM^+ = M^+, \tag{2.20}$$

$$\left(MM^{+}\right)^{*} = MM^{+}, \tag{2.21}$$

$$\left(M^{+}M\right)^{*} = M^{+}M. \tag{2.22}$$

The pseudoinverse is useful especially when matrix M is either non-square or singular.

2.1.5 Vector Norms and Matrix Norms

Norm is another important concept of vectors and matrices. It can be further developed for functions and systems as well. In this section, definitions of vector norm and matrix norm will be introduced [4]. The concept of norm can be loosely understood as a description of size or volume. A vector norm, denoted by $\|\cdot\|$, of any vector x over the field $\mathbb{C}$, must have the following properties:

1. $\|x\| > 0$, unless $x = 0$, in which case $\|x\| = 0$.
2. $\|cx\| = |c|\|x\|$ where c is any scalar in $\mathbb{C}$.
3. $\|x + y\| \leq \|x\| + \|y\|$.

Definition 2.3 Let $x = \begin{bmatrix} x_1 \\ \vdots \\ x_n \end{bmatrix}$ be a vector in $\mathbb{C}^n$. The following are norms of $\mathbb{C}^n$.

1. Vector ∞-norm: $\|x\|_\infty = \max\limits_{1 \leq i \leq n} |x_i|$.
2. Vector 1-norm: $\|x\|_1 = \sum\limits_{i=1}^{n} |x_i|$.
3. Vector 2-norm: $\|x\|_2 = \sqrt{x^* x} = \sqrt{\sum\limits_{i=1}^{n} |x_i|^2}$.
4. Vector p-norm (for $1 \leq p < \infty$): $\|x\|_p = \left(\sum\limits_{i=1}^{n} |x_i|^p\right)^{1/p}$.

In the case of matrices, a matrix norm satisfies

1. $\|A\| > 0$ unless $A = 0$, in which case $\|A\| = 0$;
2. $\|cA\| = |c|\|A\|$ where c is any scalar in $\mathbb{C}$;
3. $\|A + B\| \leq \|A\| + \|B\|$;
4. $\|AB\| \leq \|A\|\|B\|$.

Definition 2.4 Let $M = \begin{bmatrix} m_{11} & m_{12} & \cdots & m_{1n} \\ m_{21} & m_{22} & \cdots & m_{2n} \\ \vdots & \vdots & \vdots & \vdots \\ m_{m1} & m_{m2} & \cdots & m_{mn} \end{bmatrix}$ be a matrix in $\mathbb{C}^{m\times n}$. The following gives a list of different matrix norms, which will be useful for the rest of this book.

1. Matrix 1-norm (column sum): $\|M\|_1 := \max_j \sum_{i=1}^{m} |m_{ij}|$.
2. Matrix 2-norm: $\|M\|_2 := \sqrt{\lambda_{\max}(M^*M)}$.
3. Matrix ∞-norm (row sum): $\|M\|_\infty := \max_i \sum_{j=1}^{n} |m_{ij}|$.
4. Frobenius norm: $\|M\|_F := \sqrt{\text{trace}(M^*M)} = \sqrt{\sum_{i=1}^{m}\sum_{j=1}^{n} m_{ij}^* m_{ij}}$.

2.1.6 *Singular Value Decomposition*

The singular values of a matrix M are defined as

$$\sigma_i(M) := \sqrt{\lambda_i(M^*M)}. \tag{2.23}$$

The maximal singular value is denoted as

$$\overline{\sigma}(M) := \max_i (\sigma_i(M)),$$

and the minimal singular value is

$$\underline{\sigma}(M) := \min_i (\sigma_i(M)).$$

It is straightforward from the above definition that the matrix M and its complex conjugate transpose M^* have the same singular values, i.e., $\{\sigma_i(M)\} = \{\sigma_i(M^*)\}$. Let $M \in \mathbb{C}^{m\times n}$; there exist unitary matrices $U = [u_1 \ u_2 \cdots u_m] \in \mathbb{C}^{m\times m}$ and $V = [v_1 \ v_2 \cdots v_n] \in \mathbb{C}^{n\times n}$ such that

$$M = U\Sigma V^*, \tag{2.24}$$

where

$$\Sigma = \begin{bmatrix} \Sigma_1 & 0 \\ 0 & 0 \end{bmatrix}, \tag{2.25}$$

$$\Sigma_1 = \begin{bmatrix} \sigma_1 & 0 & \cdots & 0 \\ 0 & \sigma_2 & \cdots & 0 \\ \vdots & \vdots & \vdots & \vdots \\ 0 & 0 & \cdots & \sigma_r \end{bmatrix}, \tag{2.26}$$

with $\sigma_1 \geq \sigma_2 \geq \cdots \geq \sigma_r > 0$ and $r = \text{rank}(M)$. Equation (2.24) is called the singular value decomposition (SVD) of the matrix M. The matrix admits the decomposition

$$M = \sum_{i=1}^{r} \sigma_i u_i v_i{}^* = \begin{bmatrix} u_1 \; u_2 \; \cdots \; u_r \end{bmatrix} \begin{bmatrix} \sigma_1 & 0 & \cdots & 0 \\ 0 & \sigma_2 & \cdots & 0 \\ \vdots & \vdots & \vdots & \vdots \\ 0 & 0 & \cdots & \sigma_r \end{bmatrix} \begin{bmatrix} v_1 \; v_2 \; \cdots \; v_r \end{bmatrix}^*. \tag{2.27}$$

2.2 Function Spaces and Signals

Controllers or control schemes are as a matter of fact functions in the time domain or the frequency domain. Hence, the synthesis of the required controller, an optimal controller in particular, is a procedure in functional analysis. However, considering that the underlying systems in this book are mainly the linear time-invariant systems and that this book is primarily for practicing control engineers and engineering students, many mathematical definitions and deductions will not be included in order to make it more accessible to the targeted readers. Interested readers are recommended to consult relevant books, for instance [5, 6, 7, 10], for rigorous and in-depth treatment of those mathematical concepts.

2.2.1 Function Spaces

Function spaces useful for the themes introduced in this book are L_2, H_2, L_∞, and H_∞, and their orthogonal complement spaces.

The space L_p (for $1 \leq p < \infty$) consists of all Lebesgue measurable functions $w(t)$ defined in the interval $(-\infty, \infty)$ such that

$$\|w\|_p := \left(\int_{-\infty}^{\infty} |w(t)|^p \mathrm{d}t \right)^{\frac{1}{p}} < \infty. \tag{2.28}$$

The space L_∞ consists of all Lebesgue measurable functions $w(t)$ such that

$$\|w\|_\infty := \operatorname*{esssup}_{t \in R} |w(t)| < \infty. \tag{2.29}$$

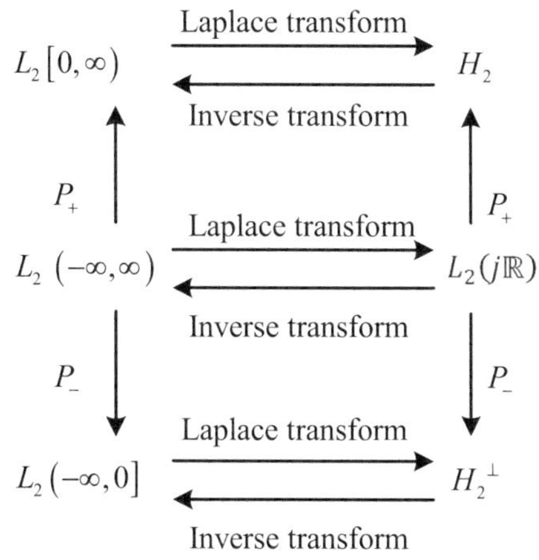

Fig. 2.1 Calculation procedures of function spaces

H_2 is the subspace of L_2 in which every function is analytic in $\mathrm{Re}(s) > 0$ (the real part of $s = \sigma + j\omega \in \mathbb{C}$), the open right-half plane, and H_∞ be a subspace of L_∞ in which every function is analytic and bounded in $\mathrm{Re}(s) > 0$. The space $H_2{}^{\perp}$ is the orthogonal complement of H_2 in L_2. If $G(s)$ is a strictly proper, stable, real, rational transfer function matrix, then $G(s) \in H_2$ implies that $G^{\sim}(s) \in H_2{}^{\perp}$, where $G^{\sim}(s) := G^{\mathrm{T}}(-s)$. The real rational subspace of H_∞ is denoted by RH_∞, which consists of all proper and real, rational, stable transfer function matrices. The relationship between spaces L_2 and H_2 is illustrated in Fig. 2.1 [3, 10].

Definition 2.5 Definitions of L_2, H_2, L_∞, and H_∞ function spaces.

1. L_2 -function space: $G(s) \in L_2$, if

$$\int_{-\infty}^{\infty} \mathrm{trace}\left[G^*(j\omega)\, G(j\omega)\right] \mathrm{d}\omega < \infty. \tag{2.30}$$

   The rational subspace of L_2, denoted by RL_2, consists of all real, rational, strictly proper transfer function matrices with no poles on the imaginary axis jR.
2. H_2-function space: $G(s) \in H_2$, if $G(s)$ is stable and

$$\|G(s)\|_2 := \sqrt{\left[\frac{1}{2\pi}\int_{-\infty}^{\infty} \mathrm{trace}\left[G^*(j\omega)\, G(j\omega)\right] \mathrm{d}\omega\right]} < \infty. \tag{2.31}$$

   Hence, the norm for H_2 can be computed just as it is done for L_2. The real rational subspace of H_2, which consists of all strictly proper and real, rational, stable transfer function matrices, is denoted by RH_2.
3. L_∞-function space: $G(s) \in L_\infty$, if

$$\|G(j\omega)\|_\infty := \operatorname*{ess\,sup}_{\omega}\ \overline{\sigma}\left[G(j\omega)\right] < \infty. \tag{2.32}$$

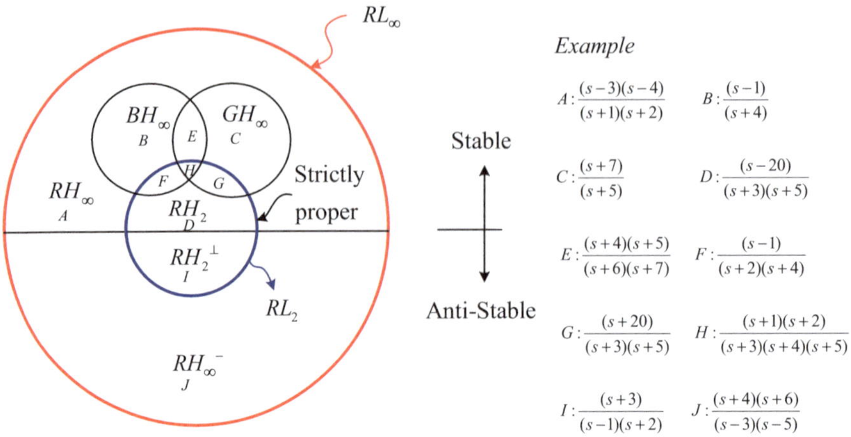

Fig. 2.2 Illustration of the relationship among different function spaces

All proper and real, rational, transfer function matrices with no poles on the imaginary axis form a subspace which is denoted by RL_∞.

4. H_∞ norm, the ∞-norm of Hardy space functions: $G \in H_\infty$, if $G(s)$ is stable and

$$\|G\|_\infty = \sup_{\mathrm{Re}(s)\geq 0} \overline{\sigma}\,[G(s)] = \sup_{\omega} \overline{\sigma}\,[G\,(j\omega)] < \infty. \tag{2.33}$$

H_∞ is a subspace of L_∞ with functions that are analytic and bounded in the open right-half plane. The real, rational subspace of H_∞ is denoted by RH_∞ which consists of all proper and real, rational, stable transfer function matrices.

This book introduces tools and concepts of optimal controller synthesis [3]. Most of the framework is set in the H_∞ function space. For the linear time-variant and causal systems, a given system $G(s) \in RH_\infty$ means the following:

(a) $G(s)$ is stable, and $\lim_{t\to\infty} \Phi(t)$ which is the impulse response of $G(s)$, is bounded.
(b) All poles of $G(s)$ are located in the open left-half plane.
(c) If $G(s)$ has a "minimal" state-space model (A, B, C, D), then the real part of all eigenvalues of the state matrix A is negative.

A state matrix is called *Hurwitz* if the real parts of all its eigenvalues are negative.

The following Fig. 2.2 shows the relationship among different function spaces, where $BH_\infty := \{F \in RH_\infty : \|F\|_\infty < 1\}$ is denoted as the set of all stable contractions and GH_∞ is the set of all units of RH_∞, i.e., if $F \in GH_\infty$, then $F \in RH_\infty$ and $F^{-1} \in RH_\infty$.

Example 2.1 Determine the function spaces for each of the following transfer functions: (1) $G_1(s) = \frac{s}{s+1}$; (2) $G_2(s) = \frac{s^2}{s+1}$; and (3) $G_3(s) = \frac{s}{(s-1)(s+2)}$.

1. It is clear that $G_1(s)$ is stable and $\sup_{\omega}|G_1(j\omega)| = \sup_{\omega}\left|\frac{j\omega}{j\omega+1}\right| = \sup_{\omega}\frac{\omega}{\sqrt{1+\omega^2}} = 1 < \infty$ for $\omega > 0$. Hence, $G_1(s) \in RH_\infty$. By decomposition of $G_1(s)$, one has $G_1(s) = \frac{s}{s+1} = 1 - \frac{1}{s+1}$. Thus,

$$\begin{aligned}
\|G\|_2 &= \sqrt{\left[\frac{1}{2\pi}\int_{-\infty}^{\infty}|G(j\omega)|^2 \mathrm{d}\omega\right]} \\
&= \sqrt{\left[\frac{1}{2\pi}\int_{-\infty}^{\infty}\left(1-\frac{1}{j\omega+1}\right)\left(1-\frac{1}{-j\omega+1}\right)\mathrm{d}\omega\right]} \\
&= \sqrt{\left[\frac{1}{2\pi}\int_{-\infty}^{\infty}\left(1-\frac{1}{j\omega+1}-\frac{1}{-j\omega+1}+\frac{1}{(j\omega+1)(-j\omega+1)}\right)\mathrm{d}\omega\right]} \\
&= \infty. \qquad (2.34)
\end{aligned}$$

   This implies $G_1(s) \notin RH_2$, which agrees with the fact that $G_1(s)$ is bi-proper.
2. By definition, $G_2(s) \notin RL_\infty$ because of $\sup_{\omega}|G_2(j\omega)| = \sup_{\omega}\left|\frac{(j\omega)^2}{j\omega+1}\right| = \infty$; $G_2(s) \notin RH_2$ because of $\int_{-\infty}^{\infty}|G(j\omega)|^2\mathrm{d}\omega = \infty$; and $G_2(s) \notin RH_\infty$ because of $\sup_{\mathrm{Re}(s)\geq 0}\left|\frac{s^2}{s+1}\right| = \infty$.
3. Apparently, $G_3(s) \notin RH_\infty$ because $G(s)$ is not analytic at $s = 1$; $G_3(s) \in RL_\infty$ because $G(s)$ is analytic on $j\omega$ axis and satisfies $\sup_{\omega}|G(j\omega)| < \infty$; and $G_3(s) \notin RH_2$ because $G(s)$ is not analytic at $s = 1$.

2.2.2 Norms for Signals and Systems

Norm symbolizes the size of a system or a function. For the control system analysis and synthesis, norm offers a direct criterion corresponding to design specifications. The detailed treatment of this topic can be found in books, such as [2, 3]. In this book, the following definitions are listed for easy reference. Note that the signal mentioned below is scalar and measurable, and the system is also scalar and linear time-invariant and causal. The vector (matrix) version of these norms can be found in, e.g., the books mentioned above.

Definition 2.6 The 1-norm of a signal $y(t)$ on $(-\infty,\infty)$ is defined as

$$\|y\|_1 := \int_{-\infty}^{\infty}|y(t)|\,\mathrm{d}t. \qquad (2.35)$$

Definition 2.7 The 2-norm of a signal $y(t)$ is defined as

$$\|y\|_2 := \sqrt{\int_{-\infty}^{\infty} y^2(t)\mathrm{d}t}. \tag{2.36}$$

Definition 2.8 The ∞-norm of a signal $y(t)$ is defined as

$$\|y\|_\infty := \sup_t |y(t)|. \tag{2.37}$$

Definition 2.9 The 1-norm of a stable system $G(s)$ is defined as

$$\|G\|_1 := \frac{1}{2\pi}\int_{-\infty}^{\infty} |G(j\omega)|\,\mathrm{d}\omega. \tag{2.38}$$

Definition 2.10 The 2-norm of a stable system $G(s)$ is defined as

$$\|G\|_2 := \sqrt{\frac{1}{2\pi}\int_{-\infty}^{\infty} |G(j\omega)|^2\mathrm{d}\omega}. \tag{2.39}$$

For a state-space described system $G(s) \overset{s}{=} \left[\begin{array}{c|c} A & B \\ \hline C & D \end{array}\right]$ defined in (2.52) below, the H_2 norm can be determined by

$$\|G\|_2 = \sqrt{\mathrm{trace}\,(B^{\mathrm{T}} P_{\mathrm{o}} B)} \tag{2.40}$$

where P_{o} is the observability gramian, which will be discussed in Chap. 7.

Definition 2.11 The ∞-norm of a stable system $G(s)$ is defined as

$$\|G\|_\infty := \sup_\omega |G(j\omega)|. \tag{2.41}$$

The $\|G\|_\infty$ equals the distance in the complex plane from the origin to the furthest point on the Nyquist plot of $G(s)$. It also appears as the peak value on the Bode magnitude plot of $G(s)$. The Hankel norm is also a representation of function size [3], especially in the design framework of H_∞ loop shaping. The Hankel norm can be exploited to determine the stability margin. Its definition is given below.

Definition 2.12 Hankel norm is used for determining the residual energy of a system before $t = 0$. For a stable system described as $y(t) = Gu(t)$, the Hankel norm is defined as

$$\|G\|_{\mathrm{H}} = \sqrt{\sup_{u \in L_2(-\infty,0)} \frac{\int_0^\infty y^{\mathrm{T}}(t)y(t)}{\int_{-\infty}^0 u^{\mathrm{T}}(t)u(t)\mathrm{d}t}}. \tag{2.42}$$

This can be determined by

$$\|G\|_{\mathrm{H}} = \sqrt{\lambda_{\max}(P_{\mathrm{c}}P_{\mathrm{o}})}, \tag{2.43}$$

where P_{c} and P_{o} are the controllability gramian and observability gramian matrices, respectively, which will be discussed in Chap. 7.

Example 2.2 Given a linear system $G(s)$ as below, determine its H_2 norm and Hankel norm.

$$\dot{x} = \begin{bmatrix} -1 & 0 \\ 0 & -2 \end{bmatrix} x + \begin{bmatrix} 1 \\ 1 \end{bmatrix} u$$

$$y = \begin{bmatrix} -1 & 2 \end{bmatrix} x. \tag{2.44}$$

The observability gramian P_{o} and controllability gramian P_{c} are

$$P_{\mathrm{o}} = \begin{bmatrix} \frac{1}{2} & -\frac{2}{3} \\ -\frac{2}{3} & 1 \end{bmatrix} \quad \text{and} \tag{2.45}$$

$$P_{\mathrm{c}} = \begin{bmatrix} \frac{1}{2} & \frac{1}{3} \\ \frac{1}{3} & \frac{1}{4} \end{bmatrix}. \tag{2.46}$$

Hence, one can obtain

$$\|G\|_2 = \sqrt{\operatorname{trace}(B^{\mathrm{T}}P_{\mathrm{o}}B)} = \frac{1}{\sqrt{6}}, \tag{2.47}$$

$$\|G\|_H = \sqrt{\lambda_{\max}(P_{\mathrm{c}}P_{\mathrm{o}})} = \frac{1}{6}. \tag{2.48}$$

2.3 Linear System Theory

The aim of this section is to introduce some basic results in linear system theory [1] that are particularly applicable to the work in the following chapters of this book. The descriptions, properties, and algebras of linear systems facilitate the development of optimal and robust control theory. These concepts preliminarily offer tools for system analysis and synthesis, and construct the main scope of modern control theory and control engineering.

2.3.1 *Linear Systems*

A finite-dimensional LTI dynamic system can be described by the following equations:

$$\begin{aligned} \dot{x} &= Ax + Bu, x(0) = x_0 \\ y &= Cx + Du, \end{aligned} \tag{2.49}$$

where $\forall\, t \geq 0$, $x(t) \in \mathbb{R}^n$ is the state vector, $u(t) \in \mathbb{R}^m$ is the input vector, and $y(t) \in \mathbb{R}^p$ is the output vector. The transfer function from u to y is defined as

$$Y(s) = G(s)U(s), \tag{2.50}$$

where $Y(s)$ and $U(s)$ are the Laplace transform of $y(t)$ and $u(t)$, respectively. It can be shown that

$$G(s) = D + C(sI - A)^{-1}B. \tag{2.51}$$

For simplicity, the state-space realization (A,B,C,D) can be written in a compact form as

$$G(s) \stackrel{s}{=} \left[\begin{array}{c|c} A & B \\ \hline C & D \end{array}\right]. \tag{2.52}$$

The state response in the time domain is

$$x(t) = \mathrm{e}^{At}x_0 + \int_0^t \mathrm{e}^{A(t-\tau)}Bu(\tau)\,\mathrm{d}\tau, \tag{2.53}$$

and the output response is

$$y(t) = C\mathrm{e}^{At}x_0 + \int_0^t C\mathrm{e}^{A(t-\tau)}Bu(\tau)\mathrm{d}\tau + Du(t). \tag{2.54}$$

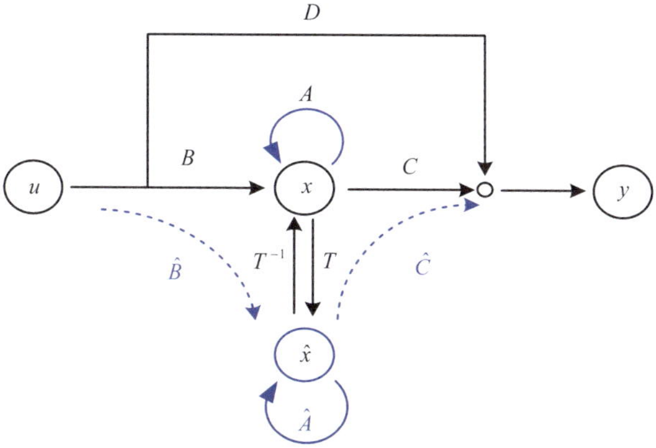

Fig. 2.3 Relationship of state-space similarity transformation

2.3.2 State Similarity Transformation

Different states can be defined to a linear time-invariant system given in (2.49) via $n \times n$ nonsingular matrix T. Let $\breve{x} = Tx$, and then the system can be described by

$$\begin{aligned} \dot{\breve{x}} &= TAT^{-1}\breve{x} + TBu, \breve{x}(0) = \breve{x}_0 = Tx_0 \\ y &= CT^{-1}\breve{x} + Du \end{aligned} \tag{2.55}$$

The transformed system is derived via the state similarity transformation of (T, T^{-1}). One has the same transfer function matrix from the input to the output, though with different state-space model:

$$G(s) \stackrel{s}{=} \left[\begin{array}{c|c} A & B \\ \hline C & D \end{array}\right] \stackrel{s}{=} \left[\begin{array}{c|c} TAT^{-1} & TB \\ \hline CT^{-1} & D \end{array}\right] = \left[\begin{array}{c|c} \breve{A} & \breve{B} \\ \hline \breve{C} & \breve{D} \end{array}\right], \tag{2.56}$$

where $\breve{A} = TAT^{-1}$, $\breve{B} = TB$, $\breve{C} = CT^{-1}$, and $\breve{D} = D$. The relationship of this transformation is illustrated in Fig. 2.3. The conjugate system $G^{\sim}(s)$ of $G(s)$ is given by

$$G^{\sim}(s) := G^T(-s) \stackrel{s}{=} \left[\begin{array}{c|c} -A^T & C^T \\ \hline -B^T & D^T \end{array}\right]. \tag{2.57}$$

Finally, if D is invertible, a state-space representation of $G(s)^{-1}$, the inverse of $G(s)$, is given by

$$G(s)^{-1} \stackrel{s}{=} \left[\begin{array}{c|c} A - BD^{-1}C & BD^{-1} \\ \hline -D^{-1}C & D^{-1} \end{array}\right]. \tag{2.58}$$

2.3.3 *Stability, Controllability, and Observability*

2.3.3.1 Stability

Stability is the most important property of a control system under study. In this section, the concepts of bounded-input-bounded-output (BIBO) stability and asymptotic stability will be discussed.

Definition 2.13 A system is BIBO stable if it generates a bounded output when it is subject to any bounded input.

For a linear system modeled by transfer function $G(s)$, it is called BIBO stable if and only if all the poles of $G(s)$ are in the open left-half plane, i.e., with negative real parts. For instance, given a $G(s) = \begin{bmatrix} \frac{1}{s+1} & \frac{1}{(s-1)(s+2)} \\ 0 & \frac{1}{s+3} \end{bmatrix}$, one can find that the poles of $G(s)$ are $\{-1, 1, -2, -3\}$. Hence, it is not BIBO stable due to that there is a positive real pole $\{1\}$. The following defines the asymptotic stability.

Definition 2.14 A system of (2.49) is called asymptotically stable if, for any given initial state x_0, the state $\|x(t)\| \to 0$, as $t \to \infty$, when $u \equiv 0$.

A necessary and sufficient condition for the system to be asymptotically stable is that the real part of all eigenvalues of A should be negative. The asymptotic stability is also called the *internal stability*, though the term is more often used in a closed-loop system setting. Asymptotic stability implies BIBO stability; however, BIBO stability does not imply the asymptotic stability. That is, asymptotically stable systems must be BIBO stable, but a BIBO stable control system may not be asymptotically stable [1]. The possible discrepancy between BIBO stability and asymptotic stability of a control system arises from whether the underlying system is completely controllable or completely observable. Controllability and observability are introduced next.

2.3.3.2 Controllability

Taking the given system in (2.49), e.g., controllability refers to the ability of the input signal u to transfer the state x from any initial state to any final state in finite time. A system is called *completely controllable* if, for any given initial state x_0 and any final state x_f, there exist a finite time T_f and an input $u(t)$, $0 \le t \le T_f$, which takes $x(0) = x_0$ to $x(T_f) = x_f$. Note that controllability of a system concerns only the matrix pair (A,B), and the state similarity transformation does not affect the controllability.

To verify the controllability and the following observability, the rank test and gramian test are the well-known methods [1]. The following summarizes these schemes.

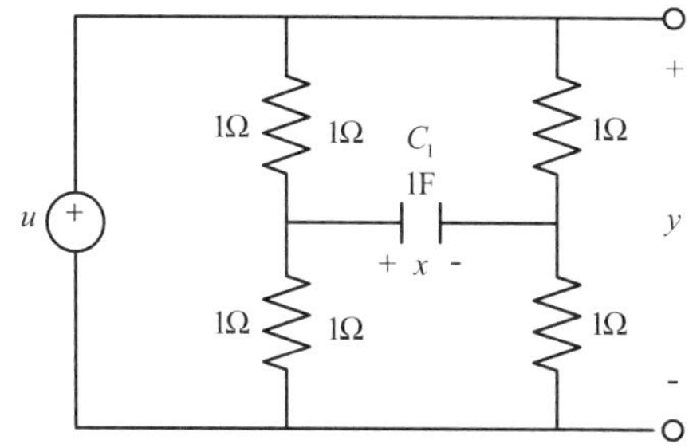

Fig. 2.4 Circuit example on observability

An n_th-order system is completely controllable if any one of the following is true:

1. The controllability matrix $\begin{bmatrix} B & AB & \cdots & A^{n-1}B \end{bmatrix}$ is of full rank.
2. The matrix $\begin{bmatrix} \lambda I - A & B \end{bmatrix}$ has full row rank at every eigenvalue of A.
3. The controllability gramian matrix

$$W_c = \int_0^t e^{A\tau} B B^T e^{A^T \tau} d\tau \tag{2.59}$$

   is nonsingular and thus positive definite for every $t > 0$.
4. All the eigenvalues of $A + BF$ can be assigned arbitrarily, where F is an appropriately chosen state feedback matrix and always exists.

A system model in (2.49) is said to be *stabilizable* if there exists a state feedback matrix F such that $A + BF$ is stable (i.e., the state matrix of the feedback system is Hurwitz).

2.3.3.3 Observability

The controllability describes the ability that the input drives the states, of which the dual concept is the observability of a system. Taking the given system in (2.49), e.g., the observability means the extent to which the system state variables are "visible" at the output. A system is called *completely observable* if, by setting the input identical to zero, any initial state $x(0)$ can be uniquely decided by the output $y(t)$, $0 \le t \le T$, for some finite T. For example, Fig. 2.4 shows that the initial voltage across the capacitor, $x(0)$, cannot be determined by the voltage output y. If no input (voltage source) u is applied to the circuit of Fig. 2.4, the initial state (voltage across the capacitor) cannot be deduced from the output y. Note that the observability concerns only the matrix pair (A,C), and the state similarity transformation does not change the observability.

The complete observability of a system can be found by using the rank test or gramian test, which are summarized as follows:

1. The observability matrix $\begin{bmatrix} C \\ CA \\ \vdots \\ CA^{n-1} \end{bmatrix}$ is of full rank.
2. The matrix $\begin{bmatrix} \lambda I - A \\ C \end{bmatrix}$ has full column rank at every eigenvalue of A.
3. The observability gramian matrix

$$W_o = \int_0^t e^{A^T \tau} C^T C e^{A\tau} d\tau \tag{2.60}$$

is nonsingular and thus positive definite for every $t > 0$.
4. All eigenvalues of $A + HC$ can be assigned arbitrarily, where H is an appropriately chosen observer gain matrix and always exists.

A system model in (2.49) is said to be *detectable* if there exists an observer gain matrix F such that $A + HC$ is stable.

2.3.4 Minimal State-Space Realization

For any given LTI system in a state-space model (2.49), an adequately chosen state similarity transfer matrix T can be applied to transform (2.52) into

$$\left[\begin{array}{c|c} TAT^{-1} & TB \\ \hline CT^{-1} & D \end{array}\right] = \left[\begin{array}{cccc|c} A_{CO} & 0 & A_{13} & 0 & B_{CO} \\ A_{21} & A_{C\bar{O}} & A_{23} & A_{24} & B_{C\bar{O}} \\ 0 & 0 & A_{\bar{C}O} & 0 & 0 \\ 0 & 0 & A_{43} & A_{\bar{C}\bar{O}} & 0 \\ \hline C_{CO} & 0 & C_{\bar{C}O} & 0 & D \end{array}\right]. \tag{2.61}$$

Representation (2.61) is the so-called canonical decomposition form (Kalman canonical decomposition). It can be easily derived that for zero initial states, the transfer function of the system is actually

$$G(s) = D + C(sI - A)^{-1}B = D + C_{CO}(sI - A_{CO})^{-1}B_{CO}, \tag{2.62}$$

which shows that the transfer function only describes the controllable and observable part of the system. Figure 2.5 shows the relation of (2.61) in a block diagram. The dynamics of the uncontrollable, unobservable, or both, if they exist in the system, will not be seen in the input/output relationship (the transfer function). That explains the possible situation of a system being BIBO stable but not asymptotically stable.

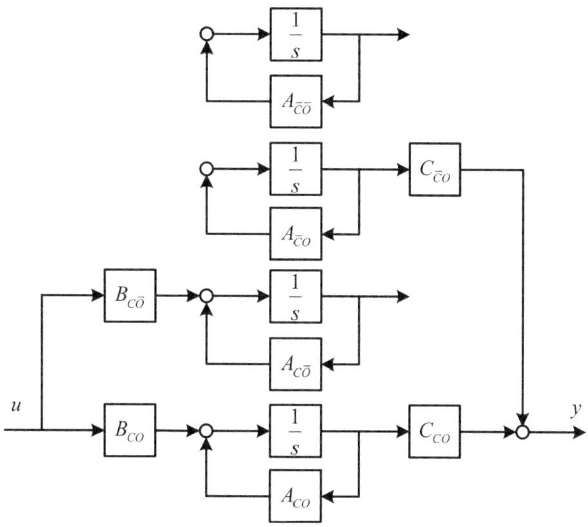

Fig. 2.5 Block diagram of canonical decomposition function

There are many state-space realizations corresponding to the same transfer function. The state-space realization (A,B,C,D) with the least dimensions of the state is called a minimal realization of the transfer function. Minimal realization (A,B,C,D) is always completely controllable and completely observable.

2.3.5 State-Space Algebra

Let state-space realizations of the systems $G_1(s)$ and $G_2(s)$ be given respectively by,

$$\begin{bmatrix} \dot{x}_1 \\ y_1 \end{bmatrix} = \begin{bmatrix} A_1 & B_1 \\ C_1 & D_1 \end{bmatrix} \begin{bmatrix} x_1 \\ u_1 \end{bmatrix} \tag{2.63}$$

and

$$\begin{bmatrix} \dot{x}_2 \\ y_2 \end{bmatrix} = \begin{bmatrix} A_2 & B_2 \\ C_2 & D_2 \end{bmatrix} \begin{bmatrix} x_2 \\ u_2 \end{bmatrix}. \tag{2.64}$$

Obviously, the system models formed from $G_1(s)$ and $G_2(s)$ could involve the variables from both systems. By augmenting (2.63) and (2.64), one obtains

$$\begin{bmatrix} \dot{x}_1 \\ x_2 \\ y_1 \end{bmatrix} = \begin{bmatrix} A_1 & 0 & B_1 \\ 0 & I & 0 \\ C_1 & 0 & D_1 \end{bmatrix} \begin{bmatrix} x_1 \\ x_2 \\ u_1 \end{bmatrix} \Rightarrow \begin{bmatrix} x_2 \\ \dot{x}_1 \\ y_1 \end{bmatrix} = \begin{bmatrix} I & 0 & 0 \\ 0 & A_1 & B_1 \\ 0 & C_1 & D_1 \end{bmatrix} \begin{bmatrix} x_2 \\ x_1 \\ u_1 \end{bmatrix} \tag{2.65}$$

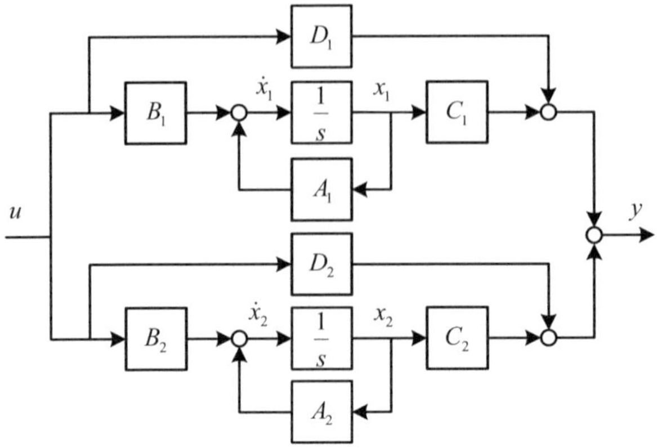

Fig. 2.6 Block diagram of a parallel system

and

$$\begin{bmatrix} \dot{x}_1 \\ \dot{x}_2 \\ y_2 \end{bmatrix} = \begin{bmatrix} I & 0 & 0 \\ 0 & A_2 & B_2 \\ 0 & C_2 & D_2 \end{bmatrix} \begin{bmatrix} \dot{x}_1 \\ x_2 \\ u_2 \end{bmatrix} \Rightarrow \begin{bmatrix} \dot{x}_2 \\ \dot{x}_1 \\ y_2 \end{bmatrix} = \begin{bmatrix} A_2 & 0 & B_2 \\ 0 & I & 0 \\ C_2 & 0 & D_2 \end{bmatrix} \begin{bmatrix} x_2 \\ \dot{x}_1 \\ u_2 \end{bmatrix}. \tag{2.66}$$

It can be seen in the following that manipulations between two control system models can be realized via the algebra of usual constant matrix operations.

2.3.6 State-Space Formula for Parallel Systems

As shown in Fig. 2.6, let $u_1 = u$ and $u_2 = u$. Since

$$y = y_1 + y_2 = \begin{bmatrix} C_1 & C_2 \end{bmatrix} \begin{bmatrix} x_1 \\ x_2 \end{bmatrix} + (D_1 + D_2)\,u, \tag{2.67}$$

a state-space realization of the transfer function from u to $y = y_1 + y_2$ can be found from (2.65) and (2.66) which have the same dimension of the total states as

$$\begin{bmatrix} \dot{x}_1 \\ \dot{x}_2 \\ y \end{bmatrix} = \begin{bmatrix} A_1 & 0 & B_1 \\ 0 & A_2 & B_2 \\ C_1 & C_2 & D_1 + D_2 \end{bmatrix} \begin{bmatrix} x_1 \\ x_2 \\ u \end{bmatrix}, \tag{2.68}$$

i.e.,

$$G_1(s) + G_2(s) \overset{s}{=} \left[\begin{array}{cc|c} A_1 & 0 & B_1 \\ 0 & A_2 & B_2 \\ \hline C_1 & C_2 & D_1 + D_2 \end{array}\right]. \tag{2.69}$$

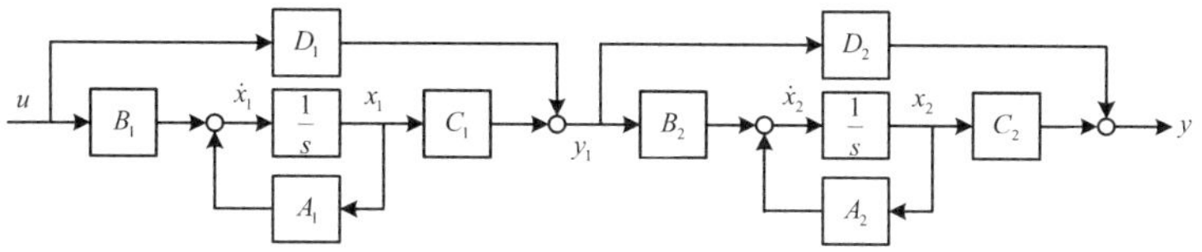

Fig. 2.7 Block diagram of a cascaded system

2.3.7 State-Space Formula for Cascaded Systems

As shown in Fig. 2.7, let $u_1 = u$, $u_2 = y_1$, and $y = y_2$. Then, one has a state-space realization of the transfer function from (2.65) and (2.66) by matrix multiplication as

$$\begin{bmatrix} \dot{x}_1 \\ \dot{x}_2 \\ y \end{bmatrix} = \begin{bmatrix} I & 0 & 0 \\ 0 & A_2 & B_2 \\ 0 & C_2 & D_2 \end{bmatrix} \begin{bmatrix} \dot{x}_1 \\ x_2 \\ u_2 \end{bmatrix} = \begin{bmatrix} I & 0 & 0 \\ 0 & A_2 & B_2 \\ 0 & C_2 & D_2 \end{bmatrix} \begin{bmatrix} A_1 & 0 & B_1 \\ 0 & I & 0 \\ C_1 & 0 & D_1 \end{bmatrix} \begin{bmatrix} x_1 \\ x_2 \\ u \end{bmatrix}$$
$$= \begin{bmatrix} A_1 & 0 & B_1 \\ B_2C_1 & A_2 & B_2D_1 \\ D_2C_1 & C_2 & D_2D_1 \end{bmatrix} \begin{bmatrix} x_1 \\ x_2 \\ u \end{bmatrix}, \tag{2.70}$$

or equivalently

$$\begin{bmatrix} \dot{x}_2 \\ \dot{x}_1 \\ y \end{bmatrix} = \begin{bmatrix} A_2 & 0 & B_2 \\ 0 & I & 0 \\ C_2 & 0 & D_2 \end{bmatrix} \begin{bmatrix} I & 0 & 0 \\ 0 & A_1 & B_1 \\ 0 & C_1 & D_1 \end{bmatrix} \begin{bmatrix} x_2 \\ x_1 \\ u \end{bmatrix} = \begin{bmatrix} A_2 & B_2C_1 & B_2D_1 \\ 0 & A_1 & B_1 \\ C_2 & D_2C_1 & D_2D_1 \end{bmatrix} \begin{bmatrix} x_2 \\ x_1 \\ u \end{bmatrix}. \tag{2.71}$$

Hence,

$$G_2(s)G_1(s) \overset{s}{=} \left[\begin{array}{cc|c} A_1 & 0 & B_1 \\ B_2C_1 & A_2 & B_2D_1 \\ \hline D_2C_1 & C_2 & D_2D_1 \end{array}\right] = \left[\begin{array}{cc|c} A_2 & B_2C_1 & B_2D_1 \\ 0 & A_1 & B_1 \\ \hline C_2 & D_2C_1 & D_2D_1 \end{array}\right]. \tag{2.72}$$

2.3.8 State-Space Formula for Similarity Transformation

Define a new state variable vector $\breve{x} = Tx$. Then one has

$$\dot{\breve{x}} = T\dot{x}, \tag{2.73}$$
$$x = T^{-1}\breve{x} \tag{2.74}$$

From (2.49), (2.73) and (2.74),

$$\dot{\breve{x}} = T\dot{x} = TAx + TBu = \left(TAT^{-1}\right)\breve{x} + (TB)u \tag{2.75}$$

and

$$\dot{\breve{x}} = y = Cx + Du = \left(CT^{-1}\right)\breve{x} + Du. \tag{2.76}$$

This implies

$$G(s) \overset{s}{=} \left[\begin{array}{c|c} A & B \\ \hline C & D \end{array}\right] = \left[\begin{array}{c|c} TAT^{-1} & TB \\ \hline CT^{-1} & D \end{array}\right]. \tag{2.77}$$

Consider the specific case that $A = \begin{bmatrix} A_{11} & A_{12} \\ A_{21} & A_{22} \end{bmatrix}$, $B = \begin{bmatrix} B_1 \\ B_2 \end{bmatrix}$, $C = \begin{bmatrix} C_1 & C_2 \end{bmatrix}$, and $T = \begin{bmatrix} I & X \\ 0 & I \end{bmatrix}$ $\left(\text{i.e., } T^{-1} = \begin{bmatrix} I & -X \\ 0 & I \end{bmatrix}\right)$ which is helpful to characterize the minimum realization of the state-space solutions later. Then,

$$T\begin{bmatrix} A & B \end{bmatrix} = \begin{bmatrix} I & X \\ 0 & I \end{bmatrix}\begin{bmatrix} A_{11} & A_{12} & B_1 \\ A_{21} & A_{22} & B_2 \end{bmatrix} = \begin{bmatrix} A_{11} + XA_{21} & A_{12} + XA_{22} & B_1 + XB_2 \\ A_{21} & A_{22} & B_2 \end{bmatrix} \tag{2.78}$$

and

$$\begin{aligned} \begin{bmatrix} TA \\ C \end{bmatrix} T^{-1} &= \begin{bmatrix} A_{11} + XA_{21} & A_{12} + XA_{22} \\ A_{21} & A_{22} \\ C_1 & C_2 \end{bmatrix}\begin{bmatrix} I & -X \\ 0 & I \end{bmatrix} \\ &= \begin{bmatrix} A_{11} + XA_{21} & -A_{11}X + XA_{21}X + A_{12} + XA_{22} \\ A_{21} & -A_{21}X + A_{22} \\ C_1 & -C_1X + C_2 \end{bmatrix}. \end{aligned} \tag{2.79}$$

This is equivalent to the matrix manipulations of

$$\begin{aligned} &\overset{-X}{\curvearrowright} \\ &\underset{X\,\circlearrowleft}{} \left[\begin{array}{cc|c} A_{11} & A_{12} & B_1 \\ A_{21} & A_{22} & B_2 \\ \hline C_1 & C_2 & D \end{array}\right] \\ &= \left[\begin{array}{cc|c} A_{11} + XA_{21} & -A_{11}X + XA_{21}X + A_{12} + XA_{22} & B_1 + XB_2 \\ A_{21} & -A_{21}X + A_{22} & B_2 \\ \hline C_1 & -C_1X + C_2 & D \end{array}\right] = \left[\begin{array}{c|c} TAT^{-1} & TB \\ \hline CT^{-1} & D \end{array}\right]. \end{aligned} \tag{2.80}$$

2.4 Linear Fractional Transformations and Chain Scattering-Matrix Description

Consider a general feedback control framework shown in Fig. 2.8, where P denotes the interconnection system of the controlled plant, namely, the standard control (or compensation) configuration (SCC) [10]. The closed-loop transfer function from w to z in Fig. 2.8 is given by

$$\mathrm{LFT_l}\left(P, K\right) = \mathrm{LFT_l}\left(\begin{bmatrix} P_{11} & P_{12} \\ P_{21} & P_{22} \end{bmatrix}, K\right) := P_{11} + P_{12}K(I - P_{22}K)^{-1}P_{21}, \tag{2.81}$$

where LFT stands for the linear fractional transformation and the subscript "l" stands for "lower." Different from the LFT, the chain scattering-matrix description (CSD) developed in the network circuits provides a straightforward interconnection in a cascaded way. The CSD transforms a LFT into a two-port network connection. Thus, many known theories which have been developed for a two-port network can then be used. The definition of CSD is briefly introduced below, while the details on background, properties, and use of CSD will be described in Chaps. 3, 4, and 5. Figure 2.9 shows the right and left CSD representations.

Define right and left CSD transformations with G and K denoted by $\mathrm{CSD_r}(G,K)$ and by $\mathrm{CSD_l}\left(\tilde{G}, K\right)$, respectively [9], as

$$\mathrm{CSD_r}\left(G, K\right) = \mathrm{CSD_r}\left(\begin{bmatrix} G_{11} & G_{12} \\ G_{21} & G_{22} \end{bmatrix}, K\right) := \left(G_{12} + G_{11}K\right)\left(G_{22} + G_{21}K\right)^{-1} \tag{2.82}$$

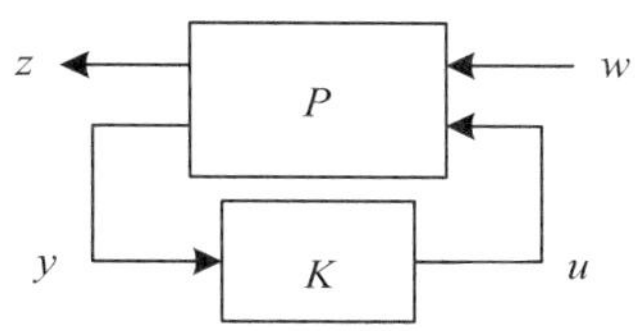

Fig. 2.8 Linear fractional transformation

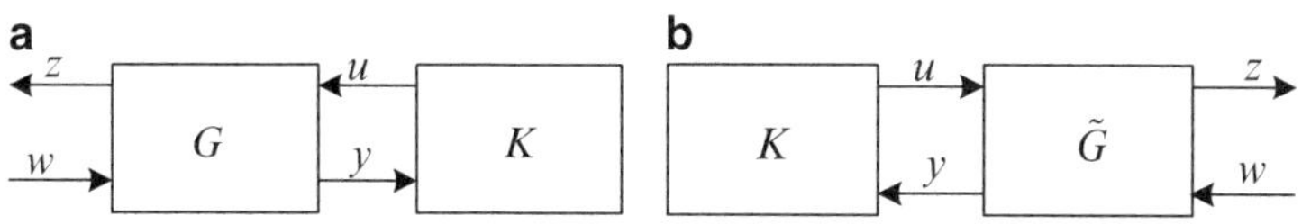

Fig. 2.9 Right and left CSD (**a**) Right CSD (**b**) Left CSD

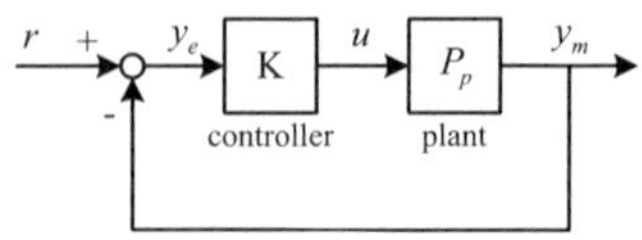

Fig. 2.10 Unity feedback control system

and

$$\mathrm{CSD}_l\left(\tilde{G}, K\right) = \mathrm{CSD}_l\left(\begin{bmatrix} \tilde{G}_{11} & \tilde{G}_{12} \\ \tilde{G}_{21} & \tilde{G}_{22} \end{bmatrix}, K\right) := -\left(\tilde{G}_{11} - K\tilde{G}_{21}\right)^{-1}\left(\tilde{G}_{12} - K\tilde{G}_{22}\right), \tag{2.83}$$

where G_{22} and $\tilde{G}_{11}$ are square and invertible. Note that, if P_{21} is invertible, the SCC matrix P can be transformed to a right CSD as

$$G = \begin{bmatrix} P_{12} - P_{11}P_{21}{}^{-1}P_{22} & P_{11}P_{21}{}^{-1} \\ -P_{21}{}^{-1}P_{22} & P_{21}{}^{-1} \end{bmatrix}. \tag{2.84}$$

Also, if P_{12} is invertible, the SCC matrix P can be transformed to a left CSD as

$$\tilde{G} = \begin{bmatrix} P_{12}{}^{-1} & P_{12}{}^{-1}P_{11} \\ P_{22}P_{12}{}^{-1} & P_{21} - P_{22}P_{12}{}^{-1}P_{11} \end{bmatrix}. \tag{2.85}$$

Example 2.3 Consider the unity feedback control system in Fig. 2.10, where P_p is a SISO-controlled plant. Find its corresponding LFT_l and CSD representations.

Let $z = \begin{pmatrix} y_e \\ u \end{pmatrix}$, $w = r$, and $y = y_e$.

From the unity feedback control system, by definition, as $u = 0$, one has $y_m = 0$; hence, $r = y_e$ from $r - P_p u = y_e$ so that $P_{11} = \dfrac{\begin{pmatrix} y_e \\ u \end{pmatrix}}{r}\Big|_{u=0} = \begin{pmatrix} 1 \\ 0 \end{pmatrix}$ and $P_{21} = \frac{y_e}{r}\big|_{u=0} = 1$. Similarly, as $r = 0$, one can also obtain

$$P_{12} = \frac{\begin{pmatrix} y_e \\ u \end{pmatrix}}{u}\Big|_{r=0} = \begin{pmatrix} -P_p \\ 1 \end{pmatrix} \text{ and } P_{22} = \frac{y_e}{u}\Big|_{r=0} = -P_p.$$

The closed-loop transfer function from r to $z = \begin{pmatrix} y_e \\ u \end{pmatrix}$ is presented as below (Fig. 2.11).

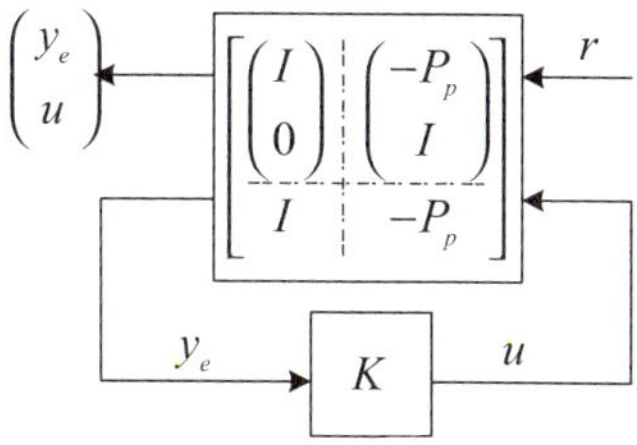

Fig. 2.11 LFT of closed-loop transfer function form

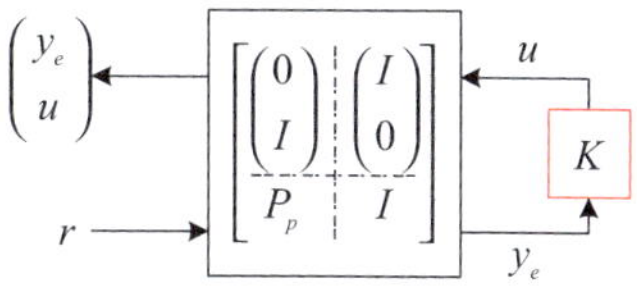

Fig. 2.12 Right CSD of closed-loop transfer function form

From

$$z = \mathrm{LFT}_l(P, K)\,w = \Big[P_{11} + P_{12}(I - KP_{22})^{-1}KP_{21}\Big]w,$$

one has

$$\begin{pmatrix} y_e \\ u \end{pmatrix} = \begin{pmatrix} 1 \\ K \end{pmatrix}(1 + P_P K)^{-1} r.$$

From the control block diagram, as $y_e = 0$, one has, by the definition of the right CSD, $G_{11} = \dfrac{\begin{pmatrix} y_e \\ u \end{pmatrix}}{u}\Big|_{y_e=0} = \begin{pmatrix} 0 \\ 1 \end{pmatrix}$ and $G_{21} = \frac{r}{u}\big|_{y_e=0} = P_p$, since $r - P_p u = y_e = 0$. As $u = 0$, one then has

$$G_{12} = \frac{\begin{pmatrix} y_e \\ u \end{pmatrix}}{y_e}\Big|_{u=0} = \begin{pmatrix} 1 \\ 0 \end{pmatrix} \text{ and from } r - P_p u = y_e.$$

Equivalently, the closed-loop transfer function from r to $z = \begin{pmatrix} y_e \\ u \end{pmatrix}$ can be represented by Fig. 2.12.

From $z = \mathrm{CSD}_r(G,K)\,w = (G_{11}K + G_{12})(G_{21}K + G_{22})^{-1}w$, one has

$$\begin{pmatrix} y_e \\ u \end{pmatrix} = \begin{pmatrix} 1 \\ K \end{pmatrix}(1 + P_p K)^{-1} r.$$

This concludes that

$$z = \mathrm{LFT_l}\,(P, K)\,w = \mathrm{CSD_r}\,(G, K)\,w.$$

Exercises

1. Prove that all the eigenvalues $\lambda(H)$ of a Hamiltonian matrix H are symmetric to the $j\omega$-axis.
2. Determine the rank of $A = \begin{bmatrix} 1 & 2 & 5 & 1 \\ 2 & 4 & -1 & 2 \\ 1 & 2 & 1 & 9 \end{bmatrix}$.
3. Let $Q = \begin{bmatrix} \frac{1}{\sqrt{2}} & \frac{1}{\sqrt{2}} \\ \frac{1}{\sqrt{2}} & -\frac{1}{\sqrt{2}} \\ 0 & 0 \end{bmatrix}$, $R = \begin{bmatrix} 1 & 1 \\ 0 & 1 \end{bmatrix}$, $b = \begin{bmatrix} 1 \\ 1 \\ 1 \end{bmatrix}$. Utilize the least square approach to solve $Ax = b$ where $A = QR$.
4. Consider the following system:

$$\begin{bmatrix} \dot{x}_1 \\ \dot{x}_2 \end{bmatrix} = \begin{bmatrix} -1 & 0 \\ 2 & -4 \end{bmatrix} \begin{bmatrix} x_1 \\ x_2 \end{bmatrix}, \quad x_0 = \begin{bmatrix} x_{10} \\ x_{20} \end{bmatrix}.$$

   Find the response of $x_1(t)$ and $x_2(t)$.
5. Sketch the state trajectories of the following system in the (x_1, x_2, x_3) plane for $x_0 = \begin{bmatrix} 2 \\ 4 \\ -4 \end{bmatrix}$ and input $u(t) = 0$. Determine the controllability of the above system

$$\dot{x} = \begin{bmatrix} 0 & 1 & 0 \\ 0 & 0 & 1 \\ -6 & -11 & -6 \end{bmatrix} x + \begin{bmatrix} 0 \\ 0 \\ 1 \end{bmatrix} u$$

$$y = \begin{bmatrix} 1 & 5 & 1 \end{bmatrix} x.$$

6. The transfer function of a linear system is given by

$$\frac{Y(s)}{U(s)} = \frac{s + a}{s^3 + 7s^2 + 14s + 8}.$$

   (a) Determine the values of a when the system is not completely controllable or not completely observable.
   (b) Define the state variables and derive the state-space model of which one of the states is unobservable.
   (c) Define the state variables and derive the state-space model of which one of the states is uncontrollable.

7. The state-space model of a third-order system is shown below:

$$\dot{x}_1 = 2x_1 + 3x_2 + 3x_3 + u$$
$$\dot{x}_2 = -2x_1 - 3x_2 - 2u$$
$$\dot{x}_3 = -2x_1 - 2x_2 - 5x_3 + 2u$$
$$y = 7x_1 + 6x_2 + 4x_3.$$

Use state similarity transformation to decouple the state-space model and explain the observability and controllability for each of the subsystems.

8. Consider the following systems and decide in which function space they belong to.

(a) $\dfrac{(s+1)}{(s+2)(s+4)}$

(b) $\dfrac{2s-1}{(s+1)(s+3)}$

(c) $\dfrac{1}{s-2}$

9. Consider the linear system below. Determine its H_2 norm, H_∞ norm, and Hankel norm.

$$\dot{x} = \begin{bmatrix} 0 & 1 \\ -1 & -2 \end{bmatrix} x + \begin{bmatrix} 0 \\ 1 \end{bmatrix} u$$
$$y = \begin{bmatrix} 1 & 0 \end{bmatrix} x.$$

References

1. Chen CT (2009) Linear system theory and design. Oxford University Press, New York
2. Doyle JC, Francis B, Tannenbaum A (1992) Feedback control theory. Macmillan Publishing Company, New York
3. Francis BA (1987) A course in H_∞ control theory, vol 88, Lecture notes in control and information sciences. Springer, Berlin
4. Golub GH, Van Loan CF (1989) Matrix computations. The Johns Hopkins University Press, London
5. Green M, Limebeer DJN (1995) Linear robust control. Prentice Hall, Englewood Cliffs
6. Helton JW, Merino O (1998) Classical control using H_∞ methods. SIAM, Philadelphia
7. Rudin W (1973) Functional analysis. McGraw-Hill, New York
8. Strang G (2004) Linear algebra and its applications, 4th edn. Academic, New York
9. Tsai MC, Tsai CS (1993) A chain scattering matrix description approach to H_∞ control. IEEE Trans Autom Control 38:1416–1421
10. Zhou K, Doyle JC, Glover K (1996) Robust and optimal control. Prentice Hall, Upper Saddle River

Chapter 3
Two-Port Networks

This chapter will briefly introduce two-port network descriptions which are closely related to that of the general control framework involving descriptions using LFT and CSD. The two-port network was developed as a common methodology to describe the relationship between inputs and outputs of an electrical circuit. For example, the impedance matrix of a two-port network can be determined by each port's voltage and current according to Ohm's law. The exposure in this book will focus on both scattering (i.e., LFT) and chain scattering (i.e., CSD) parameters as well as their applications to the modern control theory.

3.1 One-Port and Two-Port Networks

Before undertaking the study of a two-port network, it is worth reviewing the concept of a one-port network. Figure 3.1a depicts a circuit with a single terminal which forms a standard one-port network. A one-port network circuit has only one terminal in which the output current is equal to input current.

Unlike the one-port network, a two-port network circuit as illustrated in Fig. 3.1b has two terminals in which the input current is equal to the output current in each terminal. The two-terminal description of network systems offers freedom for ease of connecting several subsystems. Hence, utilizing two-port descriptions to characterize a circuit system for specific design objectives will be much more convenient than that of the one-port. For example, Thevenin and Norton circuit equivalents were developed based on one-port network theories [3]. Consider a simple circuit with two impedances Z_1 and Z_2 as illustrated in Fig. 3.2, where Z_L is the load impedance. The no-load Thevenin equivalent circuit (i.e., the open circuit without Z_L) is shown in Fig. 3.3 where the transfer function from V_1 to V_2 is given by

$$\frac{V_2}{V_1} = \frac{Z_2}{Z_1 + Z_2}. \tag{3.1}$$

M.-C. Tsai and D.-W. Gu, *Robust and Optimal Control: A Two-port Framework Approach*, Advances in Industrial Control, DOI 10.1007/978-1-4471-6257-5_3,
© Springer-Verlag London 2014

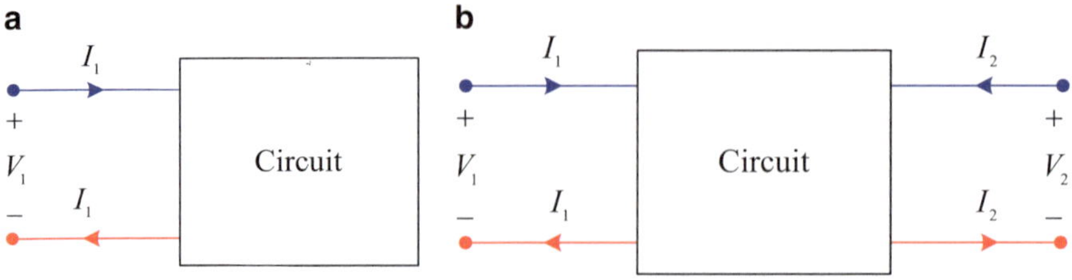

Fig. 3.1 (**a**) One-port and (**b**) two-port networks

Fig. 3.2 Two-port circuit

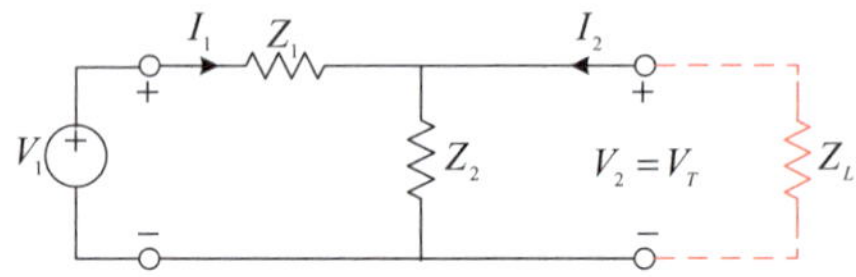

Fig. 3.3 Equivalent Thevenin circuit

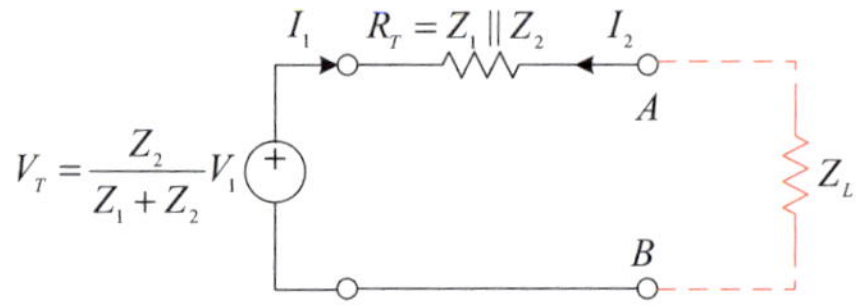

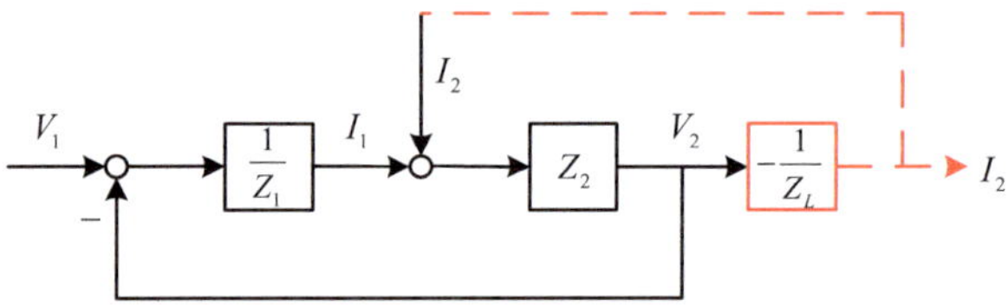

Fig. 3.4 Corresponding control block diagram

When load Z_L is included, due to the load effect, the transfer function from V_1 to V_2 can be determined as

$$\frac{V_2}{V_1} = \frac{\frac{Z_1 Z_2}{(Z_1+Z_2)}}{Z_L + \frac{Z_1 Z_2}{(Z_1+Z_2)}} = \frac{Z_2 Z_L}{Z_L\,(Z_1 + Z_2) + Z_1 Z_2}. \tag{3.2}$$

A system block diagram to describe this circuit is illustrated in Fig. 3.4. It should be noted that the signal that flows into the two-port system is $-I_2$, so that the relationship between terminal voltage and current is $Z_L = \frac{V_2}{-I_2}$. One can then apply the Mason's gain formulae to determine its transfer function from V_1 to V_2 for the cases with and without load Z_L, respectively, as

$$\frac{V_2}{V_1} = \frac{Z_2/Z_1}{1 + Z_2/Z_1} = \frac{Z_2}{Z_1 + Z_2}, \tag{3.3}$$

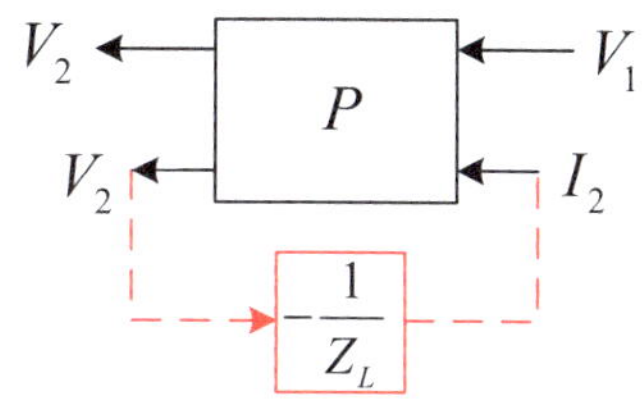

Fig. 3.5 LFT form

and

$$\frac{V_2}{V_1} = \frac{Z_2/Z_1}{1 + Z_2/Z_1 + Z_2/Z_L} = \frac{Z_2 Z_L}{Z_L\,(Z_1 + Z_2) + Z_1 Z_2}. \tag{3.4}$$

Furthermore, with the load effect, Fig. 3.4 can be formulated by a systematic framework using a two-port description as shown in Fig. 3.5. It can be determined from Fig. 3.4 by cutting around the load impedance $-\frac{1}{Z_L}$, and there are two loops in the case with load. Then by the LFT approach, one has

$$\begin{bmatrix} V_2 \\ V_2 \end{bmatrix} = \begin{bmatrix} P_{11} & P_{12} \\ P_{21} & P_{22} \end{bmatrix} \begin{bmatrix} V_1 \\ I_2 \end{bmatrix} \quad \text{and} \quad I_2 = -\frac{1}{Z_L} V_2, \tag{3.5}$$

where

$$P = \begin{bmatrix} \dfrac{Z_2}{Z_1 + Z_2} & \dfrac{Z_1 Z_2}{Z_1 + Z_2} \\ \dfrac{Z_2}{Z_1 + Z_2} & \dfrac{Z_1 Z_2}{Z_1 + Z_2} \end{bmatrix}.$$

It is reminded that the negative sign of "$-\frac{1}{Z_L}$" means that the current I_2 direction is opposite to that of I_2 as defined in the circuit in Fig. 3.2. By the definition of LFT, it can be verified that in the case of no-load,

$$\frac{V_2}{V_1} = \mathrm{LFT}_l\,(P, 0) = P_{11} = \frac{Z_2}{Z_1 + Z_2}, \tag{3.6}$$

and in the case with load,

$$\frac{V_2}{V_1} = \mathrm{LFT}_l \left(P, -\frac{1}{Z_L} \right) = P_{11} - P_{12} \frac{1}{Z_L} \left(1 + P_{22} \frac{1}{Z_L} \right)^{-1} P_{21} = \frac{Z_2 Z_L}{Z_L\,(Z_1 + Z_2) + Z_1 Z_2}. \tag{3.7}$$

Clearly, the results above are the same as (3.1)–(3.2) and (3.3)–(3.4). However, the two-port description approach is more systematic and characterizes conveniently the load effect. The system performance can be tuned easily by the load impedance as an external part. For example, engineers often use different terminated impedances to eliminate the echo problem in communication circuits. The same idea arises in control engineering. The feedback terminator of an LFT system can be chosen to achieve the desired response. Furthermore, an open loop-loop unstable system can be stabilized by a properly defined terminator in the two-port network description.

Resistors (R), inductors (L), and capacitors (C) are basic passive impedance elements of an electrical circuit. Electronic circuits are frequently needed to process electrical signals. The problem appearing in the two-port network theory is how to discover the relationship between input and output at each terminal. Based on these physical variables V_1, V_2, I_1, and I_2, there are six types of parameters which are often used for the two-port network description:

1. Impedance parameter (Z parameter)
2. Admittance parameter (Y parameter)
3. Hybrid parameter (H parameter)
4. Transmission parameter ($ABCD$ parameter)
5. Scattering parameter (S parameter)
6. Chain scattering parameter (T parameter)

Although these parameters are common in circuit design synthesis and analysis, it should be noted that some circuits may not have impedance, admittance, or transmission matrix descriptions, physically due to certain numerical constraints. For example, a circuit with transformers does not have an impedance parameter, and a simple circuit with shunt (or series) impedance does not have the two-port admittance (or impedance) matrix. The scattering-matrix description, which has its roots in microwave theory and has connections to operator theory, is then proposed to overcome problems such as the absence of physical parameters. This situation will be further discussed in the following sections.

3.2 Impedance and Admittance Parameters (Z and Y Parameters)

Figure 3.6 depicts a linear, two-port network along with appropriate voltages, currents, and its terminals. Let the matrix relationship of the impedance parameters be defined by

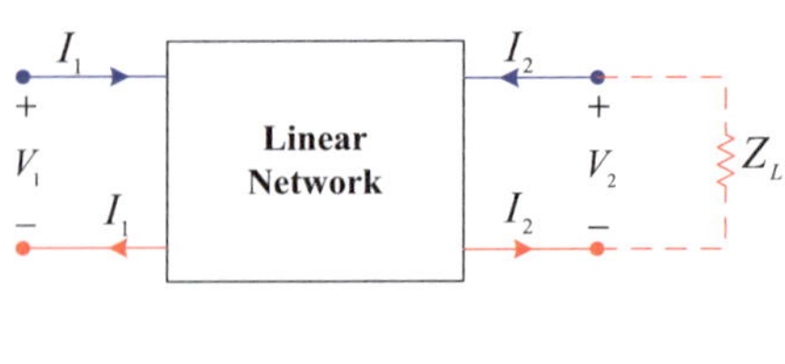

Fig. 3.6 Two-port Network

$$\begin{bmatrix} V_1 \\ V_2 \end{bmatrix} = \begin{bmatrix} Z_{11} & Z_{12} \\ Z_{21} & Z_{22} \end{bmatrix} \begin{bmatrix} I_1 \\ I_2 \end{bmatrix} \quad \text{and} \quad I_2 = -\frac{1}{Z_L} V_2, \tag{3.8}$$

where, according to the superposition principle,

$$\begin{aligned} Z_{11} &= \left.\frac{V_1}{I_1}\right|_{I_2=0} (\Omega), \quad Z_{12} = \left.\frac{V_1}{I_2}\right|_{I_1=0} (\Omega), \\ Z_{21} &= \left.\frac{V_2}{I_1}\right|_{I_2=0} (\Omega), \quad Z_{22} = \left.\frac{V_2}{I_2}\right|_{I_1=0} (\Omega). \end{aligned} \tag{3.9}$$

The impedance parameter, derived from Ohm's law, is useful for series-connected circuits. Similarly, the admittance matrix Y is defined as

$$\begin{bmatrix} I_1 \\ I_2 \end{bmatrix} = \begin{bmatrix} Y_{11} & Y_{12} \\ Y_{21} & Y_{22} \end{bmatrix} \begin{bmatrix} V_1 \\ V_2 \end{bmatrix}, \tag{3.10}$$

where

$$\begin{aligned} Y_{11} &= \left.\frac{I_1}{V_1}\right|_{V_2=0} \left(\frac{1}{\Omega}\right), \quad Y_{12} = \left.\frac{I_1}{V_2}\right|_{V_1=0} \left(\frac{1}{\Omega}\right), \\ Y_{21} &= \left.\frac{I_2}{V_1}\right|_{V_2=0} \left(\frac{1}{\Omega}\right), \quad Y_{22} = \left.\frac{I_2}{V_2}\right|_{V_1=0} \left(\frac{1}{\Omega}\right). \end{aligned} \tag{3.11}$$

Note that the admittance parameter description is useful for parallel-connected circuits. Apparently, it can be observed that

$$\begin{bmatrix} Y_{11} & Y_{12} \\ Y_{21} & Y_{22} \end{bmatrix} = \begin{bmatrix} Z_{11} & Z_{12} \\ Z_{21} & Z_{22} \end{bmatrix}^{-1}. \tag{3.12}$$

In addition, the parameters of matrices Z and Y are for physical units. One can easily examine the load (Z_L) effect for a given two-port impedance (or admittance) matrix Z (or Y) by the LFT description as illustrated in Fig. 3.7.

Now recall the circuit presented in Fig. 3.2. According to (3.8) and (3.10), the two-port impedance matrix Z and admittance matrix Y can be determined, respectively, as

$$Z = \begin{bmatrix} Z_{11} & Z_{12} \\ Z_{21} & Z_{22} \end{bmatrix} = \begin{bmatrix} Z_1 + Z_2 & Z_2 \\ Z_2 & Z_2 \end{bmatrix}, \tag{3.13}$$

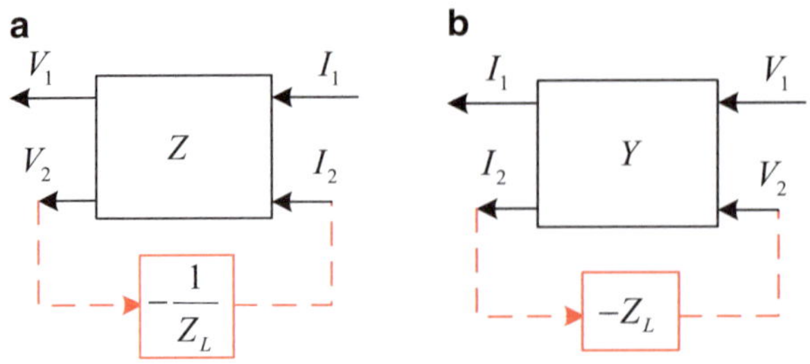

Fig. 3.7 LFT forms of Z and Y parameters: (**a**) LFT form of Z and (**b**) LFT form of Y

$$Y = \begin{bmatrix} Y_{11} & Y_{12} \\ Y_{21} & Y_{22} \end{bmatrix} = \begin{bmatrix} \frac{1}{Z_1} & -\dfrac{1}{Z_1} \\ -\frac{1}{Z_1} & \dfrac{Z_1 + Z_2}{Z_1 Z_2} \end{bmatrix}. \tag{3.14}$$

In fact, it can be verified that

$$Y = Z^{-1} = \begin{bmatrix} Z_1 + Z_2 & Z_2 \\ Z_2 & Z_2 \end{bmatrix}^{-1} = \begin{bmatrix} \dfrac{1}{Z_1} & -\dfrac{1}{Z_1} \\ -\dfrac{1}{Z_1} & \dfrac{Z_1 + Z_2}{Z_1 Z_2} \end{bmatrix}. \tag{3.15}$$

It can be realized from (3.14) that for the series short circuit where $Z_1 = 0$, Fig. 3.2 does not have the two-port Y parameter description. It can also be found from (3.14) that for the shunt open circuit where Y parameter is equal to zero, and the circuit shown in Fig. 3.2 does not have the two-port Z parameter description. Equations (3.8) and (3.9) can be used to determine the relationship (transfer function) between the currents and voltages by exploring the LFT. For instance, the overall input impedance which is the transfer function from I_1 to V_1 is given by (3.13) and by $I_2 = -\frac{1}{Z_L} V_2$ when closing the loop,

$$\begin{aligned} \frac{V_1}{I_1} &= \mathrm{LFT}_l\left(Z, -\frac{1}{Z_L}\right) = (Z_1 + Z_2) + Z_2\left(\frac{1}{Z_L}\right)\left(1 + \frac{Z_2}{Z_L}\right)^{-1} Z_L \\ &= Z_1 + (Z_2 \,||\, Z_L) = \frac{Z_1 (Z_2 + Z_L) + Z_2 Z_L}{Z_2 + Z_L}. \end{aligned} \tag{3.16}$$

Moreover, the overall input admittance which is the transfer function from V_1 to I_1 is given by (3.16) and by $V_2 = -Z_L I_2$ when closing the loop,

$$\frac{I_1}{V_1} = \text{LFT}_l\left(Y, -Z_L\right) = \frac{1}{Z_1} - \frac{Z_L}{Z_1}\left(1 + \frac{Z_L\left(Z_1 + Z_2\right)}{Z_1 Z_2}\right)^{-1}$$

$$\frac{1}{Z_1} = \frac{Z_2 + Z_L}{Z_1\left(Z_2 + Z_L\right) + Z_2 Z_L}. \tag{3.17}$$

3.3 Hybrid Parameters (*H* Parameters)

Now consider the hybrid parameter of a two-port network defined as

$$\begin{bmatrix} V_1 \\ I_2 \end{bmatrix} = \begin{bmatrix} H_{11} & H_{12} \\ H_{21} & H_{22} \end{bmatrix} \begin{bmatrix} I_1 \\ V_2 \end{bmatrix}, \tag{3.18}$$

where $H_{11} = \left.\frac{V_1}{I_1}\right|_{V_2=0} (\Omega)$ is the short-circuit input *impedance*, $H_{12} = \left.\frac{V_1}{V_2}\right|_{I_1=0}$ the open-circuit reverse *voltage gain*, $H_{21} = \left.\frac{I_2}{I_1}\right|_{V_2=0}$ the short-circuit forward *current gain*, and $H_{22} = \left.\frac{I_2}{V_2}\right|_{I_1=0} \left(\frac{1}{\Omega}\right)$ the open-circuit output *admittance*. The hybrid parameter H is commonly seen in the analysis of transistor circuits. For the circuit of Fig. 3.2, one has, by (3.18),

$$H = \begin{bmatrix} Z_1 & 1 \\ -1 & \frac{1}{Z_2} \end{bmatrix}. \tag{3.19}$$

Here, the overall input impedance can be determined as in Fig. 3.8 and is given by

$$\frac{V_1}{I_1} = \text{LFT}_l\left(H, -Z_L\right) = Z_1 + Z_L\left(1 + \frac{Z_L}{Z_2}\right)^{-1} = \frac{Z_1\left(Z_2 + Z_L\right) + Z_2 Z_L}{Z_2 + Z_L}, \tag{3.20}$$

which is the same as (3.16).

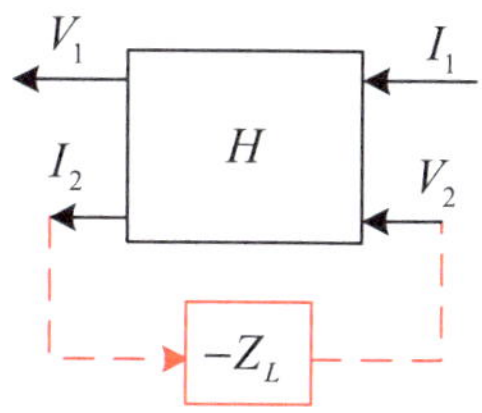

Fig. 3.8 LFT form of H parameter

3.4 Transmission Parameters (*ABCD* Parameters)

The above cases have shown how to find input/output relations by using the LFT structure. This section shows how the transmission parameters can be used to derive those relations by directly considering two-port network chains. It will be seen that the two-port network chain description is an alternative to that of LFT. It is, however, more appealing to electrical engineers and communication engineers, due to its direct connection to the two-port network structure.

The transmission parameter matrix description can connect several two-port network circuits in a series as illustrated in Fig. 3.9. The transmission parameter matrix is defined by

$$\begin{bmatrix} V_1 \\ I_1 \end{bmatrix} = \begin{bmatrix} A & B \\ C & D \end{bmatrix} \begin{bmatrix} V_2 \\ -I_2 \end{bmatrix}, \tag{3.21}$$

where $A = \left.\frac{V_1}{V_2}\right|_{I_2=0}$ denotes the open-circuit reverse voltage gain, $B = \left.\frac{V_1}{-I_2}\right|_{V_2=0}$ (Ω) the short-circuit transfer impedance, $C = \left.\frac{I_1}{V_2}\right|_{I_2=0}$ $\left(\frac{1}{\Omega}\right)$ the open-circuit transfer admittance, and $D = \left.\frac{I_1}{-I_2}\right|_{V_2=0}$ the short-circuit reverse current gain. The transmission parameters are often called the *ABCD* parameters in the electrical engineering community. Figure 3.10 shows the two-port transmission parameter description with load Z_L.

For the circuit in Fig. 3.2, the transmission parameters in (3.21) can be found as

$$\begin{bmatrix} A & B \\ C & D \end{bmatrix} = \begin{bmatrix} 1 + \frac{Z_1}{Z_2} & Z_1 \\ \frac{1}{Z_2} & 1 \end{bmatrix}. \tag{3.22}$$

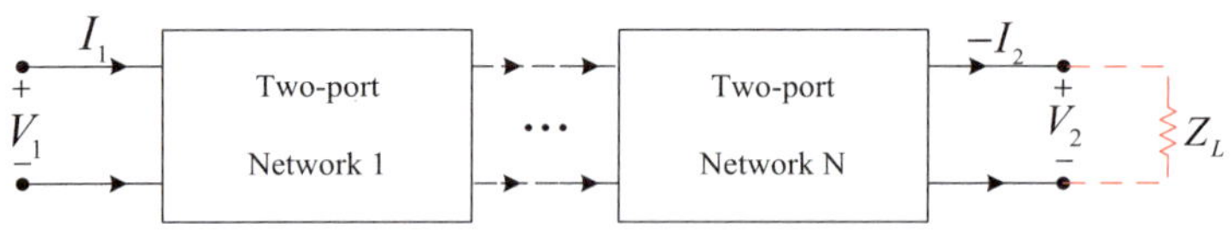

Fig. 3.9 Two-port network chain

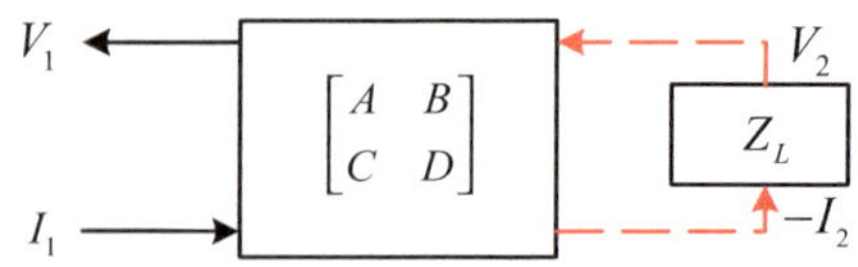

Fig. 3.10 Transmission parameter chain description

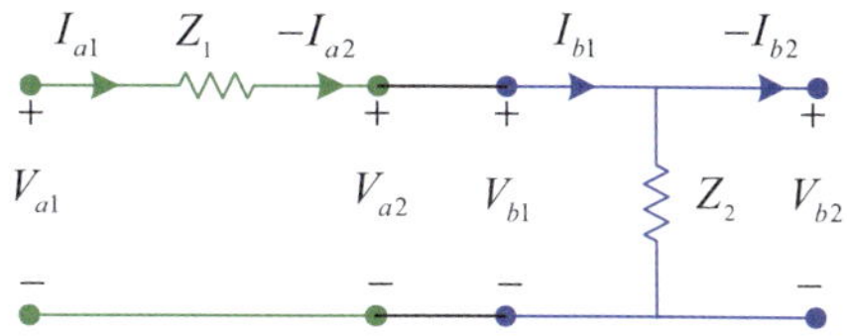

Fig. 3.11 Two-port circuit

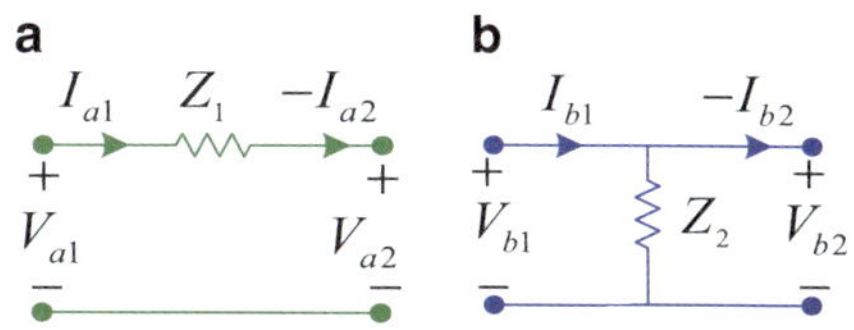

Fig. 3.12 Partition of the two-port circuit: (**a**) Left sub-circuit and (**b**) Right sub-circuit

One can then obtain the overall input impedance via

$$\frac{V_1}{I_1} = \mathrm{CSD_r}\left(\begin{bmatrix} A & B \\ C & D \end{bmatrix}, Z_L\right) = (B + AZ_L)(D + CZ_L)^{-1} = \frac{Z_1(Z_2+Z_L) + Z_2 Z_L}{Z_2 + Z_L}. \tag{3.23}$$

As expected, the result is the same as (3.16) and (3.20). The transmission parameter is useful for chaining several two-port networks in a series, which is the same as matrix multiplication. Hence, transmission parameters are also called *chain matrices*. Consider the circuit in Fig. 3.2 again and redraw it in Fig. 3.11. It is further decomposed with two sub-circuits as shown in Fig. 3.12.

The sub-circuit of Fig. 3.12a gives

$$\begin{bmatrix} V_{a1} \\ I_{a1} \end{bmatrix} = \begin{bmatrix} A_a & B_a \\ C_a & D_a \end{bmatrix} \begin{bmatrix} V_{a2} \\ -I_{a2} \end{bmatrix} \tag{3.24}$$

with

$$\begin{bmatrix} A_a & B_a \\ C_a & D_a \end{bmatrix} = \begin{bmatrix} 1 & Z_1 \\ 0 & 1 \end{bmatrix}. \tag{3.25}$$

Similarly, the sub-circuit of Fig. 3.12b gives

$$\begin{bmatrix} V_{b1} \\ I_{b1} \end{bmatrix} = \begin{bmatrix} A_b & B_b \\ C_b & D_b \end{bmatrix} \begin{bmatrix} V_{b2} \\ -I_{b2} \end{bmatrix} \tag{3.26}$$

with

$$\begin{bmatrix} A_b & B_b \\ C_b & D_b \end{bmatrix} = \begin{bmatrix} 1 & 0 \\ \frac{1}{Z_2} & 1 \end{bmatrix}. \tag{3.27}$$

Finally, the transmission parameter can be chained up (i.e., multiplying two matrices) as

$$\begin{aligned}\begin{bmatrix} V_1 \\ I_1 \end{bmatrix} = \begin{bmatrix} V_{a1} \\ I_{a1} \end{bmatrix} &= \begin{bmatrix} A_a & B_a \\ C_a & D_a \end{bmatrix} \begin{bmatrix} V_{a2} \\ -I_{a2} \end{bmatrix} \\ &= \begin{bmatrix} A_a & B_a \\ C_a & D_a \end{bmatrix} \begin{bmatrix} A_b & B_b \\ C_b & D_b \end{bmatrix} \begin{bmatrix} V_{b2} \\ -I_{b2} \end{bmatrix} = \begin{bmatrix} A & B \\ C & D \end{bmatrix} \begin{bmatrix} V_2 \\ I_2 \end{bmatrix}\end{aligned} \tag{3.28}$$

with

$$\begin{bmatrix} A & B \\ C & D \end{bmatrix} = \begin{bmatrix} 1 & Z_1 \\ 0 & 1 \end{bmatrix} \begin{bmatrix} 1 & 0 \\ \frac{1}{Z_2} & 1 \end{bmatrix} = \begin{bmatrix} 1 + \frac{Z_1}{Z_2} & Z_1 \\ \frac{1}{Z_2} & 1 \end{bmatrix}. \tag{3.29}$$

This concludes the same result as in (3.22).

In summary, the impedance, admittance, hybrid, and chain parameters in two-port network circuits are defined, respectively, based on the relationship between the input and output (voltages and currents) as

$$\begin{bmatrix} V_1 \\ V_2 \end{bmatrix} = \begin{bmatrix} Z_{11} & Z_{12} \\ Z_{21} & Z_{22} \end{bmatrix} \begin{bmatrix} I_1 \\ I_2 \end{bmatrix}, \tag{3.30}$$

$$\begin{bmatrix} I_1 \\ I_2 \end{bmatrix} = \begin{bmatrix} Y_{11} & Y_{12} \\ Y_{21} & Y_{22} \end{bmatrix} \begin{bmatrix} V_1 \\ V_2 \end{bmatrix}, \tag{3.31}$$

$$\begin{bmatrix} V_1 \\ I_2 \end{bmatrix} = \begin{bmatrix} H_{11} & H_{12} \\ H_{21} & H_{22} \end{bmatrix} \begin{bmatrix} I_1 \\ V_2 \end{bmatrix}, \tag{3.32}$$

$$\begin{bmatrix} V_1 \\ I_1 \end{bmatrix} = \begin{bmatrix} A & B \\ C & D \end{bmatrix} \begin{bmatrix} V_2 \\ -I_2 \end{bmatrix}. \tag{3.33}$$

Additionally, the overall input impedance can be equivalently represented by

$$\begin{aligned}\mathrm{LFT_l}\left(Z, -\frac{1}{Z_L}\right) &= (\mathrm{LFT_l}\,(Y, -Z_L))^{-1} = \mathrm{LFT_l}\,(H, -Z_L) \\ &= \mathrm{CSD_r}\left(\begin{bmatrix} A & B \\ C & D \end{bmatrix}, Z_L\right).\end{aligned} \tag{3.34}$$

The equivalence naturally leads to exploration of conversion formulae between these parameters. In the following, an example is used to illustrate how the parameter H from Z is determined.

Let the impedance matrix described in (3.8) be augmented as

$$\begin{bmatrix} V_1 \\ V_2 \\ I_1 \\ I_2 \end{bmatrix} = \begin{bmatrix} Z_{11} & Z_{12} \\ Z_{21} & Z_{22} \\ I & 0 \\ 0 & I \end{bmatrix} \begin{bmatrix} I_1 \\ I_2 \end{bmatrix} \Rightarrow \begin{bmatrix} V_1 \\ I_2 \\ I_1 \\ V_2 \end{bmatrix} = \begin{bmatrix} Z_{11} & Z_{12} \\ 0 & I \\ I & 0 \\ Z_{21} & Z_{22} \end{bmatrix} \begin{bmatrix} I_1 \\ I_2 \end{bmatrix}. \tag{3.35}$$

If Z_{22} is invertible, then

$$\begin{bmatrix} V_1 \\ I_2 \end{bmatrix} = \begin{bmatrix} Z_{11} & Z_{12} \\ 0 & I \end{bmatrix} \begin{bmatrix} I & 0 \\ Z_{21} & Z_{22} \end{bmatrix}^{-1} \begin{bmatrix} I_1 \\ V_2 \end{bmatrix} = \begin{bmatrix} H_{11} & H_{12} \\ H_{21} & H_{22} \end{bmatrix} \begin{bmatrix} I_1 \\ V_2 \end{bmatrix}, \tag{3.36}$$

where, by the matrix inversion,

$$\begin{bmatrix} H_{11} & H_{12} \\ H_{21} & H_{22} \end{bmatrix} = \begin{bmatrix} Z_{11} - Z_{12}Z_{22}^{-1}Z_{21} & Z_{12}Z_{22}^{-1} \\ -Z_{22}^{-1}Z_{21} & Z_{22}^{-1} \end{bmatrix}. \tag{3.37}$$

In the analogy, the Z parameter matrix can be converted into the chain parameters. Let

$$\begin{bmatrix} V_1 \\ V_2 \\ I_1 \\ -I_2 \end{bmatrix} = \begin{bmatrix} Z_{11} & -Z_{12} \\ Z_{21} & -Z_{22} \\ I & 0 \\ 0 & I \end{bmatrix} \begin{bmatrix} I_1 \\ -I_2 \end{bmatrix} \Rightarrow \begin{bmatrix} V_1 \\ I_1 \\ V_2 \\ -I_2 \end{bmatrix} = \begin{bmatrix} Z_{11} & -Z_{12} \\ I & 0 \\ Z_{21} & -Z_{22} \\ 0 & I \end{bmatrix} \begin{bmatrix} I_1 \\ -I_2 \end{bmatrix}. \tag{3.38}$$

If Z_{21} is invertible, then one can obtain that

$$\begin{bmatrix} V_1 \\ I_2 \end{bmatrix} = \begin{bmatrix} Z_{11} & -Z_{12} \\ I & 0 \end{bmatrix} \begin{bmatrix} Z_{21} & -Z_{22} \\ 0 & I \end{bmatrix}^{-1} \begin{bmatrix} V_2 \\ -I_2 \end{bmatrix} = \begin{bmatrix} A & B \\ C & D \end{bmatrix} \begin{bmatrix} V_2 \\ -I_2 \end{bmatrix}, \tag{3.39}$$

where

$$\begin{bmatrix} A & B \\ C & D \end{bmatrix} = \begin{bmatrix} Z_{11}Z_{21}^{-1} & Z_{11}Z_{21}^{-1}Z_{22} - Z_{12} \\ Z_{21}^{-1} & Z_{21}^{-1}Z_{22} \end{bmatrix}. \tag{3.40}$$

Note that the matrix conversion between any two parameter descriptions can be carried out by the same methodology.

3.5 Scattering Parameters (*S* Parameters)

In this section, a scattering parameter, namely, *S* parameter [5], will be discussed. Consider a transmission line as illustrated in Fig. 3.13, where *Z* denotes its equivalent impedance which is composed of four basic passive elements (R, L, G, C). Let Z_o be the characteristic impedance defined by

$$Z_0 = \sqrt{\frac{R + Ls}{G + Cs}}. \tag{3.41}$$

A transmission line which is only composed of (L, C) is called lossless. Then, the characteristic impedance (3.41) becomes $Z_0 = \sqrt{\frac{L}{C}}$. The impedance Z of a lossless transmission line contains only the imaginary part so that signals in the lossless transmission line do not consume any real power.

From the power wave theory [5], engineers often utilize *S* parameter to describe the injection and reflection phenomena of high-frequency microwave circuits and to determine phenomena such as standing wave, echo, and impedance matching problems throughout the entire communication procedure. Define

$$\begin{aligned} a_1 &= \frac{V_1 + Z_0 I_1}{2\sqrt{Z_0}} \left(\sqrt{\text{watt}}\right),\ a_2 = \frac{V_2 + Z_0 I_2}{2\sqrt{Z_0}} \left(\sqrt{\text{watt}}\right), \\ b_1 &= \frac{V_1 - Z_0 I_1}{2\sqrt{Z_0}} \left(\sqrt{\text{watt}}\right),\ b_2 = \frac{V_2 - Z_0 I_2}{2\sqrt{Z_0}} \left(\sqrt{\text{watt}}\right), \end{aligned} \tag{3.42}$$

where a_i denotes the incident wave (signal) at port i and b_i represents the reflected wave (signal) at port i. Let

$$\begin{bmatrix} b_1 \\ b_2 \end{bmatrix} = \begin{bmatrix} S_{11} & S_{12} \\ S_{21} & S_{22} \end{bmatrix} \begin{bmatrix} a_1 \\ a_2 \end{bmatrix}, \tag{3.43}$$

where

$$S_{11} = \left.\frac{b_1}{a_1}\right|_{a_2=0},\quad S_{12} = \left.\frac{b_2}{a_1}\right|_{a_2=0},\quad S_{21} = \left.\frac{b_1}{a_2}\right|_{a_1=0},\quad S_{22} = \left.\frac{b_2}{a_2}\right|_{a_1=0}. \tag{3.44}$$

The two-port *S* parameter description is illustrated, in LFT and CSD, in Fig. 3.14.

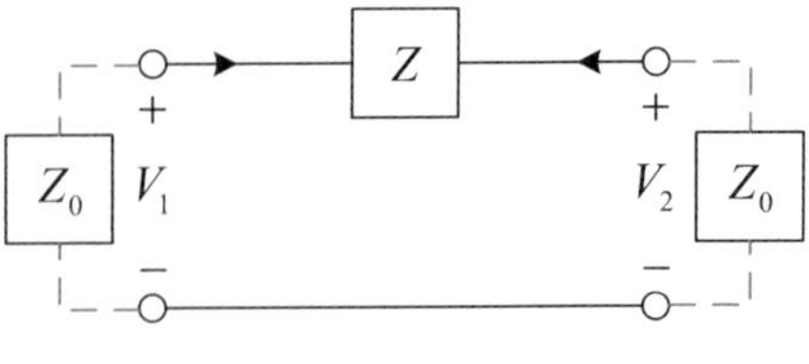

Fig. 3.13 Transmission line circuit

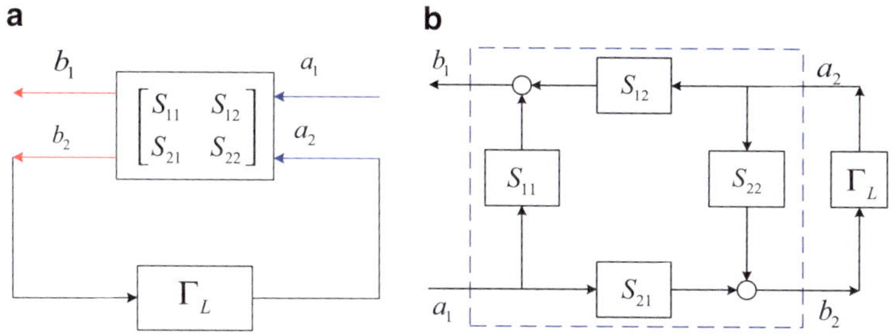

Fig. 3.14 LFT form of *S* parameter and its block description in CSD: (**a**) LFT form of *S* parameters and (**b**) Block description of *S* in CSD

Generally speaking, two-port networks can be interconnected by a terminator in a lower LFT or in a right CSD. The two-port *S* parameter description is different from those discussed in the previous sections in that matrix *S* is unit-free. This scattering description is devoted to utilizing the signals information of (a_1, b_1) and (a_2, b_2). However, it will be seen next that *S* parameter is in fact a transformation of variables from voltage/current to incident/reflected waves. Rewriting (3.42) yields

$$\begin{bmatrix} b_1 \\ a_1 \end{bmatrix} = \begin{bmatrix} \dfrac{V_1 - Z_0 I_1}{2\sqrt{Z_0}} \\ \dfrac{V_1 + Z_0 I_1}{2\sqrt{Z_0}} \end{bmatrix} = \Pi \begin{bmatrix} V_1 \\ I_1 \end{bmatrix}, \tag{3.45}$$

$$\begin{bmatrix} a_2 \\ b_2 \end{bmatrix} = \begin{bmatrix} \dfrac{V_2 + Z_0 I_2}{2\sqrt{Z_0}} \\ \dfrac{V_2 - Z_0 I_2}{2\sqrt{Z_0}} \end{bmatrix} = \Pi \begin{bmatrix} V_2 \\ -I_2 \end{bmatrix}, \tag{3.46}$$

where

$$\Pi = \frac{1}{2\sqrt{Z_0}} \begin{bmatrix} 1 & -Z_0 \\ 1 & Z_0 \end{bmatrix}. \tag{3.47}$$

Note that Π is a transformation matrix between voltage/current and power wave at a port.

Example 3.1 Find the two-port *S* parameter description for the circuit in Fig. 3.13.

As shown in Fig. 3.15, a voltage source V_s is added in the input (left) port, and a load of the characteristic impedance Z_o is connected in the output (right) port.

From

$$I_1 = -I_2,\ I_1 = \frac{V_s}{Z + Z_0},\ V_2 = \frac{Z_0}{Z + Z_0} V_s,\ V_1 = V_s, \tag{3.48}$$

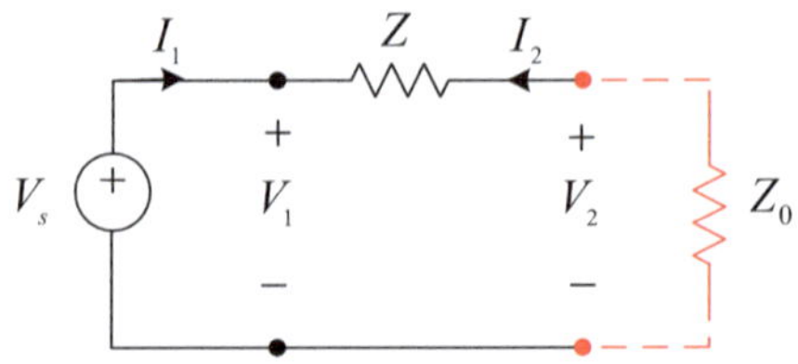

Fig. 3.15 Transmission line circuit

one gathers, by the superposition principle,

$$a_1 = \frac{V_1 + Z_0 I_1}{2\sqrt{Z_0}} = \frac{1 + \frac{Z_0}{Z+Z_0}}{2\sqrt{Z_0}} V_s = \frac{\frac{Z+2Z_0}{Z+Z_0}}{2\sqrt{Z_0}} V_s, \tag{3.49}$$

$$b_1 = \frac{V_1 - Z_0 I_1}{2\sqrt{Z_0}} = \frac{1 - \frac{Z_0}{Z+Z_0}}{2\sqrt{Z_0}} V_s = \frac{\frac{Z}{Z+Z_0}}{2\sqrt{Z_0}} V_s, \tag{3.50}$$

$$b_2 = \frac{V_2 - Z_0 I_2}{2\sqrt{Z_0}} = \frac{\frac{Z_0}{Z+Z_0} + \frac{Z_0}{Z+Z_0}}{2\sqrt{Z_0}} V_s = \frac{\frac{2Z_0}{Z+Z_0}}{2\sqrt{Z_0}} V_s, \tag{3.51}$$

$$a_2 = \frac{V_2 + Z_0 I_2}{2\sqrt{Z_0}} = \frac{\frac{Z_0}{Z+Z_0} - \frac{Z_0}{Z+Z_0}}{2\sqrt{Z_0}} V_s = 0. \tag{3.52}$$

Therefore, as $a_2 = 0$ in (3.52), the transfer functions from a_1 to b_1 and b_2, respectively, can be found from the ratio of (3.49) and (3.50), and the ratio of (3.49) and (3.51) as

$$S_{11} = \left.\frac{b_1}{a_1}\right|_{a_2=0} = \frac{Z}{Z + 2Z_0}, \tag{3.53}$$

$$S_{21} = \left.\frac{b_2}{a_1}\right|_{a_2=0} = \frac{2Z_0}{Z + 2Z_0}. \tag{3.54}$$

Similarly, utilizing the circuit shown in Fig. 3.16 results in

$$I_1 = -I_2, \quad I_2 = \frac{V_s}{Z + Z_0}, \quad V_1 = \frac{Z_0}{Z + Z_0} V_s, \quad V_2 = V_s, \tag{3.55}$$

and then

$$a_2 = \frac{V_2 + Z_0 I_2}{2\sqrt{Z_0}} = \frac{1 + \frac{Z_0}{Z+Z_0}}{2\sqrt{Z_0}} V_s = \frac{\frac{Z+2Z_0}{Z+Z_0}}{2\sqrt{Z_0}} V_s, \tag{3.56}$$

$$b_2 = \frac{V_2 - Z_0 I_2}{2\sqrt{Z_0}} = \frac{1 - \frac{Z_0}{Z+Z_0}}{2\sqrt{Z_0}} V_s = \frac{\frac{Z}{Z+Z_0}}{2\sqrt{Z_0}} V_s, \tag{3.57}$$

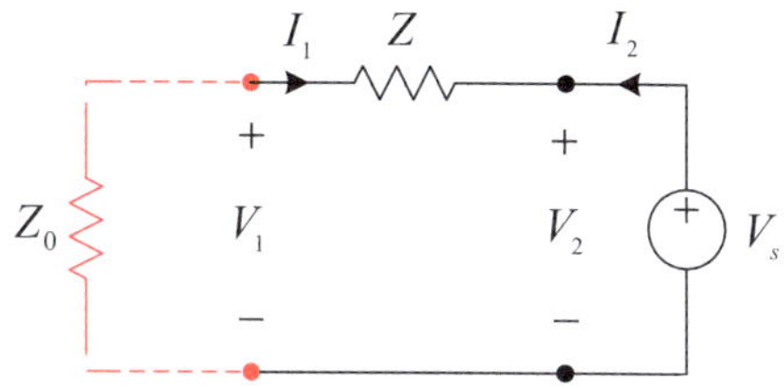

Fig. 3.16 Transmission line circuit

$$b_1 = \frac{V_1 - Z_0 I_1}{2\sqrt{Z_0}} = \frac{\frac{Z_0}{Z+Z_0} + \frac{Z_0}{Z+Z_0}}{2\sqrt{Z_0}} V_s = \frac{\frac{2Z_0}{Z+Z_0}}{2\sqrt{Z_0}} V_s, \tag{3.58}$$

$$a_1 = \frac{V_1 + Z_0 I_1}{2\sqrt{Z_0}} = \frac{\frac{Z_0}{Z+Z_0} - \frac{Z_0}{Z+Z_0}}{2\sqrt{Z_0}} V_s = 0. \tag{3.59}$$

Therefore, as $a_1 = 0$ in (3.59), the transfer functions from a_2 to b_1 and b_2, respectively, can be found from the ratio of (3.56) and (3.57), and the ratio of (3.56) and (3.58) as

$$S_{22} = \left.\frac{b_2}{a_2}\right|_{a_1=0} = \frac{Z}{Z+2Z_0}, \tag{3.60}$$

$$S_{12} = \left.\frac{b_1}{a_2}\right|_{a_1=0} = \frac{2Z_0}{Z+2Z_0}. \tag{3.61}$$

This concludes

$$\begin{bmatrix} b_1 \\ b_2 \end{bmatrix} = \begin{bmatrix} S_{11} & S_{12} \\ S_{21} & S_{22} \end{bmatrix} \begin{bmatrix} a_1 \\ a_2 \end{bmatrix}, \tag{3.62}$$

where

$$\begin{bmatrix} S_{11} & S_{12} \\ S_{21} & S_{22} \end{bmatrix} = \begin{bmatrix} \frac{Z}{Z+2Z_0} & \frac{2Z_0}{Z+2Z_0} \\ \frac{2Z_0}{Z+2Z_0} & \frac{Z}{Z+2Z_0} \end{bmatrix}. \tag{3.63}$$

3.6 Chain Scattering Parameters (*T* Parameters)

As shown in Fig. 3.17, the chain scattering parameter of a two-port network is defined as

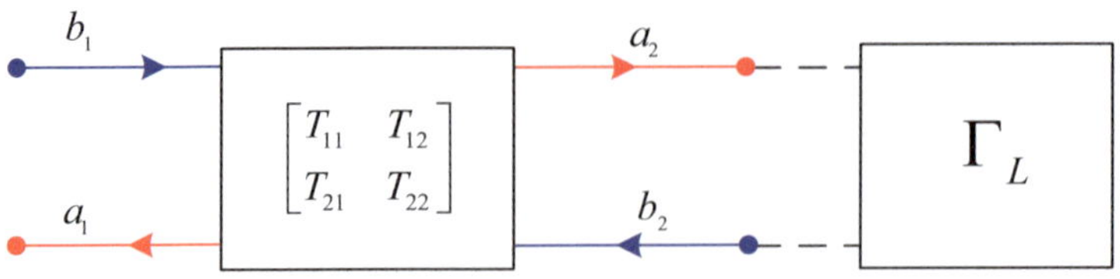

Fig. 3.17 Chain description of T parameter

$$\begin{bmatrix} b_1 \\ a_1 \end{bmatrix} = \begin{bmatrix} T_{11} & T_{12} \\ T_{21} & T_{22} \end{bmatrix} \begin{bmatrix} a_2 \\ b_2 \end{bmatrix}, \tag{3.64}$$

where

$$T_{11} = \left.\frac{b_1}{a_2}\right|_{b_2=0}, \quad T_{12} = \left.\frac{b_1}{b_2}\right|_{a_2=0}, \quad T_{21} = \left.\frac{a_1}{a_2}\right|_{b_2=0}, \quad T_{22} = \left.\frac{a_1}{b_2}\right|_{a_2=0}. \tag{3.65}$$

Apparently, the T parameter matrix is also unit-free like the S parameter.

The chain scattering parameter is often denoted as the *T parameter* in two-port network circuits. Recall that the scattering parameter was proposed to describe incident and reflected waves. It should be noted that the chain scattering matrix T was introduced to easily cascade several networks. The relationship between a scattering matrix (in LFT) and a chain scattering matrix (in CSD) can be determined as follows. Rearranging signals in T yields

$$\begin{bmatrix} b_1 \\ a_1 \\ a_2 \\ b_2 \end{bmatrix} = \begin{bmatrix} T_{11} & T_{12} \\ T_{21} & T_{22} \\ I & 0 \\ 0 & I \end{bmatrix} \begin{bmatrix} a_2 \\ b_2 \end{bmatrix} \Rightarrow \begin{bmatrix} b_1 \\ b_2 \\ a_1 \\ a_2 \end{bmatrix} = \begin{bmatrix} T_{11} & T_{12} \\ 0 & I \\ T_{21} & T_{22} \\ I & 0 \end{bmatrix} \begin{bmatrix} a_2 \\ b_2 \end{bmatrix}. \tag{3.66}$$

Thus, the S parameter is given by

$$\begin{aligned} \begin{bmatrix} b_1 \\ b_2 \end{bmatrix} &= \begin{bmatrix} T_{11} & T_{22} \\ 0 & I \end{bmatrix} \begin{bmatrix} a_2 \\ b_2 \end{bmatrix} = \begin{bmatrix} T_{11} & T_{12} \\ 0 & I \end{bmatrix} \begin{bmatrix} T_{21} & T_{22} \\ I & 0 \end{bmatrix}^{-1} \begin{bmatrix} a_1 \\ a_2 \end{bmatrix} \\ &= \begin{bmatrix} S_{11} & S_{12} \\ S_{21} & S_{22} \end{bmatrix} \begin{bmatrix} a_1 \\ a_2 \end{bmatrix}, \end{aligned} \tag{3.67}$$

where

$$\begin{bmatrix} S_{11} & S_{12} \\ S_{21} & S_{22} \end{bmatrix} = \begin{bmatrix} T_{11} & T_{12} \\ 0 & I \end{bmatrix} \begin{bmatrix} T_{21} & T_{22} \\ I & 0 \end{bmatrix}^{-1}. \tag{3.68}$$

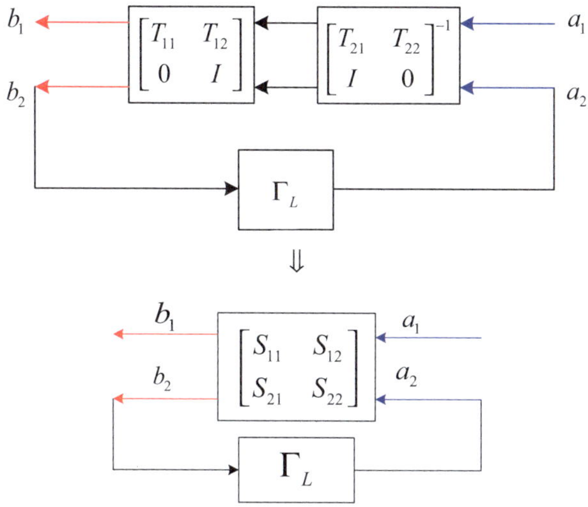

Fig. 3.18 Parameter conversion from *T* to *S*

Note that here, T_{22} should be invertible. Then, the S parameter as illustrated in Fig. 3.14 can be obtained. Figure 3.18 shows the corresponding manipulations.

Similarly, one can also derive parameter T from S,

$$\begin{bmatrix} b_1 \\ b_2 \\ a_1 \\ a_2 \end{bmatrix} = \begin{bmatrix} S_{11} & S_{12} \\ S_{21} & S_{22} \\ I & 0 \\ 0 & I \end{bmatrix} \begin{bmatrix} a_1 \\ a_2 \end{bmatrix} \Rightarrow \begin{bmatrix} b_1 \\ a_1 \\ a_2 \\ b_2 \end{bmatrix} = \begin{bmatrix} S_{11} & S_{12} \\ I & 0 \\ 0 & I \\ S_{21} & S_{22} \end{bmatrix} \begin{bmatrix} a_1 \\ a_2 \end{bmatrix}. \tag{3.69}$$

If S_{21} is invertible, then T parameter matrix is given by

$$\begin{bmatrix} b_1 \\ a_1 \end{bmatrix} = \begin{bmatrix} S_{11} & S_{12} \\ I & 0 \end{bmatrix} \begin{bmatrix} 0 & I \\ S_{21} & S_{22} \end{bmatrix}^{-1} \begin{bmatrix} a_2 \\ b_2 \end{bmatrix} = \begin{bmatrix} T_{11} & T_{12} \\ T_{21} & T_{22} \end{bmatrix} \begin{bmatrix} a_2 \\ b_2 \end{bmatrix}, \tag{3.70}$$

where

$$\begin{bmatrix} T_{11} & T_{12} \\ T_{21} & T_{22} \end{bmatrix} = \begin{bmatrix} S_{11} & S_{12} \\ I & 0 \end{bmatrix} \begin{bmatrix} 0 & I \\ S_{21} & S_{22} \end{bmatrix}^{-1}. \tag{3.71}$$

3.7 Conversions Between (*ABCD*) and (*S*, *T*) Matrix Parameters

In this section, the conversions from impedance, admittance, chain, and hybrid matrices to the scattering and chain scattering matrices will be discussed. Firstly, as shown in Fig. 3.19, the conversion from transmission parameters *ABCD* to *T* parameters will be taken as an example.

Recall (3.42), (3.45), and (3.46). Then, by (3.21), one has

$$\begin{bmatrix} V_1 \\ I_1 \end{bmatrix} = \begin{bmatrix} A & B \\ C & D \end{bmatrix}\begin{bmatrix} V_2 \\ -I_2 \end{bmatrix} \Rightarrow \Pi\begin{bmatrix} V_1 \\ I_1 \end{bmatrix} = \Pi\begin{bmatrix} A & B \\ C & D \end{bmatrix}\Pi^{-1}\Pi\begin{bmatrix} V_2 \\ -I_2 \end{bmatrix}. \tag{3.72}$$

Hence, from (3.64),

$$\begin{bmatrix} b_1 \\ a_1 \end{bmatrix} = \Pi\begin{bmatrix} A & B \\ C & D \end{bmatrix}\Pi^{-1}\begin{bmatrix} a_2 \\ b_2 \end{bmatrix} = \begin{bmatrix} T_{11} & T_{12} \\ T_{21} & T_{22} \end{bmatrix}\begin{bmatrix} a_2 \\ b_2 \end{bmatrix}. \tag{3.73}$$

Further, by (3.47),

$$\begin{bmatrix} T_{11} & T_{12} \\ T_{21} & T_{22} \end{bmatrix} = \begin{bmatrix} 1 & -Z_0 \\ 1 & Z_0 \end{bmatrix}\begin{bmatrix} A & B \\ C & D \end{bmatrix}\begin{bmatrix} 1 & -Z_0 \\ 1 & Z_0 \end{bmatrix}^{-1}. \tag{3.74}$$

As illustrated in Fig. 3.20, the function of Π is in fact to transform the system states (V_i, I_i) into (a_i, b_i). This mapping is also called the "Möbius transformation" [4, 6], mathematically speaking. In addition, readers can easily convert *T* parameter into the *S* parameter according to (3.68), as shown in Fig. 3.18, which reveals that the *S* parameter interconnects with Γ_L. Other conversions can be carried out by the same techniques.

Example 3.2 For the circuit of Fig. 3.15, determine its *ABCD*, *T*, and *S* parameters.

The chain matrix can be determined as

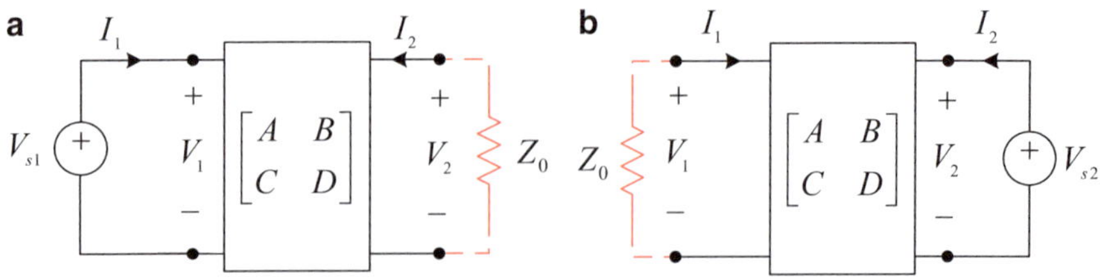

Fig. 3.19 Two-port transmission circuits: (**a**) With left voltage source and (**b**) With right voltage source

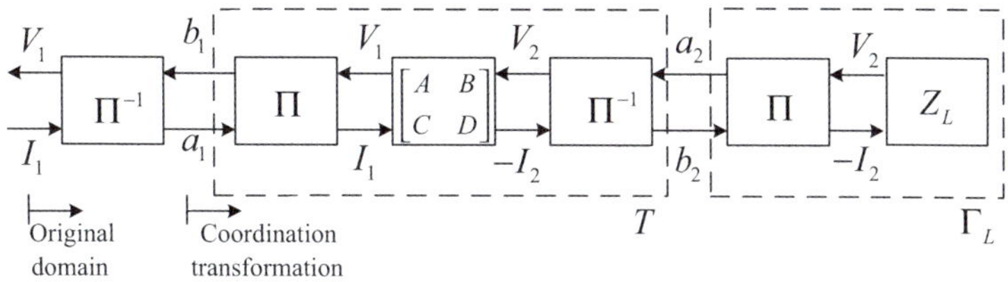

Fig. 3.20 Effect of Π on coordination transformation

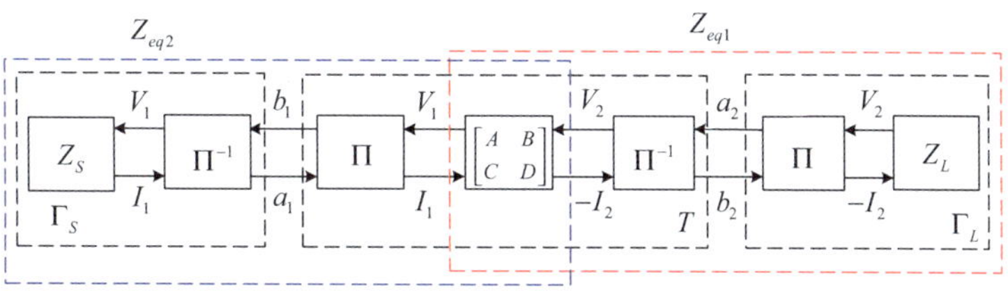

Fig. 3.21 Two-port circuit with source and load impedances

$$\begin{bmatrix} V_1 \\ I_1 \end{bmatrix} = \begin{bmatrix} A & B \\ C & D \end{bmatrix} \begin{bmatrix} V_2 \\ -I_2 \end{bmatrix} = \begin{bmatrix} 1 & Z \\ 0 & 1 \end{bmatrix} \begin{bmatrix} V_2 \\ -I_2 \end{bmatrix}. \tag{3.75}$$

Then T can be determined as

$$\begin{bmatrix} T_{11} & T_{12} \\ T_{21} & T_{22} \end{bmatrix} = \Pi \begin{bmatrix} A & B \\ C & D \end{bmatrix} \Pi^{-1} = \begin{bmatrix} 1 - \dfrac{Z}{2Z_0} & \dfrac{Z}{2Z_0} \\ -\dfrac{Z}{2Z_0} & 1 + \dfrac{Z}{2Z_0} \end{bmatrix}. \tag{3.76}$$

Furthermore, by (3.68), the S parameter is given by

$$\begin{bmatrix} S_{11} & S_{12} \\ S_{21} & S_{22} \end{bmatrix} = \begin{bmatrix} \dfrac{Z}{2Z_0 + Z} & \dfrac{2Z_0}{2Z_0 + Z} \\ \dfrac{2Z_0}{2Z_0 + Z} & \dfrac{Z}{2Z_0 + Z} \end{bmatrix}. \tag{3.77}$$

3.8 Lossless Networks

As shown in Fig. 3.21, define the input and output reflection coefficient (Γ) as

$$\Gamma_1 = \frac{b_1}{a_1} = \mathrm{CSD_r}\left(\Pi, Z_{\mathrm{eq1}}\right) = \frac{Z_{\mathrm{eq1}} - Z_0}{Z_{\mathrm{eq1}} + Z_0}, \tag{3.78}$$

$$\Gamma_2 = \frac{b_2}{a_2} = \mathrm{CSD}_l\left(\Pi^{-1}, Z_{eq2}\right) = \frac{Z_{eq2} - Z_0}{Z_{eq2} + Z_0}, \tag{3.79}$$

where $Z_{eq1} = \mathrm{CSD_r}\left(\begin{bmatrix} A & B \\ C & D \end{bmatrix}, Z_L\right)$ is the equivalent impedance looking from port 1, $Z_{eq2} = \mathrm{CSD_l}\left(\begin{bmatrix} A & B \\ C & D \end{bmatrix}, Z_S\right)$ is the equivalent impedance looking from port 2, and Z_o is the characteristic impedance. Clearly, if $Z_{eq1} = Z_o$ (i.e., the equivalent impedance matches the characteristic impedance), then $\Gamma_1 = 0$ by (3.78). This indicates that the incident wave from port 1 will fully come out from port 2 and will not cause any reflection in port 1. Also $\Gamma_1 = 0$ means that the incident power wave from port 1 does not reflect a power wave in the output of port 1 that causes an echo. Similarly, if $Z_{eq2} = Z_o$, $\Gamma_2 = 0$. This means that the incident power wave from port 2 does not reflect a power wave in the output of port 2 that produces an echo. Alternatively, all the output power waves come from another port of a two-port network. Such a condition is called *all-pass* in microwave theory [2]. If the released power is not attenuated through the propagation of two-port networks, this circuit is considered *lossless*. In addition, a no-reflection echo two-port network is also called the *matched network*.

Define the average delivered power at each port as

$$P_{av1} = \frac{1}{2}\left(a_1{}^* a_1 - b_1{}^* b_1\right), \tag{3.80}$$

$$P_{av2} = \frac{1}{2}\left(a_2{}^* a_2 - b_2{}^* b_2\right). \tag{3.81}$$

Consequently, the total delivered power is

$$P_{av} = P_{av1} + P_{av2}. \tag{3.82}$$

Then, one has, by (3.43),

$$P_{av} = \frac{1}{2}\left\{\begin{bmatrix} a_1 \\ a_2 \end{bmatrix}^* \begin{bmatrix} a_1 \\ a_2 \end{bmatrix} - \begin{bmatrix} b_1 \\ b_2 \end{bmatrix}^* \begin{bmatrix} b_1 \\ b_2 \end{bmatrix}\right\} = \frac{1}{2}\left\{\begin{bmatrix} a_1 \\ a_2 \end{bmatrix}^* \begin{bmatrix} a_1 \\ a_2 \end{bmatrix} - \begin{bmatrix} a_1 \\ a_2 \end{bmatrix}^* S^*S \begin{bmatrix} a_1 \\ a_2 \end{bmatrix}\right\} = \frac{1}{2}\left\{\begin{bmatrix} a_1 \\ a_2 \end{bmatrix}^* [I - S^*S] \begin{bmatrix} a_1 \\ a_2 \end{bmatrix}\right\}. \tag{3.83}$$

A two-port network which consumes no power (i.e., $P_{av} = 0$) is called lossless. From (3.83), a lossless network has the property $S^*S = I$. Generally speaking, the energy will not be attenuated in this propagation. For a lossless network, all of the power (energy) directed to any one port has to be accounted for by summing

Fig. 3.22 Transmission line circuit

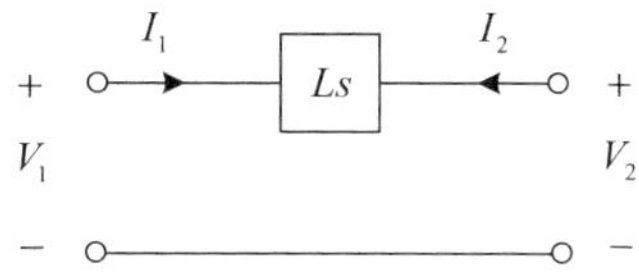

the power output at the other ports with the power reflected at the incident port. From $\begin{bmatrix} S_{11} & S_{12} \\ S_{21} & S_{22} \end{bmatrix} = \begin{bmatrix} T_{11} & T_{12} \\ 0 & I \end{bmatrix} \begin{bmatrix} T_{21} & T_{22} \\ I & 0 \end{bmatrix}^{-1}$ and $S^*S = I$, one can obtain, for $J := \begin{bmatrix} I & 0 \\ 0 & -I \end{bmatrix}$,

$$T^*JT = \begin{bmatrix} T_{11} & T_{12} \\ T_{21} & T_{22} \end{bmatrix}^* \begin{bmatrix} I & 0 \\ 0 & -I \end{bmatrix} \begin{bmatrix} T_{11} & T_{12} \\ T_{21} & T_{22} \end{bmatrix} = \begin{bmatrix} I & 0 \\ 0 & -I \end{bmatrix} = J. \quad (3.84)$$

Therefore, a lossless two-port network has a *unitary* S matrix or a *J-unitary* T matrix.

Example 3.3 Prove the S parameter is a unitary matrix in Fig. 3.22 with characteristic impedance $Z_o = 1(\Omega)$.

Consider the transmission line in Fig. 3.22, where the impedance is an inductance which does not contain any real part. Then by (3.63) with $Z = sL$ and $Z_o = R_o(\Omega)$, one has

$$S = \begin{bmatrix} \frac{Ls}{Ls + R_o} & \frac{R_o}{Ls + R_o} \\ \frac{R_o}{Ls + R_o} & \frac{Ls}{Ls + R_o} \end{bmatrix}, \quad (3.85)$$

or $sL\,(= jwL) = jX$

$$S = \begin{bmatrix} \frac{jX}{jX + R_o} & \frac{R_o}{jX + R_o} \\ \frac{R_o}{jX + R_o} & \frac{jX}{jX + R_o} \end{bmatrix}, \quad (3.86)$$

and hence,

$$S^*S = \begin{bmatrix} \frac{X^2}{R_o^2 + X^2} + \frac{R_o^2}{R_o^2 + X^2} & \frac{jR_oX}{R_o^2 + X^2} - \frac{jR_oX}{R_o^2 + X^2} \\ -\frac{jR_oX}{R_o^2 + X^2} - \frac{jR_oX}{R_o^2 + X^2} & \frac{R_o^2}{R_o^2 + X^2} + \frac{X^2}{R_o^2 + X^2} \end{bmatrix} = I. \quad (3.87)$$

This concludes that the S matrix is unitary. Consequently, this is a lossless system.

As mentioned in Sect. 3.7, the transformation of Π acts as a transformation coordinator. The following examples will show the effect of function transformation via the operation of Π. First, two definitions are given below.

Definition 3.1 A real rational function $Z(s)$ is called positive real if the following conditions are satisfied [1]:

1. $Z(s)$ is analytic in $\text{Re}[s] > 0$.
2. $Z(s)$ is real when s is positive and real.
3. $Z^*(s) + Z(s) > 0$ for $\text{Re}[s] > 0$.

Definition 3.2 A real rational function $\Gamma(s)$ is called bounded real if the following conditions are satisfied:

1. $\Gamma(s)$ is analytic in $\text{Re}[s] > 0$.
2. $\Gamma(s)$ is real for real positive s.
3. $I - \Gamma^*(s)\Gamma(s) \geq 0$ for $\text{Re}[s] > 0$.

Recall that $\Pi = \frac{1}{2\sqrt{Z_0}}\begin{bmatrix} 1 & -Z_o \\ 1 & Z_o \end{bmatrix}$ defined in (3.47), which is a transformation matrix between voltage/current and power wave at a port. From Fig. 3.21, one can obtain that

$$\Gamma_L = \frac{a_2}{b_2} = \text{CSD}_\text{r}\left(\Pi, Z_L\right) = \frac{Z_L - Z_o}{Z_L + Z_o}, \tag{3.88}$$

$$\Gamma_S = \frac{a_1}{b_1} = \text{CSD}_\text{l}\left(\Pi^{-1}, Z_S\right) = \frac{Z_S - Z_o}{Z_S + Z_o} \left(= \text{CSD}_\text{r}\left(\Pi, Z_S\right)\right). \tag{3.89}$$

In fact, these are called the bilinear transformations. For Π with real Z_o, the positive real (PR) Z_L and Z_S such as (R,L,C) can be transformed to the bounded real (BR) functions via Π. For example, if $Z_o = 1(\Omega)$ and $Z_L = sL$, then $\Gamma_L = \text{CSD}_\text{r}\left(\Pi, Z_L\right) = \frac{\text{sL}-1}{\text{sL}+1}$, or if $Z_L = \frac{1}{Cs}$, then $\Gamma_L = \text{CSD}_\text{r}\left(\Pi, Z_L\right) = \frac{\frac{1}{Cs}-1}{\frac{1}{Cs}+1} = \frac{1-Cs}{1+Cs}$. Figure 3.23 reveals this transformation.

The T parameter description is convenient for network cascading. However, the representation of T matrix may not always be a proper function. Recall the series circuit in Fig. 3.24, which gives

From Fig. 3.21, one has, for $Z = sL$,

$$T = \Pi \begin{bmatrix} 1 & Z \\ 0 & 1 \end{bmatrix} \Pi^{-1} = \begin{bmatrix} 1 - \dfrac{Z}{2Z_o} & \dfrac{Z}{2Z_o} \\ -\dfrac{Z}{2Z_o} & 1 + \dfrac{Z}{2Z_o} \end{bmatrix}. \tag{3.90}$$

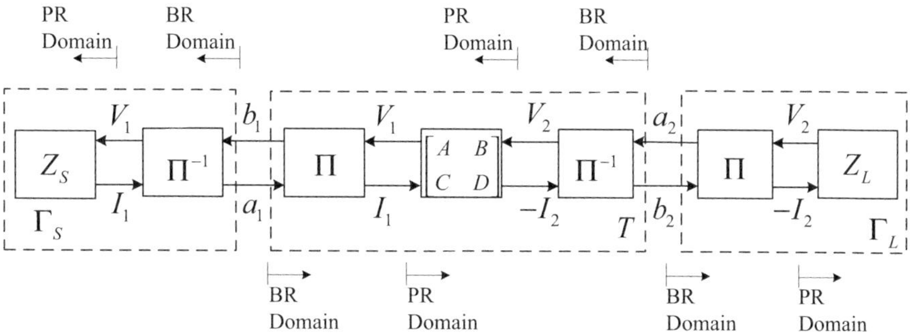

Fig. 3.23 The coordination transformation properties

Fig. 3.24 Transmission line with series circuit

$$\begin{bmatrix} A & B \\ C & D \end{bmatrix} = \begin{bmatrix} 1 & Z \\ 0 & 1 \end{bmatrix}.$$

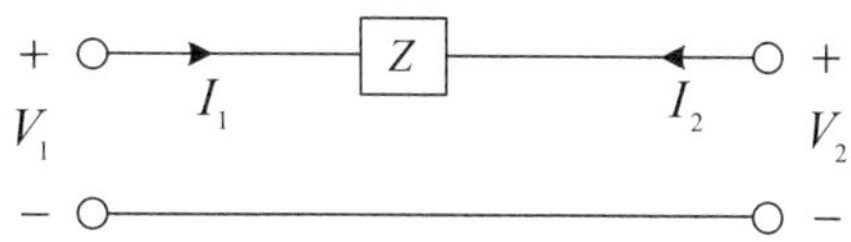

Fig. 3.25 Transmission line with shunt circuit

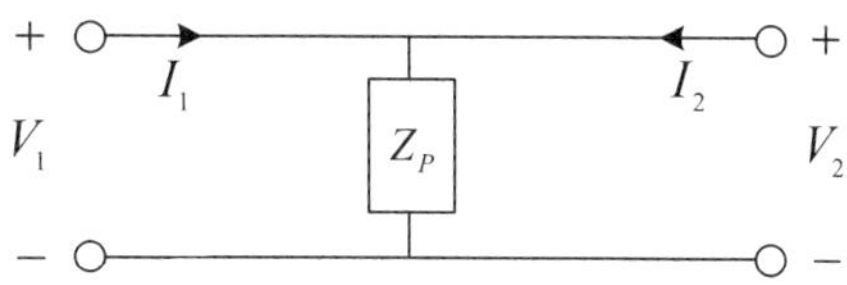

For the case that $Z_o = 1(\Omega)$ and $L = 2$, $T = \begin{bmatrix} 1-s & s \\ -s & 1+s \end{bmatrix}$ becomes an improper function matrix. However, by terminating the load $Z_L = 3(\Omega)$, one has $\Gamma_L = \text{CSD}_r(\Pi, 3) = \frac{3-1}{3+1} = 0.5$. From (3.88), the transfer function is given by

$$\frac{b_1}{a_1} = \text{CSD}_r\left(\Pi \begin{bmatrix} A & B \\ C & D \end{bmatrix}, Z_L\right) = \text{CSD}_r(T, \Gamma_L) = \frac{2\Gamma_L Z_o + \Gamma_L Z - Z}{\Gamma_L Z - 2Z_o - Z} = \frac{s-1}{s+2}, \tag{3.91}$$

which is bounded real.

From the shunt circuit depicted in Fig. 3.25, one also obtains

$$\begin{bmatrix} A & B \\ C & D \end{bmatrix} = \begin{bmatrix} 1 & 0 \\ \frac{1}{Z_P} & 1 \end{bmatrix}, \tag{3.92}$$

Fig. 3.26 RLC circuit

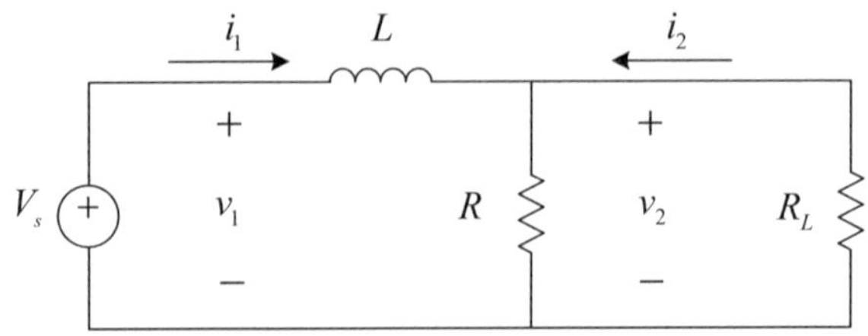

and

$$T = \Pi \begin{bmatrix} 1 & 0 \\ \frac{1}{Z_P} & 1 \end{bmatrix} \Pi^{-1} = \begin{bmatrix} 1 - \dfrac{Z_o}{2Z_P} & -\dfrac{Z_o}{2Z_P} \\ \dfrac{Z_o}{2Z_P} & 1 + \dfrac{Z_o}{2Z_P} \end{bmatrix}. \tag{3.93}$$

For a capacitive circuit, the T matrix is improper. For the load Γ_L, the transfer function is

$$\frac{b_1}{a_1} = \mathrm{CSD_r}\left(\Pi \begin{bmatrix} A & B \\ C & D \end{bmatrix}, Z_L\right) = \mathrm{CSD_r}(T, \Gamma_L) = \frac{2\Gamma_L Z_P + \Gamma_L Z_0 - Z_0}{\Gamma_L Z_0 - 2Z_P - Z_0}. \tag{3.94}$$

Converting the system description to LFT as shown in Fig. 3.18, one has

$$S_P = \begin{bmatrix} \dfrac{-Z_0}{2Z_P + Z_0} & \dfrac{2Z_P}{2Z_P + Z_0} \\ \dfrac{2Z_P}{2Z_P + Z_0} & \dfrac{-Z_0}{2Z_P + Z_0} \end{bmatrix} = \begin{bmatrix} S_{P_{11}} & S_{P_{12}} \\ S_{P_{21}} & S_{P_{22}} \end{bmatrix} \tag{3.95}$$

which is always adequate with real Z_o. Apparently,

$$\frac{b_1}{a_1} = \mathrm{LFT_l}(S_P, \Gamma_L) = \frac{2\Gamma_L Z_P + \Gamma_L Z_0 - Z_0}{\Gamma_L Z_0 - 2Z_P - Z_0}. \tag{3.96}$$

Example 3.4 Consider an RLC circuit illustrated in Fig. 3.26. Derive its *ABCD* parameter and T parameters when $R = 10$ and $L = 1$.

The *ABCD* parameter of Fig. 3.26 can be derived from simple matrix manipulation

$$G = \begin{bmatrix} A & B \\ C & D \end{bmatrix} = \begin{bmatrix} 1 & Ls \\ 0 & 1 \end{bmatrix} \begin{bmatrix} 1 & 0 \\ \dfrac{1}{R} & 1 \end{bmatrix} = \begin{bmatrix} 1 + \dfrac{Ls}{R} & Ls \\ \dfrac{1}{R} & 1 \end{bmatrix} = \begin{bmatrix} 1 + \dfrac{s}{10} & s \\ 0.1 & 1 \end{bmatrix}$$

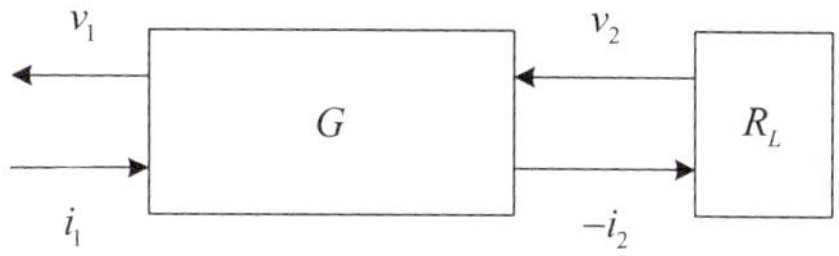

Fig. 3.27 Two-port description

The overall input impedance can be calculated from (Fig. 3.27),

$$\frac{V_1}{i_1} = \mathrm{CSD_r}\,(G, R_L) = \frac{(R + R_L)\,Ls + RR_L}{R_L + R} = \frac{(10 + R_L)\,s + 10R_L}{R_L + 10}$$

which is the positive real for any positive real R_L. However, it is an improper function. Let $\Pi = \frac{1}{2}\begin{bmatrix} 1 & -1 \\ 1 & 1 \end{bmatrix}$, and one gets

$$T = \Pi G \Pi^{-1} = \frac{1}{2}\begin{bmatrix} \dfrac{19-9s}{10} & \dfrac{11s-1}{10} \\ \dfrac{1-9s}{10} & \dfrac{11s+21}{10} \end{bmatrix}$$

and $\Gamma_L = \mathrm{CSD_r}\,(\Pi, R_L) = \frac{R_L-1}{R_L+1}$. From Fig. 3.23, one can obtain

$$\frac{b_1}{a_1} = \mathrm{CSD_r}\,(T, \Gamma_L) = \frac{(11 - 9\Gamma_L)\,s + (19\Gamma_L - 1)}{(11 - 9\Gamma_L)\,s + (21 + \Gamma_L)}.$$

Evidently, since $\Gamma_L = \mathrm{CSD_r}\,(\Pi, R_L) = \frac{R_L-1}{R_L+1} < 1$, for any positive real $R_L > 0$, the norm of b_1 is always less or equal to the norm of a_1. For example, when assuming $\Gamma_L = 0.7$, then $\frac{b_1}{a_1} = \mathrm{CSD_r}\,(T, \Gamma_L) = \frac{4.7s+12.3}{4.7s+21.7} < 1, \forall s > 0$; one can conclude that $\frac{b_1}{a_1} \in BH_\infty$.

Exercises

1. Determine the transfer function from V_1 to V_2 of the following circuit, with and without the load R_L. Convert this circuit into the control block diagram and verify $\frac{V_2}{V_1}$ via Mason's gain formula and LFT, respectively.

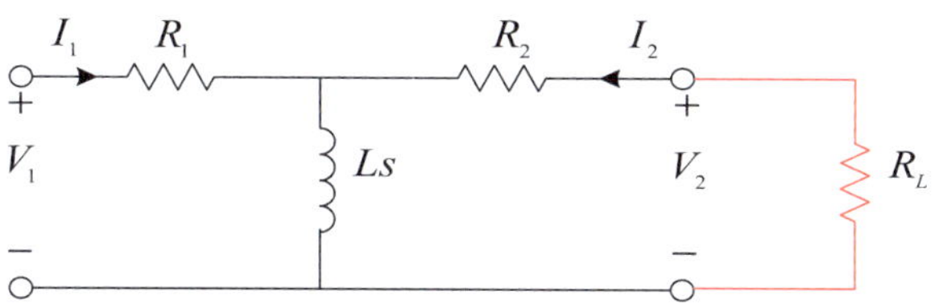

2. Find the Y parameter of the following two-port circuit.

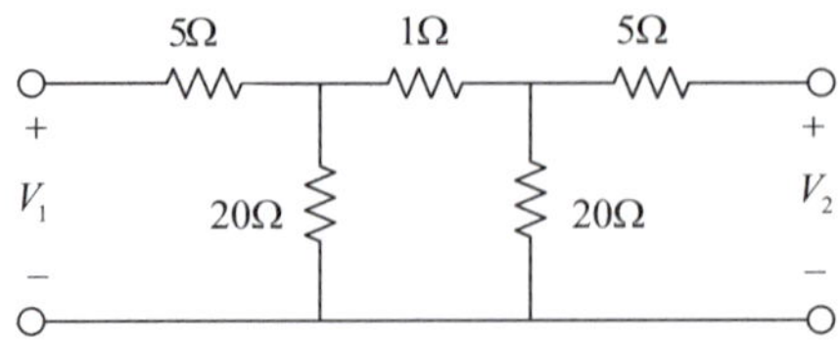

3. Determine the $ABCD$ and H parameters for the following two-port circuit.

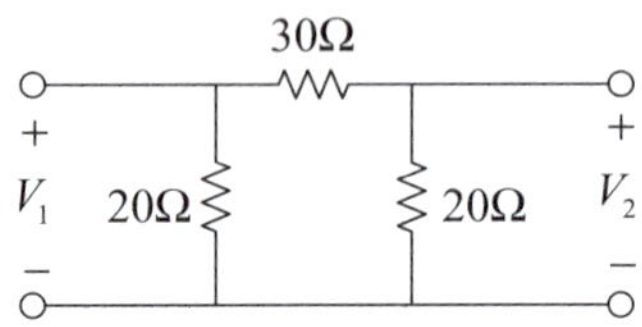

4. Determine the S parameter of the following two-port circuit with the characteristic impedance $Z_o = 64\ \Omega$.

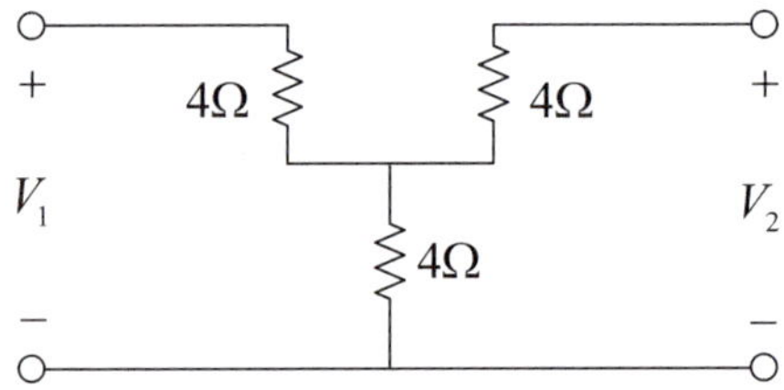

5. Determine the S parameter of the following two-port circuit with the characteristic impedance $Z_o = 10\ \Omega$.

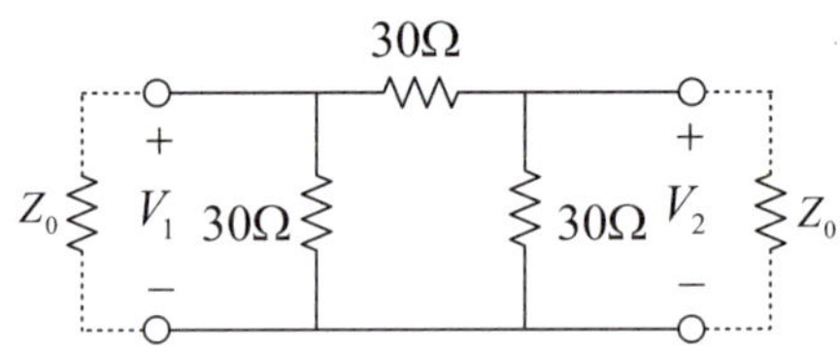

6. Determine the $ABCD$ parameter of the given circuit and then derive the T parameter using the $ABCD$ parameter.

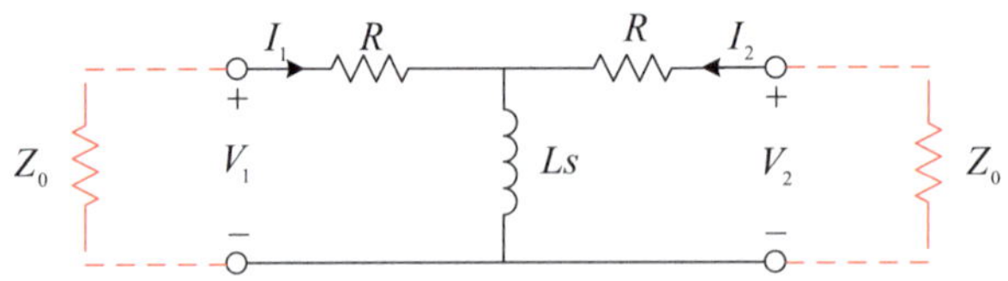

7. Check whether the S parameters of the following circuit is unitary (lossless) or not, then transfer S to Y. Does the real part of Y equal zero?

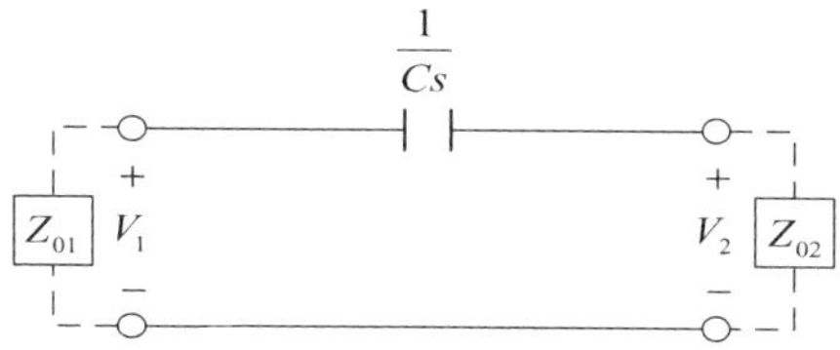

References

1. Anderson BDO, Vongpanitlerd S (1973) Network analysis and synthesis: a modern systems theory approach. Prentice-Hall Inc, Englewood Cliffs
2. Cheng DK (1992) Fundamentals of engineering electromagnetics. Addison-Wesley, New York
3. Franco S (1995) Electric circuits fundamentals. Saunders College Publishing, Orlando
4. Knopp K (1952) Elements of the theory of functions. Dover, New York
5. Misra DK (2001) Radio-frequency and microwave communication circuits. Wiley, New York
6. Needham T (2000) Möbius transformations and inversion. Clarendon, New York

Chapter 4
Linear Fractional Transformations

This chapter introduces the linear fractional transformation (LFT), which is a convenient and powerful formulation in control system analysis and controller synthesis. The LFT formulation employs a two-port matrix description linked by a terminator to represent a closed-loop feedback system with two individual open-loop systems. This representation is inherently suitable for MIMO systems. Several examples are given to show how to locate the interconnected transfer function for a given system by using LFT and also how to formulate a control design problem into LFT. Additionally, in order to understand the benefit of utilizing LFT, the relationship between Mason's gain formulae and LFT will be discussed in this chapter. Inner and co-inner systems are relevant to various aspects of control theory, especially H_∞ control. Definitions of inner and co-inner functions are thus introduced in the last section of this chapter.

4.1 Linear Fractional Transformations

Figure 4.1 illustrates the framework of an LFT, which includes two parts, i.e., a two-port matrix P and one-port feedback terminator K. Without the feedback termination $y \mapsto u$, the two-port matrix P is an open-loop system of $\begin{bmatrix} r \\ u \end{bmatrix} \mapsto \begin{bmatrix} w \\ y \end{bmatrix}$. Thus, the matrix representation can be characterized as

$$\begin{cases} w = P_{11} r + P_{12} u \\ y = P_{21} r + P_{22} u, \end{cases} \tag{4.1}$$

where

$$P_{11} = \left.\frac{w}{r}\right|_{u=0}, \quad P_{12} = \left.\frac{w}{u}\right|_{r=0}, \quad P_{21} = \left.\frac{y}{r}\right|_{u=0}, \quad P_{22} = \left.\frac{y}{u}\right|_{r=0}. \tag{4.2}$$

M.-C. Tsai and D.-W. Gu, *Robust and Optimal Control: A Two-port Framework Approach*, Advances in Industrial Control, DOI 10.1007/978-1-4471-6257-5_4,
© Springer-Verlag London 2014

Fig. 4.1 LFT form

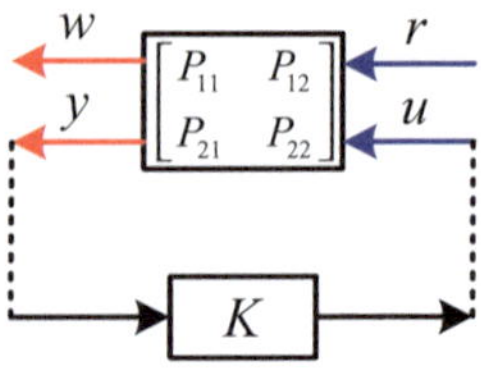

Note that each entrant in $\begin{bmatrix} w \\ y \end{bmatrix}$ and $\begin{bmatrix} r \\ u \end{bmatrix}$ can be vector-valued signals. Here, (4.2) symbolizes its input/output relations in the MIMO cases, not the actual mathematical "division." It can be said as, e.g., P_{11} represents how w is dependent on r when $u=0$. In the feedback control, the terminator K (to be designed) encloses the open-loop system (4.1) via the feedback part

$$u = Ky. \tag{4.3}$$

Apparently, the LFT formulation can be regarded as two individual open-loop systems but linked by the terminator to represent the closed-loop transfer function of $r \mapsto w$. From (4.1) and (4.3), one has

$$\begin{cases} w = P_{11}r + P_{12}u = P_{11}r + P_{12}Ky \\ y = P_{21}r + P_{22}u = P_{21}r + P_{22}Ky, \end{cases} \tag{4.4}$$

therefore,

$$y = (I - P_{22}K)^{-1}P_{21}r. \tag{4.5}$$

Substituting (4.5) into (4.4) yields

$$w = \mathrm{LFT_l}\,(P, K)\,r, \tag{4.6}$$

where

$$\mathrm{LFT_l}\,(P, K) := P_{11} + P_{12}K(I - P_{22}K)^{-1}P_{21}. \tag{4.7}$$

In general, the LFT exploits the matrix form to describe linear systems in which the LFT allows the terminator to be placed either below or above (interconnected) plant P. Herein, the subscript l in (4.7) indicates that this is a lower LFT, i.e., terminator K is below plant P.

Similarly, Fig. 4.2 shows the upper LFT form that

$$\begin{bmatrix} y \\ w \end{bmatrix} = \begin{bmatrix} P_{11} & P_{12} \\ P_{21} & P_{22} \end{bmatrix} \begin{bmatrix} u \\ r \end{bmatrix} \tag{4.8}$$

Fig. 4.2 Upper LFT form

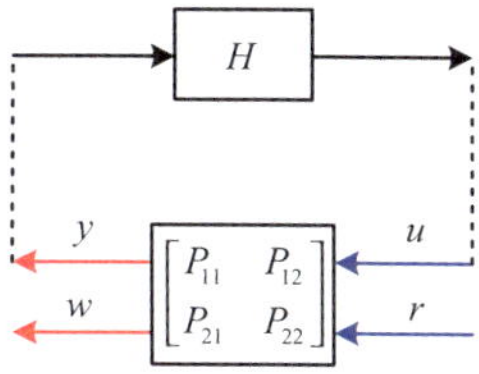

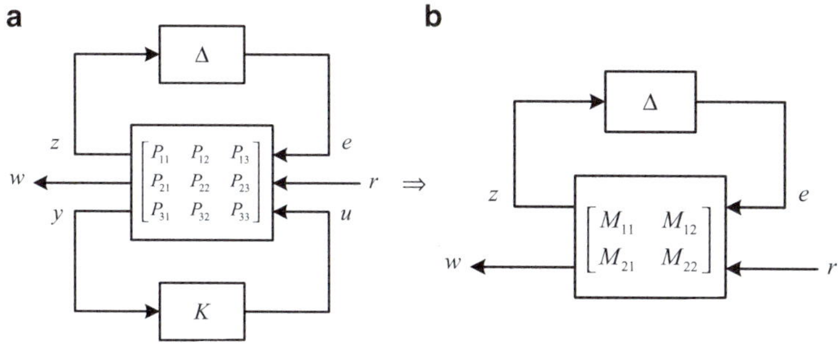

Fig. 4.3 LFT forms: (**a**) (3×3) LFT form and (**b**) (2×2) LFT form

and

$$u = Hy. \tag{4.9}$$

By the same manipulations as (4.4), (4.5) and (4.6), one can then obtain $w = \mathrm{LFT_u}(P,H)r$ where

$$\mathrm{LFT_u}(P, H) := P_{22} + P_{21}H(I - P_{11}H)^{-1}P_{12}. \tag{4.10}$$

Note that in (4.7) and (4.10), the terminators have implicitly been assumed in the above such that $(I - P_{22}K)$ and $(I - P_{11}H)$ are invertible. For $\mathrm{LFT_l}(P,K)$ (or $\mathrm{LFT_u}(P,H)$), if $(I - P_{22}K)$ (or $(I - P_{11}H)$) is invertible, this system is called *well-defined* (or *well-posed*). In practice, the invertibility condition in almost all feedback control systems design should be satisfied for the existence of a controller.

Additionally, Fig. 4.3a shows the full LFT form with the terminators that appear both in the upper and lower presentations. For example, Fig. 4.3a can be employed to describe a feedback control system which suffers from perturbed dynamics Δ. In this case, P consists of, accordingly, 9 (3 × 3) sub-matrices, i.e.,

$$\begin{bmatrix} \begin{pmatrix} z \\ w \end{pmatrix} \\ y \end{bmatrix} = \left[\begin{array}{cc|c} P_{11} & P_{12} & P_{13} \\ P_{21} & P_{22} & P_{23} \\ \hline P_{31} & P_{32} & P_{33} \end{array}\right] \begin{bmatrix} \begin{pmatrix} e \\ r \end{pmatrix} \\ u \end{bmatrix}. \tag{4.11}$$

Closing the feedback loop by $u = Ky$ will yield, by a straightforward manipulation,

$$\begin{bmatrix} z \\ w \end{bmatrix} = \begin{bmatrix} M_{11} & M_{12} \\ M_{21} & M_{22} \end{bmatrix} \begin{bmatrix} e \\ r \end{bmatrix}, \tag{4.12}$$

where

$$\begin{bmatrix} M_{11} & M_{12} \\ M_{21} & M_{22} \end{bmatrix} = LFT_l\left(\left[\begin{array}{cc|c} P_{11} & P_{12} & P_{13} \\ P_{21} & P_{22} & P_{23} \\ \hline P_{31} & P_{32} & P_{33} \end{array}\right], K \right) = \begin{bmatrix} P_{11} & P_{12} \\ P_{21} & P_{22} \end{bmatrix} + \begin{bmatrix} P_{13} \\ P_{23} \end{bmatrix} K\left(I - P_{33}K\right)^{-1} \left[P_{31}\ P_{32}\right]. \tag{4.13}$$

This shows that the full (3×3) LFT form of Fig. 4.3a can be reduced by the closed-loop $u = Ky$ into a (2×2) LFT form depicted in Fig. 4.3b, which includes an uncertain dynamics appealing for robustness consideration [2].

Example 4.1 Consider the upper LFT form of (4.8) and use MATLAB to process the LFT manipulations.

The LFT determination can be carried out in many ways and be formed as an S-function. The following MATLAB code is an example included here for readers' reference.

```
clc;
clear;
disp('LFTl method')
syms P11 P12 P21 P22 K S;
syms p11 p12 p21 p22 k s;
P11=input('z => u (P11) ?');
P12=input('z => y (P12) ?');
P21=input('w => u (P21) ?');
P22=input('w => y (P22) ?');
K=input('Please input K:');
if (eye(size(P22))-P22*K==0);
    disp('The transfer function is singular !');
    disp('Plz enter Ctrl c !');
    pause;
else
Transfer_function_w_to_z=simplify(P11+P12*K*inv(eye
(size(P22))-P22*K)*P21)
end
```

4.2 Application of LFT in State-Space Realizations

It is known that the output of a linear time-invariant (LTI) system is a convolution of the system's impulse response and input functions [4]. Inherently, there is no simple analytic way to compute the convolution, especially for MIMO matrix cases. The state-space description is a useful approach to represent the system's input/output relationship. Consider an nth-order $(m \times r)$ MIMO system and a state-space description given by

$$\begin{aligned} \dot{x}(t) &= Ax(t) + Bu(t), x(t_0) = x_0 \\ y(t) &= Cx(t) + Du(t), \end{aligned} \tag{4.14}$$

where $x(t_0)$ denotes the initial state and $A \in \mathbb{R}^{n \times n}$, $B \in \mathbb{R}^{n \times r}$, $C \in \mathbb{R}^{m \times n}$, $D \in \mathbb{R}^{m \times r}$ are system matrices. The transfer matrix of the input/output relationship for the zero initial state (i.e., $x(t_0) = 0$) can be written as

$$G(s) = \frac{Y(s)}{U(s)}, \quad i.e., \quad Y(s) = G(s)U(s), \tag{4.15}$$

where $G(s)$ is the transfer function (matrix) of the system, $Y(s)$ and $U(s)$ are the Laplace transforms of the output and input functions, respectively. If the state and output equations of (4.14) are known, $G(s)$ can be uniquely computed as

$$G(s) = C(sI - A)^{-1}B + D. \tag{4.16}$$

In this book, the state-space realization of a given system $G(s)$ is denoted as

$$\begin{bmatrix} \dot{x} \\ y \end{bmatrix} = \begin{bmatrix} A & B \\ C & D \end{bmatrix} \begin{bmatrix} x \\ u \end{bmatrix}, \text{ or } G(s) \overset{s}{=} \left[\begin{array}{c|c} A & B \\ \hline C & D \end{array}\right]. \tag{4.17}$$

In the former, the manipulation on the right-hand side of the equation is the usual multiplication between a matrix and a vector. Figure 4.4 below shows the corresponding block diagram of (4.17) where the *integrator* (i.e., $\frac{1}{s}$) is an operator to characterize the relationship of state variables x and $\dot{x}$. As an example, it is shown next that the LFT manipulation can be used to determine the state-space realization. Briefly speaking, through the LFT method, one can obtain the state-space model of a system by cutting off the connections around the *integrator*.

To determine the matrix P in the LFT formulation, the first step is to properly cut off all the internal loops such that no feedback loop is left and the isolated part will be the terminator. It can be seen from the block diagram in Fig. 4.4 that if the integrator is truncated, there are no internal loops. Thus, a two-port matrix for

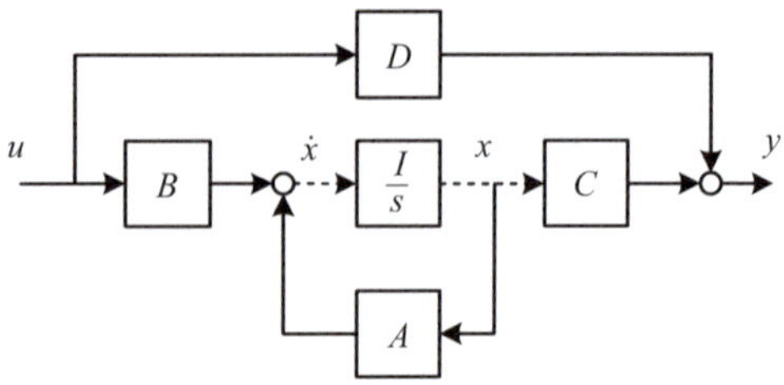

Fig. 4.4 State-space model

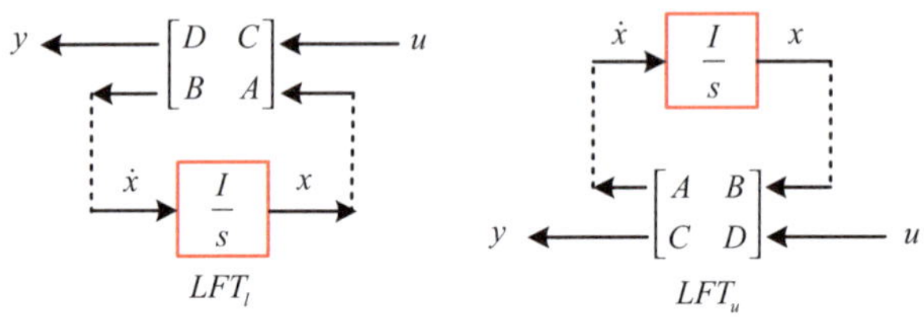

Fig. 4.5 LFT representations

the relationship $\begin{bmatrix} u \\ x \end{bmatrix} \mapsto \begin{bmatrix} y \\ \dot{x} \end{bmatrix}$ can be found directly as $\begin{bmatrix} D & C \\ B & A \end{bmatrix}$ or alternatively $\begin{bmatrix} A & B \\ C & D \end{bmatrix}$ for the relationship $\begin{bmatrix} x \\ u \end{bmatrix} \mapsto \begin{bmatrix} \dot{x} \\ y \end{bmatrix}$. Using $\frac{I}{s}$ as the terminator that encloses $\dot{x} \mapsto x$ will yield the transfer matrix $G(s)$ from $u \mapsto y$. Figure 4.5 shows both lower and upper LFT representations where

$$G(s) = \text{LFT}_\text{l}\left(\begin{bmatrix} D & C \\ B & A \end{bmatrix}, \frac{I}{s}\right) = \text{LFT}_\text{u}\left(\begin{bmatrix} A & B \\ C & D \end{bmatrix}, \frac{I}{s}\right) = D + C(sI - A)^{-1}B. \tag{4.18}$$

A transfer function matrix $G(s)$ is said to be realizable if there exists a finite-dimensional state-space realization. A transfer function matrix $G(s)$ is uniquely determined from the input/output relationship for a given initial state; however, there exist infinitely many state-space realizations for a given $G(s)$. As an introduction dealing with the realization problem, one can first consider the SISO, third-order LTI system below

$$Y(s) = G(s)U(s) = \left(\frac{b_2s^2 + b_1s + b_0}{s^3 + a_2s^2 + a_1s + a_0} + d_0\right)U(s). \tag{4.19}$$

As can be seen in (4.19), the transfer function $G(s)$ is purposely formed as the summation of a strict proper function and a constant $d_0(=G(\infty))$. Here, (4.19) can be rewritten as

$$Y(s) = \left(\frac{\frac{1}{s}b_2 + \frac{1}{s^2}b_1 + \frac{1}{s^3}b_0}{1 + \frac{1}{s}a_2 + \frac{1}{s^2}a_1 + \frac{1}{s^3}a_0} + d_0 \right) U(s) = \left(\frac{n\left(\frac{1}{s}\right)}{d\left(\frac{1}{s}\right)} + d_0 \right) U(s) \quad (4.20)$$

or

$$Y(s) = \left[\frac{1}{d\left(\frac{1}{s}\right)} \left(\frac{1}{s}b_2 + \frac{1}{s^2}b_1 + \frac{1}{s^3}b_0 \right) + d_0 \right] U(s). \quad (4.21)$$

Furthermore, defining

$$X(s) = \frac{1}{d\left(\frac{1}{s}\right)} U(s) = \frac{1}{1 + \frac{1}{s}a_2 + \frac{1}{s^2}a_1 + \frac{1}{s^3}a_0} U(s) \quad (4.22)$$

yields

$$\left(1 + \frac{1}{s}a_2 + \frac{1}{s^2}a_1 + \frac{1}{s^3}a_0 \right) X(s) = U(s) \quad (4.23)$$

and

$$X(s) = U(s) - \frac{1}{s}a_2 X(s) - \frac{1}{s^2}a_1 X(s) - \frac{1}{s^3}a_0 X(s). \quad (4.24)$$

Then,

$$Y(s) = n\left(\frac{1}{s}\right) X(s) + d_0 U(s) = \left(\frac{1}{s}b_2 + \frac{1}{s^2}b_1 + \frac{1}{s^3}b_0 \right) X(s) + d_0 U(s). \quad (4.25)$$

The corresponding system block diagram is depicted in Fig. 4.6, where only the negative sign, if exists, is noted as "–" in all of the summing points.

Now, the LFT approach is adopted to find the state-space representation, and the integrators can be cut off as illustrated in Fig. 4.7. If the state variables are defined as shown in Fig. 4.7, then the realization matrix $P = \begin{bmatrix} A & B \\ C & D \end{bmatrix}$ can be found from $\begin{bmatrix} x \\ u \end{bmatrix} \mapsto \begin{bmatrix} \dot{x} \\ y \end{bmatrix}$ directly as

$$\begin{bmatrix} \dot{x}_1 \\ \dot{x}_2 \\ \dot{x}_3 \\ \hline y \end{bmatrix} = \left[\begin{array}{ccc|c} 0 & 1 & 0 & 0 \\ 0 & 0 & 1 & 0 \\ -a_0 & -a_1 & -a_2 & 1 \\ \hline b_0 & b_1 & b_2 & d_0 \end{array}\right] \begin{bmatrix} x_1 \\ x_2 \\ x_3 \\ \hline u \end{bmatrix} = \begin{bmatrix} A & B \\ C & D \end{bmatrix} \begin{bmatrix} x_1 \\ x_2 \\ x_3 \\ \hline u \end{bmatrix}, \quad (4.26)$$

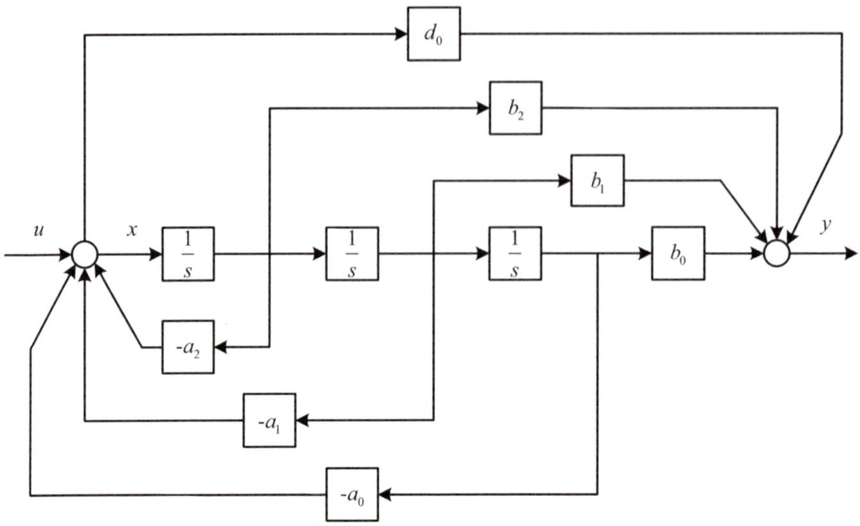

Fig. 4.6 Controller canonical form

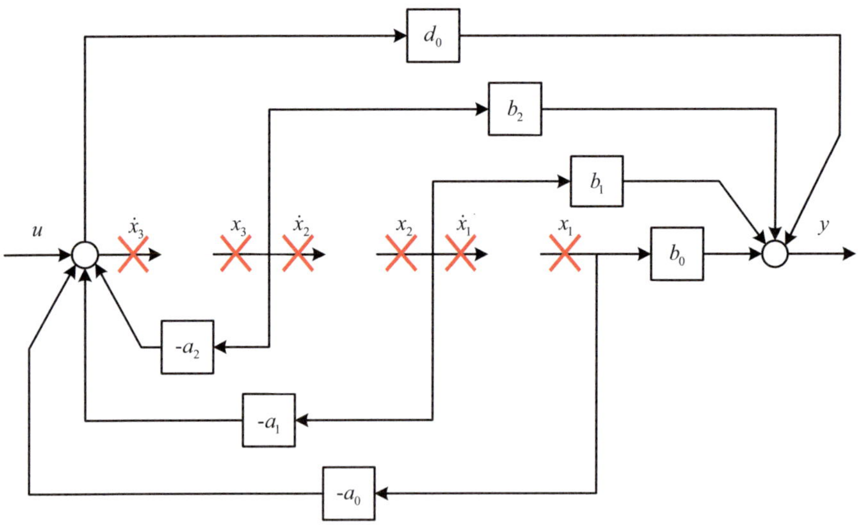

Fig. 4.7 Cut off the integrators

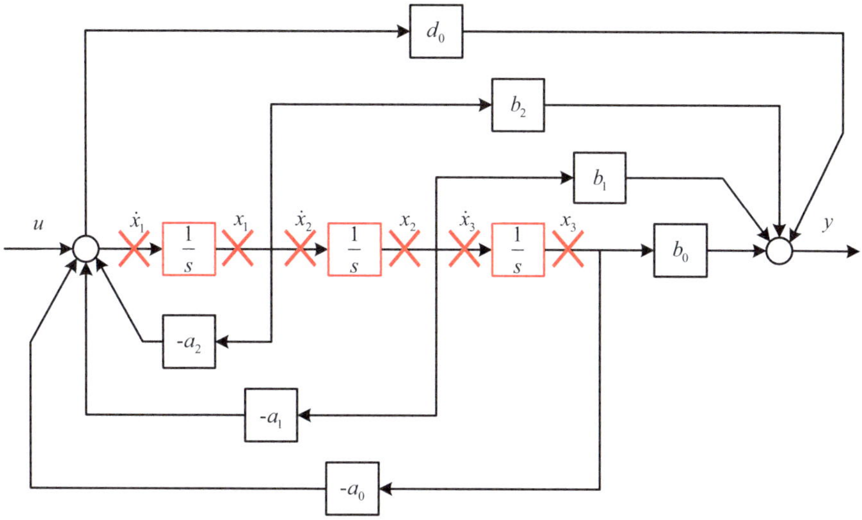

Fig. 4.8 Define state variables

i.e.,

$$\begin{bmatrix} \dot{x}_1 \\ \dot{x}_2 \\ \dot{x}_3 \end{bmatrix} = \begin{bmatrix} 0 & 1 & 0 \\ 0 & 0 & 1 \\ -a_0 & -a_1 & -a_2 \end{bmatrix} \begin{bmatrix} x_1 \\ x_2 \\ x_3 \end{bmatrix} + \begin{bmatrix} 0 \\ 0 \\ 1 \end{bmatrix} u,$$

$$y = \begin{bmatrix} b_0 & b_1 & b_2 \end{bmatrix} \begin{bmatrix} x_1 \\ x_2 \\ x_3 \end{bmatrix} + d_0 u. \tag{4.27}$$

Obviously, if the state variables are defined in a different sequence as represented in Fig. 4.8, another state-space realization is derived as

$$\begin{bmatrix} \dot{x}_1 \\ \dot{x}_2 \\ \dot{x}_3 \\ \hline y \end{bmatrix} = \left[\begin{array}{ccc:c} -a_2 & -a_1 & -a_0 & 1 \\ 1 & 0 & 0 & 0 \\ 0 & 1 & 0 & 0 \\ \hdashline b_2 & b_1 & b_0 & d_0 \end{array}\right] \begin{bmatrix} x_1 \\ x_2 \\ x_3 \\ \hline u \end{bmatrix} = \begin{bmatrix} \hat{A} & \hat{B} \\ \hat{C} & \hat{D} \end{bmatrix} \begin{bmatrix} x_1 \\ x_2 \\ x_3 \\ \hline u \end{bmatrix}. \tag{4.28}$$

Apparently, (4.26) and (4.28) indicate that these two different state-space realizations result from the same third-order transfer function. In fact, for two different expressions with the same dimension, there always exists a similarity transformation that links them together. For this example,

$$T = \begin{bmatrix} 0 & 0 & 1 \\ 0 & 1 & 0 \\ 1 & 0 & 0 \end{bmatrix}, \tag{4.29}$$

which gives

$$\begin{bmatrix} \widehat{A} & \widehat{B} \\ \widehat{C} & \widehat{D} \end{bmatrix} = \begin{bmatrix} T^{-1} & 0 \\ 0 & I \end{bmatrix} \begin{bmatrix} A & B \\ C & D \end{bmatrix} \begin{bmatrix} T & 0 \\ 0 & I \end{bmatrix}. \tag{4.30}$$

In summary, for a general nth-order system of

$$G(s) = \frac{b_{n-1}s^{n-1} + \cdots + b_1 s + b_0}{s^n + a_{n-1}s^{n-1} + \cdots + a_1 s + a_0} + d_0, \tag{4.31}$$

the state-space form given by

$$\begin{bmatrix} \dot{x}_1 \\ \dot{x}_2 \\ \vdots \\ \vdots \\ \dot{x}_n \\ \hline y \end{bmatrix} = \left[\begin{array}{ccccc|c} 0 & 1 & 0 & 0 & 0 & 0 \\ 0 & 0 & 1 & 0 & 0 & 0 \\ \vdots & \vdots & \ddots & \vdots & \vdots & \vdots \\ 0 & \cdots & \cdots & \cdots & 1 & \vdots \\ -a_0 & -a_1 & \cdots & \cdots & -a_{n-1} & 1 \\ \hline b_0 & b_1 & \cdots & \cdots & b_{n-1} & d_0 \end{array}\right] \begin{bmatrix} x_1 \\ x_2 \\ \vdots \\ \vdots \\ x_n \\ \hline u \end{bmatrix} \Rightarrow \begin{bmatrix} \dot{x} \\ u \end{bmatrix} = \begin{bmatrix} A & B \\ C & D \end{bmatrix} \begin{bmatrix} x \\ u \end{bmatrix} \tag{4.32}$$

is called the controllable canonical form [1]. The observable canonical realization is dual in the form of (4.28).

4.3 Examples of Determining LFT Matrices

In order to enhance the understanding of LFT manipulations, this section provides several examples for exercise. These exercises are aimed at assisting readers in developing knowledge and skills in selecting the cutoff points in LFT formulations.

Example 4.2 The following shows three different state-space expressions which are from the same transfer function $G(s) = \frac{3s+4}{s^2+3s+2}$.

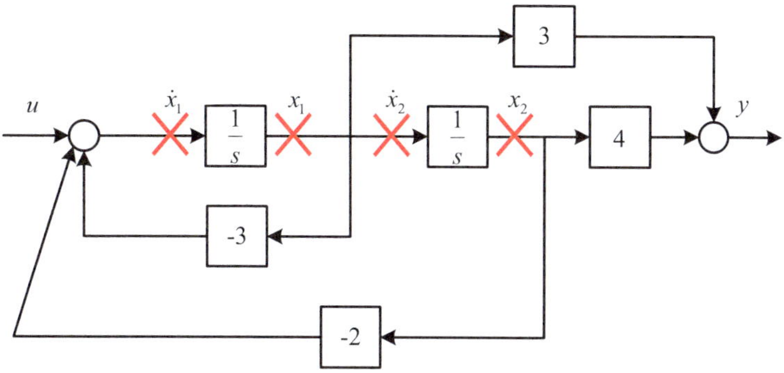

Fig. 4.9 Realization in a canonical form

4.3.1 Canonical Form

The transfer function of the input/output relationship can be written as

$$G(s) = \frac{3s + 4}{s^2 + 3s + 2} = \frac{Y(s)}{U(s)} \tag{4.33}$$

or equivalently,

$$Y(s) = \frac{3\frac{1}{s^1} + 4\frac{1}{s^2}}{1 + 3\frac{1}{s^1} + 2\frac{1}{s^2}} U(s). \tag{4.34}$$

Applying the procedures developed above to obtain (4.24) and (4.25) will yield the block diagram of Fig. 4.9. Then, the integrators can be separated to obtain a two-port matrix, of which the state-space realization is

$$\begin{bmatrix} \begin{pmatrix} \dot{x}_1 \\ \dot{x}_2 \end{pmatrix} \\ y \end{bmatrix} = \left[\begin{array}{cc|c} -3 & -2 & 1 \\ 1 & 0 & 0 \\ \hline 3 & 4 & 0 \end{array}\right] \begin{bmatrix} \begin{pmatrix} x_1 \\ x_2 \end{pmatrix} \\ u \end{bmatrix} \Rightarrow \begin{bmatrix} \dot{x} \\ y \end{bmatrix} = \left[\begin{array}{c|c} A_1 & B_1 \\ \hline C_1 & D_1 \end{array}\right] \begin{bmatrix} x \\ u \end{bmatrix}, \tag{4.35}$$

which can be verified by

$$G(s) = LFT_u\left(\left[\begin{array}{cc|c} -3 & -2 & 1 \\ 1 & 0 & 0 \\ \hline 3 & 4 & 0 \end{array}\right], \frac{I}{s} \right) = \frac{3s+4}{s^2+3s+2}. \tag{4.36}$$

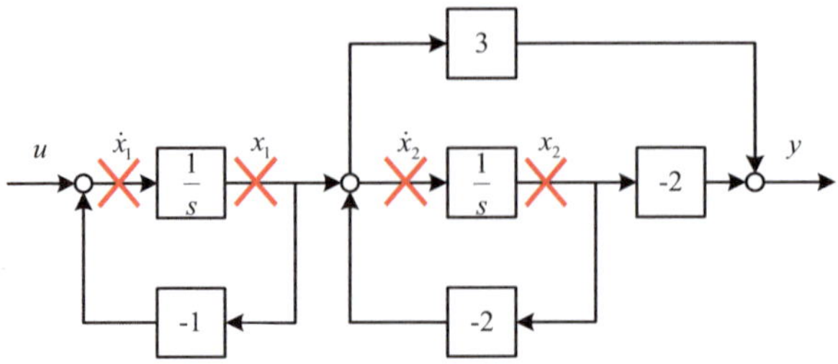

Fig. 4.10 Realization in a cascade form

4.3.2 *Cascade Form*

Let

$$G(s) = \frac{3s+4}{s^2+3s+2} = \left(\frac{1}{s+1}\right)\left(3 - \frac{2}{s+2}\right) = \left(\frac{\frac{1}{s}}{1+\frac{1}{s}}\right)\left(3 + \frac{-2\frac{1}{s}}{1+2\frac{1}{s}}\right) \tag{4.37}$$

which will result in the block diagram of Fig. 4.10. The corresponding two-port matrix becomes

$$\begin{bmatrix} \begin{pmatrix} \dot{x}_1 \\ \dot{x}_2 \end{pmatrix} \\ y \end{bmatrix} = \left[\begin{array}{cc|c} -1 & 0 & 1 \\ 1 & -2 & 0 \\ \hline 3 & -2 & 0 \end{array}\right] \begin{bmatrix} \begin{pmatrix} \dot{x}_1 \\ \dot{x}_2 \end{pmatrix} \\ u \end{bmatrix} \Rightarrow \begin{bmatrix} \dot{x} \\ y \end{bmatrix} = \left[\begin{array}{c|c} A_2 & B_2 \\ \hline C_2 & D_2 \end{array}\right] \begin{bmatrix} x \\ u \end{bmatrix}, \tag{4.38}$$

where the A-matrix of this expression is low triangular. It can be seen that

$$G(s) = LFT_u\left(\left[\begin{array}{cc|c} -1 & 0 & 1 \\ 1 & -2 & 0 \\ \hline 3 & -2 & 0 \end{array}\right], \frac{I}{s}\right) = \frac{3s+4}{s^2+3s+2}. \tag{4.39}$$

4.3.3 *Parallel Form*

From

$$G(s) = \frac{3s+4}{s^2+3s+2} = \frac{1}{s+1} + \frac{2}{s+2} = \frac{\frac{1}{s}}{1+\frac{1}{s}} + \frac{\frac{2}{s}}{1+\frac{2}{s}}, \tag{4.40}$$

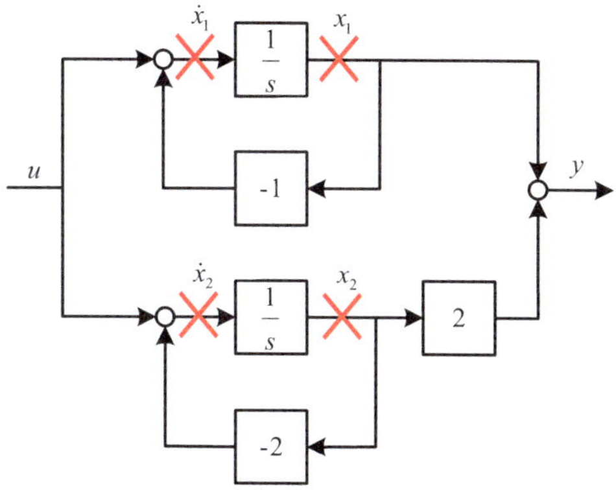

Fig. 4.11 Realization in a parallel form

one has the block diagram of Fig. 4.11. Then the corresponding two-port matrix becomes

$$\begin{bmatrix} \begin{pmatrix} \dot{x}_1 \\ \dot{x}_2 \end{pmatrix} \\ y \end{bmatrix} = \left[\begin{array}{cc|c} -1 & 0 & 1 \\ 0 & -2 & 1 \\ \hline 1 & 2 & 0 \end{array}\right] \begin{bmatrix} \begin{pmatrix} x_1 \\ x_2 \end{pmatrix} \\ u \end{bmatrix} \Rightarrow \begin{bmatrix} \dot{x} \\ y \end{bmatrix} = \left[\begin{array}{c|c} A_3 & B_3 \\ \hline C_3 & D_3 \end{array}\right] \begin{bmatrix} x \\ u \end{bmatrix}, \tag{4.41}$$

where the A-matrix of this realization is diagonal. It can be found that

$$G(s) = LFT_u\left(\left[\begin{array}{cc|c} -1 & 0 & 1 \\ 0 & -2 & 1 \\ \hline 1 & 2 & 0 \end{array}\right], \frac{I}{s} \right) = \frac{3s+4}{s^2+3s+2}. \tag{4.42}$$

From this example, it can be seen that the LFT state-space representation of a given transfer is not unique. However, from (4.35), (4.38), and (4.41),

$$C_1 A_1^i B_1 = C_2 A_2^i B_2 = C_3 A_3^i B_3, \quad \forall i = 0, 1, \ldots, n. \tag{4.43}$$

The subscripts 1, 2, and 3 denote the (A, B, C) matrices from (4.35), (4.38), and (4.41), respectively. This implies that their input/output mappings are the same. Different state-space realizations are theoretically equivalent in that they lead to the same transfer function. However, in practice, how to choose an appropriate expression is still important due to the numerical stability and sensitivity as well as manipulation convenience, which will always appear in physical implementations.

The following summarizes the use of LFT.

1. The LFT structure represents the interconnection of a two-port, open-loop matrix and a feedback terminator.

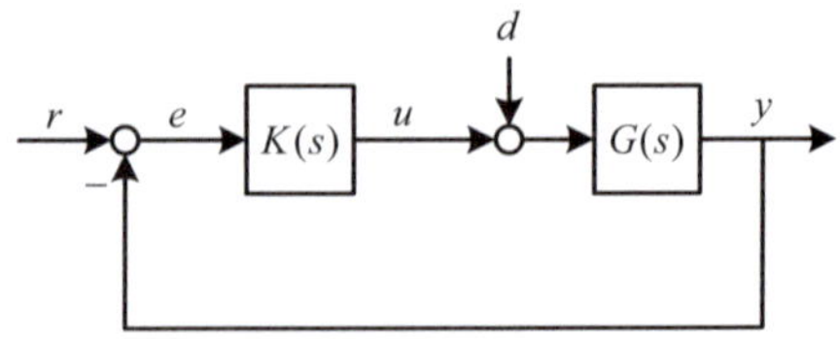

Fig. 4.12 Feedback control case

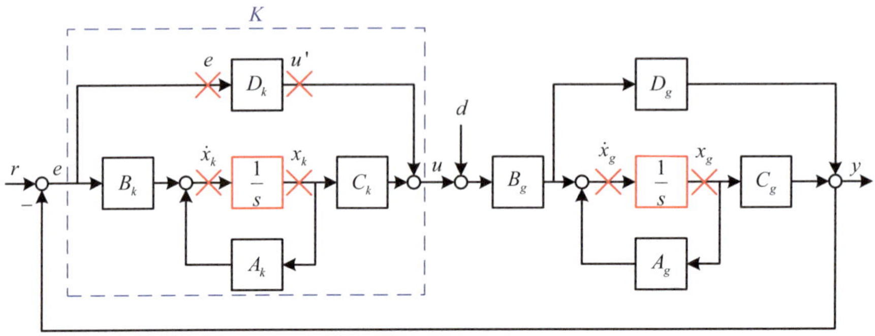

Fig. 4.13 State-space realization of a unity feedback system

2. In determining the LFT form of a feedback connection system, one should break all the inner feedback loops to obtain the two-port, open-loop matrix.
3. Different cutoffs will lead to a different terminator and different two-port matrices, but the input/output mappings are the same.

Recall the unity feedback system, depicted again in Fig. 4.12 below, where $G(s)$ is a controlled plant and $K(s)$ is the controller. The closed-loop transfer (2×2) matrix of $\begin{bmatrix} d \\ r \end{bmatrix} \mapsto \begin{bmatrix} e \\ y \end{bmatrix}$ can be found as

$$T = \begin{bmatrix} KG(I + KG)^{-1} & (I + GK)^{-1} \\ G(I + KG)^{-1} & GK(I + GK)^{-1} \end{bmatrix}. \tag{4.44}$$

Let $G(s) \overset{s}{=} \left[\begin{array}{c|c} A_g & B_g \\ \hline C_g & D_g \end{array}\right]$ and $K(s) \overset{s}{=} \left[\begin{array}{c|c} A_k & B_k \\ \hline C_k & D_k \end{array}\right]$ denote the state-space realizations. In order to obtain the state-space expression of (4.44) in the full LFT formulation, as depicted in Fig. 4.3a, one may cut off all of the integrators in Fig. 4.13 and also eliminate D_k (or D_g) such that there are no inner loops. Then, the full LFT matrix P can be found directly from the assigned state variables as

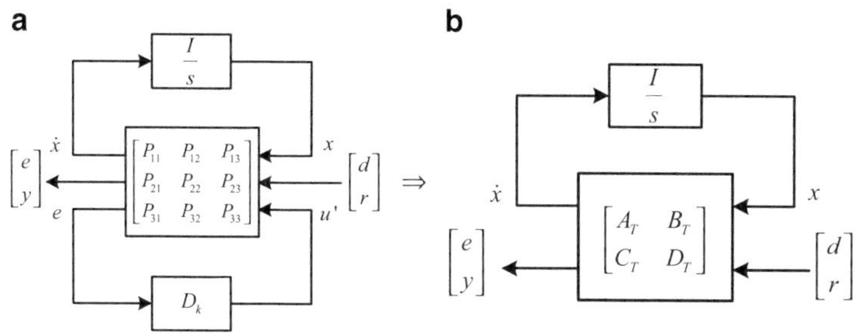

Fig. 4.14 LFT forms: (**a**) (3×3) LFT form and (**b**) (2×2) LFT form

$$\begin{bmatrix} \dot{x}_k \\ \dot{x}_g \\ \hline e \\ y \\ \hline e \end{bmatrix} = P \begin{bmatrix} x_k \\ x_g \\ \hline d \\ r \\ \hline u' \end{bmatrix} = \left[\begin{array}{cc|cc|c} A_k - B_k D_g & -B_k C_g & -B_k D_g & B_k & -B_k D_g \\ B_g C_k & A_g & B_g & 0 & B_g \\ \hline -D_g C_k & -C_g & -D_g & 0 & -D_g \\ D_g C_k & C_g & D_g & I & D_g \\ \hline -D_g C_k & -C_g & -D_g & 0 & -D_g \end{array}\right] \begin{bmatrix} x_k \\ x_g \\ \hline d \\ r \\ \hline u' \end{bmatrix} \tag{4.45}$$

Suppose that $I + D_g D_k \neq 0$, which reflects the well-posedness condition of a feedback system [4, 11]. Let the state-space realization of the closed-loop transfer matrix of $\begin{bmatrix} d \\ r \end{bmatrix} \mapsto \begin{bmatrix} e \\ y \end{bmatrix}$ in (4.44) be denoted by

$$\begin{bmatrix} \dot{x}_k \\ \dot{x}_g \\ \hline e \\ y \end{bmatrix} = \begin{bmatrix} A_T & B_T \\ C_T & D_T \end{bmatrix} \begin{bmatrix} x_k \\ x_g \\ \hline d \\ r \end{bmatrix} \tag{4.46}$$

where

$$\begin{bmatrix} A_T & B_T \\ C_T & D_T \end{bmatrix} = LFT_l \left(\left[\begin{array}{cc|c} P_{11} & P_{12} & P_{13} \\ P_{21} & P_{22} & P_{23} \\ \hline P_{31} & P_{32} & P_{33} \end{array}\right], D_k \right). \tag{4.47}$$

Thus, one has $T(s) = \mathrm{LFT_l}(P, D_k) = \mathrm{LFT_u}\left(\begin{bmatrix} A_T & B_T \\ C_T & D_T \end{bmatrix}, \frac{I}{s} \right)$. Obviously, the feedback system is always well-posed for the case of either $D_k = 0$ or $D_g = 0$, i.e., the strictly proper case. Then, for $D_k = 0$, one can find from (4.45) that the state-space realization of the feedback system is given by Fig. 4.14b

$$T(s) \stackrel{s}{=} \left[\begin{array}{c|c} A_T & B_T \\ \hline C_T & D_T \end{array}\right] = \left[\begin{array}{cc|cc} A_k - B_k D_g & -B_k C_g & -B_k D_g & B_k \\ B_g C_k & A_g & B_g & 0 \\ \hline -D_g C_k & -C_g & -D_g & 0 \\ D_g C_k & C_g & D_g & I \end{array}\right]. \tag{4.48}$$

4.4 Relationship Between Mason's Gain Formulae and LFT

Mason's approach [7, 8] has been a useful tool to determine the transfer function for a given system for many years. A substantial number of practicing control engineers are familiar with it. However, this approach involves a set of rules that can be difficult for a new learner to remember.

This section reveals a relationship between the classic Mason's gain formulae and the LFT approach to characterize the transfer function from input to output. It is well-known that the Mason's gain formulae are a set of rules for determining the transfer function of a single-input and single-output (SISO) system. From the discussions in previous sections, one knows that the LFT formulation can generalize the determination of transfer functions in MIMO cases. In this section, several examples are given to show the differences and similarities between these two approaches. Through the following examples, one can observe that these two approaches reveal their individual benefits in determining the transfer function based on differently structured control systems.

The Mason's gain formulae [6] are as follows:

1. $M = \dfrac{\sum_j M_j \Delta_j}{\Delta}$
2. M = transfer function or gain of the system
3. M_j = gain of jth forward path
4. j = an integer representing the forward paths in the system
5. $\Delta = 1 - \sum$ (all different loop gains)

   $+\sum$(sum of the gain products of all combinations of 2 nontouching loop gains)

   $-\sum$(sum of the gain products of all combinations of 3 nontouching loop gains)

   $+ \cdots$
6. Δ_j = value of Δ for that part of the block diagram that does not touch the jth forward path. (4.49)

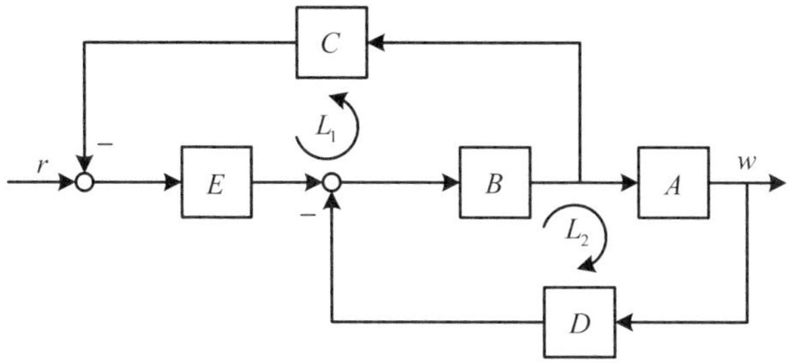

Fig. 4.15 System block diagram

Example 4.3 Determine the transfer function of the system block diagram depicted in Fig. 4.15 from r to w using Mason's gain formulae and the LFT formulation, respectively.

Firstly, the Mason's gain formulae (i.e., the following steps (1) to (5)) are applied.

Mason's Approach

Step 1: Find all of the feedback loops and their gains.
In this example, there are two feedback loops: $L_1 = -CBE$ and $L_2 = -DAB$.

Step 2: Find the forward paths and their gains.
A forward path is the path from r to w that does not cross the same point more than once. In this example, there is only one forward path, i.e., $j = 1$ and $M_1 = ABE$.

Step 3: Find Δ.
In this example, there are no non-touching loop pairs so that

$$\Delta = 1 - \sum \text{loop gains} = 1 + CBE + DAB. \tag{4.50}$$

Step 4: Find Δ_j.
From the block diagram of Fig. 4.15, one can see that $L_1 = -CBE$ and $L_2 = -DAB$ touch the forward path $M_1 = ABE$ so that $\Delta_1 = 1$.

Step 5: Final solution.
Using the Mason's gain formulae (4.49), the transfer function from r to w is given by

$$\frac{w}{r} = M = \frac{ABE}{1 + ECB + DAB}. \tag{4.51}$$

LFT Approach

To find the transfer function of a feedback control system by using the LFT approach, the truncated termination part can always be chosen to be an identity. Note that the cutting place should break all of the internal closed loops. In this example, the minimum cutting point is 1. Figure 4.16 shows an alternative cutting selection. As can be seen in this figure, the number of cutting points is 2. Figure 4.17

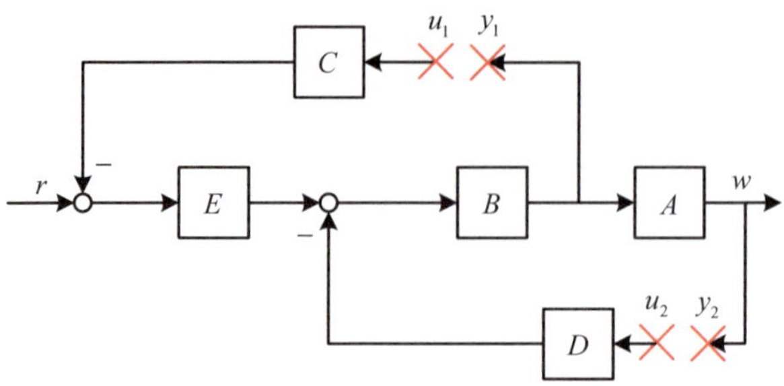

Fig. 4.16 System block diagram with two cutting points

Fig. 4.17 LFT representation

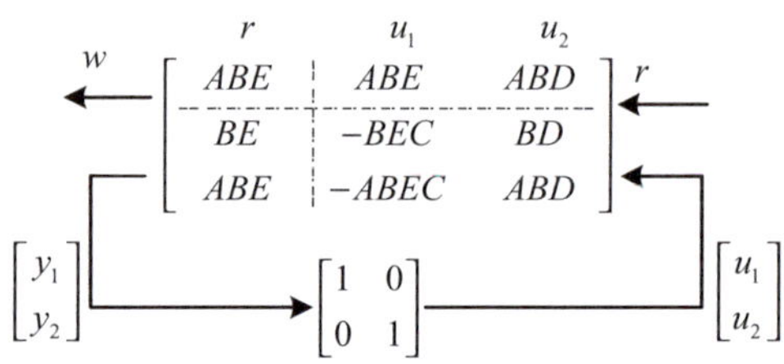

illustrates its corresponding LFT formulation. The transfer function from r to w is given by

$$\begin{aligned} & LFT_l\left(\left[\begin{array}{c|cc} ABE & ABE & ABD \\ \hline BE & -BEC & BD \\ ABE & -ABEC & ABD \end{array}\right], \begin{bmatrix} 1 & 0 \\ 0 & 1 \end{bmatrix}\right) \\ & = ABE + \begin{bmatrix} ABE & ABD \end{bmatrix} \begin{bmatrix} 1 & 0 \\ 0 & 1 \end{bmatrix} \left(\begin{bmatrix} 1 & 0 \\ 0 & 1 \end{bmatrix} - \begin{bmatrix} -BEC & BD \\ -ABEC & ABD \end{bmatrix} \begin{bmatrix} 1 & 0 \\ 0 & 1 \end{bmatrix} \right)^{-1} \begin{bmatrix} BE \\ ABE \end{bmatrix} \\ & = \frac{ABE}{1+(DA+EC)B}. \end{aligned} \tag{4.52}$$

Now, consider the truncation as depicted in Fig. 4.18, in which there is only one break point. Clearly, one has the terminator $K = 1$, and the two-port matrix is reduced to

$$\begin{bmatrix} w \\ y \end{bmatrix} = \left[\begin{array}{c|c} 0 & AB \\ \hline E & -(DA+EC)B \end{array}\right] \begin{bmatrix} r \\ u \end{bmatrix} = \begin{bmatrix} P_{11} & P_{12} \\ P_{21} & P_{22} \end{bmatrix} \begin{bmatrix} r \\ u \end{bmatrix} = P \begin{bmatrix} r \\ u \end{bmatrix}. \tag{4.53}$$

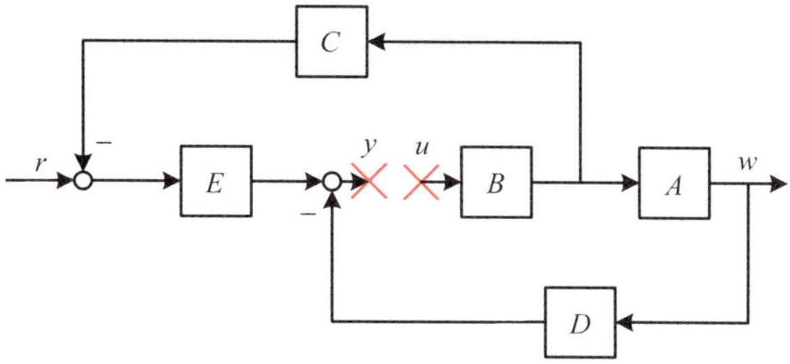

Fig. 4.18 System block diagram with one cutting point

Fig. 4.19 Reduced LFT form

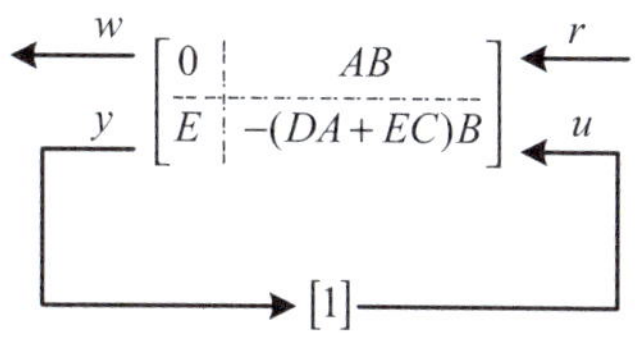

Figure 4.19 illustrates its corresponding LFT formulation. By the definition of $\text{LFT}_l(P,K)$, the transfer function from r to w here is illustrated by

$$LFT_l\left(\left[\begin{array}{c|c} 0 & AB \\ \hline E & -(DA+EC)B \end{array}\right],1\right) = 0 + AB\left[1+(DA+EC)B\right]^{-1}E$$
$$= \frac{ABE}{1+(DA+EC)B}. \tag{4.54}$$

Apparently, the results shown in Eqs. (4.51), (4.52), and (4.54) are the same. Here, one can observe that $(I - P_{22}K) = 1 + (DA + EC)B$ in the above example is equal to $\Delta = 1 + (DA + EC)B$ in (4.50). The *well-posedness* condition of the LFT formulation is thus related to requirement of nonzero Δ in Mason's gain formulae.

Example 4.4 Consider the following control block diagram, as shown in Fig. 4.20, and determine the transfer function from r to w by using Mason's gain formulae and LFT, respectively.

Mason's Approach

Step 1: Find all the loops and their gains.
Individual loops:

$$L_1 = -H_1G_2, \quad L_2 = -H_2G_5, \quad L_3 = -G_5G_3G_2G_1, \quad L_4 = -G_5G_4G_2G_1. \tag{4.55}$$

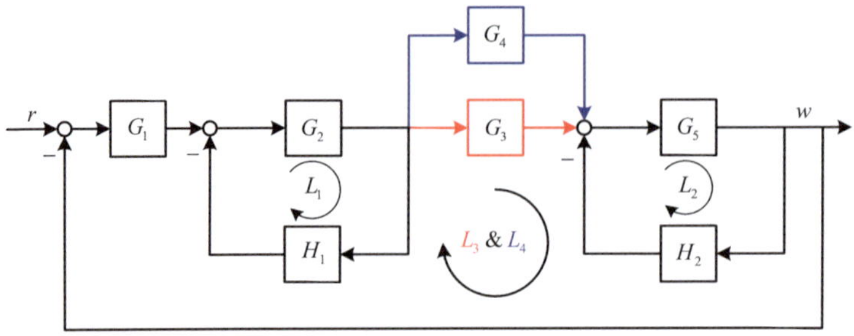

Fig. 4.20 System block diagram

A pair of non-touching loops:

$$\text{Loop 1–Loop 2} \Rightarrow H_1G_2H_2G_5. \tag{4.56}$$

Step 2: Find the forward paths and their gains.
In this example, there are two forward paths, so one has $j = 1$ and 2 as

$$M_1 = G_5G_3G_2G_1 \quad \text{and} \quad M_2 = G_5G_4G_2G_1. \tag{4.57}$$

Step 3: Find Δ.

$$\begin{aligned}\Delta &= 1 - \sum \text{loop gains} + \sum \text{nontouching loop gains taken two at a time} \\ &= 1 - [(-H_1G_2) + (-H_2G_5) + (-G_5G_3G_2G_1) + (-G_5G_4G_2G_1)] \\ &\quad + H_1G_2H_2G_5.\end{aligned} \tag{4.58}$$

Step 4: Find Δ_j.

$$\Delta_1 = 1 \quad \text{and} \quad \Delta_2 = 1. \tag{4.59}$$

Step 5: Finalsolution.
By (4.49),

$$M = \frac{G_5(G_3 + G_4)G_2G_1}{1 + H_1G_2 + H_2G_5 + G_5(G_3 + G_4)G_2G_1 + H_1G_2H_2G_5}. \tag{4.60}$$

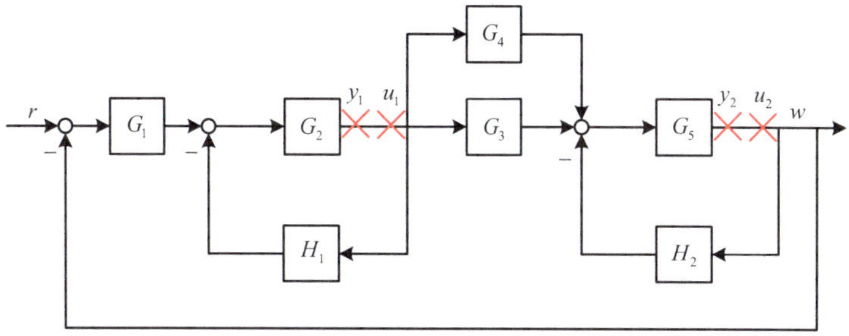

Fig. 4.21 System block diagram with chosen cutting points

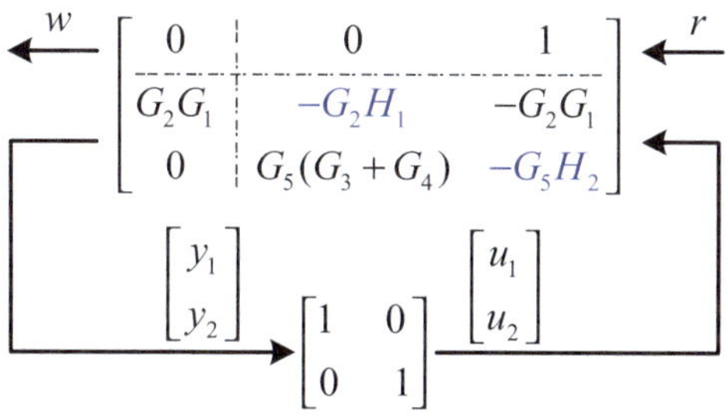

Fig. 4.22 LFT formulation

LFT Approach

Breaking the system as depicted in Fig. 4.21, its corresponding LFT form is as shown in Fig. 4.22 where the two-port matrix is given by

$$P = \left[\begin{array}{c|cc} 0 & 0 & 1 \\ \hline G_2G_1 & -G_2H_1 & -G_2G_1 \\ 0 & G_5(G_3+G_4) & -G_5H_2 \end{array}\right]. \tag{4.61}$$

By definition,

$$\begin{aligned} \mathrm{LFT_l}\,(P,K) &= \begin{bmatrix} 0 & 1 \end{bmatrix} \begin{bmatrix} 1+G_2H_1 & G_2G_1 \\ -G_5\,(G_3+G_4) & 1+G_5H_2 \end{bmatrix}^{-1} \begin{bmatrix} G_2G_1 \\ 0 \end{bmatrix} \\ &= \frac{G_1G_2\,(G_3+G_4)\,G_5}{1+H_1G_2+H_2G_5+G_5\,(G_3+G_4)\,G_2G_1+H_1G_2H_2G_5}. \end{aligned} \tag{4.62}$$

Fig. 4.23 RC circuit

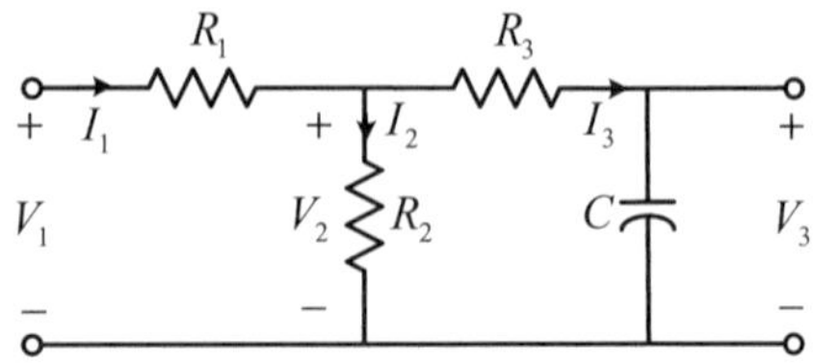

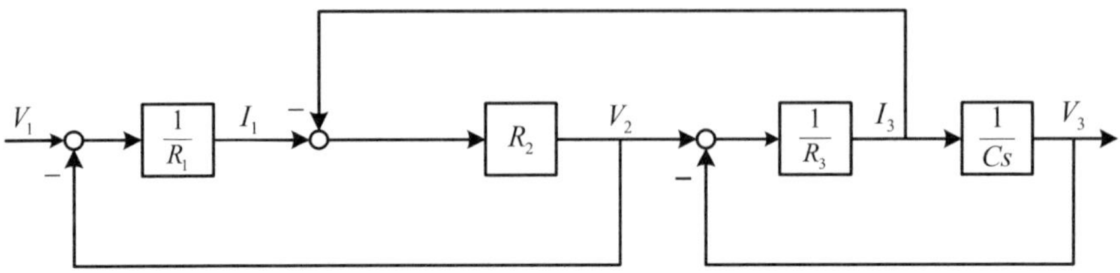

Fig. 4.24 Block diagram of RC circuit

Consequently, the LFT yields the same results as Mason's gain formula. It can be seen from (4.61) that P_{22} is 2×2 with non-touching loops in the main diagonal elements, additionally, $\det(I - P_{22}) = \Delta$, where

$$\Delta = 1 + H_1G_2 + H_2G_5 + G_5(G_3 + G_4)G_2G_1 + H_1G_2H_2G_5. \tag{4.63}$$

In the following, engineering circuit problem is utilized for advanced comprehension. Readers can compare the differences between these three analysis methodologies shown below.

Example 4.5 Determine the transfer function from V_1 to V_3 of the circuit in Fig. 4.23 by using the LFT formulation.

From the given circuit, one can find the gain relationship between V_1 and V_3, denoted by $T(s)$, using Kirchhoff's law [3], as

$$\frac{V_3}{V_1} = T(s) = \frac{\left[R_2 \,||\, \left(R_3 + \frac{1}{Cs}\right)\right] \frac{\frac{1}{Cs}}{R_3 + \frac{1}{Cs}}}{R_1 + R_2 \,||\, \left(R_3 + \frac{1}{Cs}\right)}, \tag{4.64}$$

where $R_1 \,||\, R_2 := \frac{R_1R_2}{R_2+R_1}$ denotes the equivalent impedance in a parallel circuit. In the special case that $R_1 = R_2 = R_3 = R$, then $T(s) = \frac{1}{3RCs+2}$.

From the viewpoint of feedback control, one can draw a block diagram as shown in Fig. 4.24 to represent this circuit. By properly choosing the cutting points to break the two inner feedback loops, as shown in Fig. 4.25, the transfer function from V_1 to V_3 can be found from an LFT formulation, where matrix P in Fig. 4.26 can be determined as

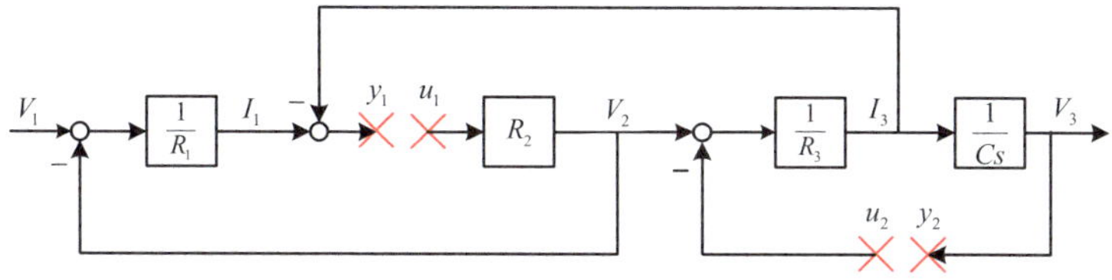

Fig. 4.25 Chosen cutting points

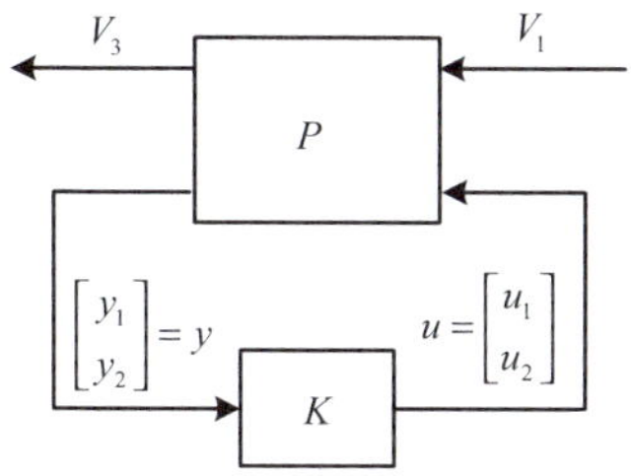

Fig. 4.26 LFT formulation of RC circuit

$$\begin{bmatrix} V_3 \\ \begin{pmatrix} y_1 \\ y_2 \end{pmatrix} \end{bmatrix} = P \begin{bmatrix} V_1 \\ \begin{pmatrix} u_1 \\ u_2 \end{pmatrix} \end{bmatrix} = \left[\begin{array}{c|c} P_{11} & P_{12} \\ \hline P_{21} & P_{22} \end{array}\right] \begin{bmatrix} V_1 \\ \begin{pmatrix} u_1 \\ u_2 \end{pmatrix} \end{bmatrix} = \left[\begin{array}{c|cc} 0 & \dfrac{R_2}{R_3Cs} & \dfrac{-1}{R_3Cs} \\ \hline \dfrac{1}{R_1} & -\dfrac{R_2}{R_3} - \dfrac{R_2}{R_1} & \dfrac{1}{R_3} \\ 0 & \dfrac{R_2}{R_3Cs} & \dfrac{-1}{R_3Cs} \end{array}\right] \begin{bmatrix} V_1 \\ \begin{pmatrix} u_1 \\ u_2 \end{pmatrix} \end{bmatrix}. \quad (4.65)$$

For the case $R_1 = R_2 = R_3 = R$, the implementing termination $K = \begin{bmatrix} 1 & 0 \\ 0 & 1 \end{bmatrix}$ yields

$$\begin{aligned} \frac{V_3}{V_1} &= \mathrm{LFT}_l\,(P, K) = P_{11} + P_{12}K\,(I - P_{22}K)^{-1} P_{21} \\ &= 0 + \begin{bmatrix} \dfrac{1}{Cs} & \dfrac{-1}{RCs} \end{bmatrix} \left(I - \begin{bmatrix} -2 & \frac{1}{R} \\ \frac{1}{Cs} & \frac{-1}{RCs} \end{bmatrix} \right)^{-1} \begin{bmatrix} \frac{1}{R} \\ 0 \end{bmatrix} = \frac{1}{3RCs + 2}. \end{aligned} \quad (4.66)$$

Apparently, the LFT approach of Fig. 4.26 yields the same result as that by using Kirchhoff's law.

It should be noted that the cutting places which break the internal feedback loops are not unique. In this example, one can also choose the other cutting points, e.g.,

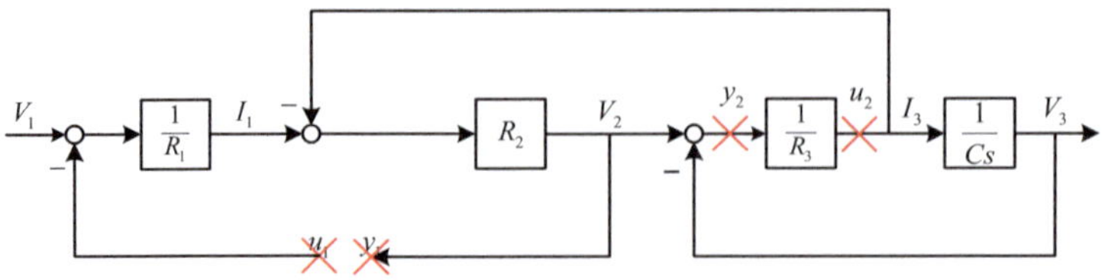

Fig. 4.27 Another cutting points of LFT

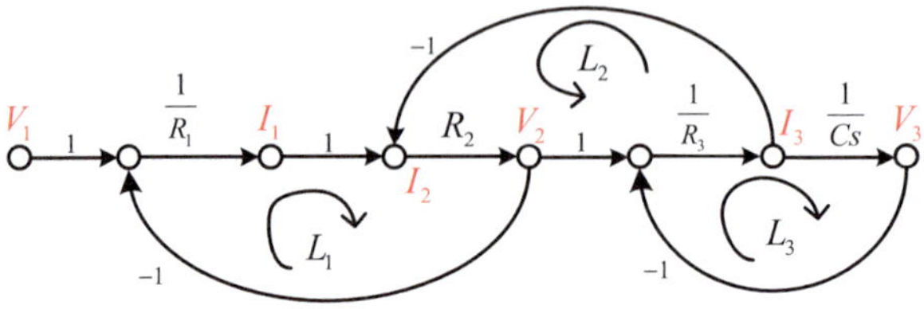

Fig. 4.28 Signal flow diagram of RC circuit

as shown in Fig. 4.27, in which one has the termination $K = \begin{bmatrix} 1 & 0 \\ 0 & \frac{1}{R_3} \end{bmatrix}$ and the corresponding two-port matrix P given by

$$\begin{bmatrix} P_{11} & P_{12} \\ P_{21} & P_{22} \end{bmatrix} = \left[\begin{array}{c|cc} 0 & 0 & \dfrac{1}{Cs} \\ \hline \dfrac{R_2}{R_1} & -\dfrac{R_2}{R_1} & -R_2 \\ \dfrac{R_2}{R_1} & -\dfrac{R_2}{R_1} & -\dfrac{1}{Cs} - R_2 \end{array}\right]. \tag{4.67}$$

For the case $R_1 = R_2 = R_3 = R$, this yields

$$\begin{aligned} \mathrm{LFT}_l(P, K) &= P_{11} + P_{12}K(1 - P_{22}K)^{-1}P_{21} \\ &= 0 + \begin{bmatrix} 0 & \frac{1}{Cs} \end{bmatrix} \begin{bmatrix} 1 & 0 \\ 0 & \frac{1}{R} \end{bmatrix} \left(I - \begin{bmatrix} -1 & -R \\ -1 & -\dfrac{1}{Cs} - R \end{bmatrix} \begin{bmatrix} 1 & 0 \\ 0 & \dfrac{1}{R} \end{bmatrix} \right)^{-1} \begin{bmatrix} 1 \\ 1 \end{bmatrix} \\ &= \frac{1}{3RCs + 2}. \end{aligned} \tag{4.68}$$

Here, one can also exploit Mason's gain formulae for further verifications. Convert this system control block diagram into the signal flow graph as shown in Fig. 4.28.

$$T(s) = \frac{V_3}{V_1} = \frac{\dfrac{R_2}{R_1 R_3 C s}}{1 - (L_1 + L_2 + L_3) + L_1 L_3}, \tag{4.69}$$

where

$$L_1 = -\frac{R_2}{R_1}, \quad L_2 = -\frac{R_2}{R_3}, \quad L_3 = -\frac{1}{R_3 C s}. \tag{4.70}$$

For the case $R_1 = R_2 = R_3 = R$, one can obtain

$$T(s) = \frac{1}{3RCs + 2}. \tag{4.71}$$

Evidently, the results of (4.64), (4.66), and (4.71) are the same. The enhanced discussions between LFT and Mason's gain formulae will be revealed in the next section.

From the examples above, using the LFT approach to determine the transfer function is much more straightforward than that by Mason's gain formulae. The only key point in using the LFT approach is how to choose the places to break all of the feedback loops. To summarize, for determining the closed-loop transfer function of a feedback system, the more loops the control system has, the more complicated Mason's approach becomes. While in the LFT approach, it only needs to properly choose the breaking points (even for MIMO systems), and then the rest of the procedure is systematically manipulated. Moreover, some topological behaviors such as non-touching loop gains can be observed in the two-port matrix of the LFT formulation.

4.5 LFT Description and Feedback Controllers

Most of the feedback systems can be described in the standard control configuration (SCC) of Fig. 4.1, in which the LFT approach is used to synthesize and analyze the control problem. In this section, the feedback control design using the LFT approach is firstly introduced. The LFT formulation is now mature, and subroutines are available in MATLAB toolboxes. In the SCC of Fig. 4.1, LFT matrix P denotes a generalized plant that represents an open-loop system with two sets of inputs stacked as $\begin{bmatrix} w \\ u \end{bmatrix}$ and two sets of outputs $\begin{bmatrix} r \\ y \end{bmatrix}$, where signals can be vector-valued functions of time. Controller K (to be designed) is to form the closed loop of $y \mapsto u$ as a terminator in the SCC. If a given system is "stabilizable" via the feedback controller K, all internally stabilizing controllers can be found by the Youla parameterization [10, 11], in which a generator for the stabilizing controller K is also given by an

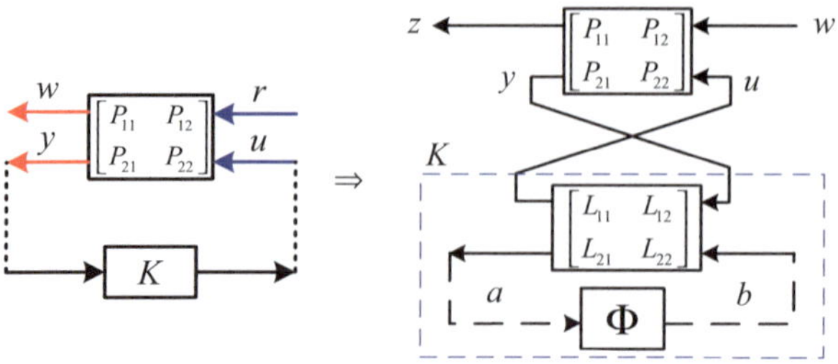

Fig. 4.29 Star product

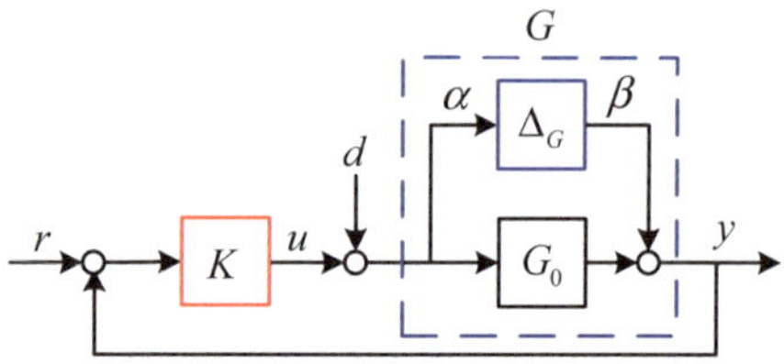

Fig. 4.30 Control system with disturbance and model uncertainty

LFT in $K = \text{LFT}_l(\Pi,\Phi)$. Then, the interconnection of two LFTs will include a product connection, namely, the *star product* [9], which is very complicated in the representation. Figure 4.29 shows the star product of two LFT matrices.

Modeluncertainties are almost inevitable in real-world control systems, and therefore, robustness consideration plays an important role in control system theory and control system design. Model uncertainties usually arise from unavailable information on the plant dynamics or represent part of the dynamics which are purposely left out for the sake of simplicity and ease in system analysis and controller design. As a consequence, the controller, which is designed based on the mathematical model, has to be "robust" in the sense that it must achieve the design specifications for the real plant dynamics containing the unmodeled, uncertain dynamic component. In fact, almost all control systems, in practice, suffer from disturbances/noises and model uncertainties. Basically, a disturbance is an external signal which is neither controllable nor dependent on internal variables of the plant, such as signal d in Fig. 4.12. However, the model uncertainty describes inconsistencies in dynamics between the model and the real plant. As shown in Fig. 4.30 below, it is assumed that the actual plant dynamics G is the sum of nominal model G_o and model uncertainty Δ_G. The so-called additive uncertainty, i.e., $G = G_o + \Delta_G$, includes therefore uncertain dynamics [2]. In fact, uncertainties also arise in the model reduction issues, which simplify calculations or avoid difficulties from the high complexity of a complete model. It is only *known* that some discrepancies exist between the simulated model and the real plant. For instance, the neglect of some high-order terms or some nonlinear phenomena will be summed up in a global model error, and the model error block Δ_G can be a full transfer function matrix.

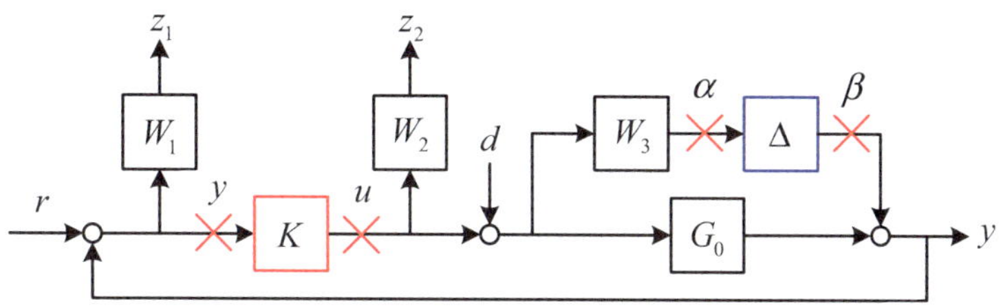

Fig. 4.31 A controller synthesis scheme

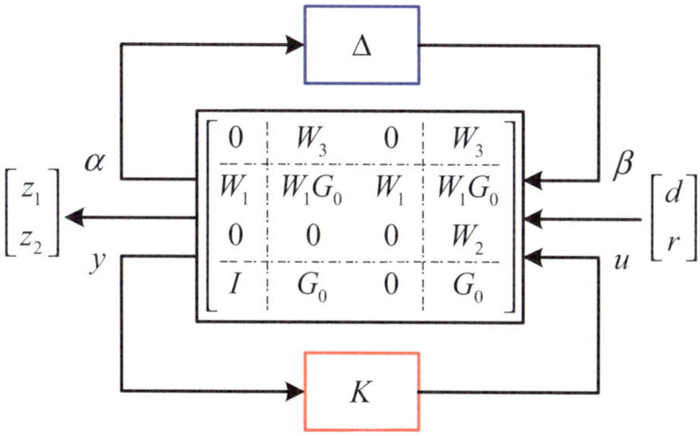

Fig. 4.32 LFT form

The synthesis of controller K would have to consider how to design a robust controller against modeling errors and at the same time achieving engineering requirements. Consider the controller synthesis scheme of Fig. 4.31, where W_1, W_2, and W_3 are weighting functions with stable dynamics to reflect the corresponding design specifications and to characterize the uncertainty profile, i.e., $\Delta_G = \Delta W_3$. Then, taking this synthesis scheme into the LFT form as shown in Fig. 4.32, the controller synthesis problem can be represented in LFT form as

$$\begin{bmatrix} \alpha \\ \hline z_1 \\ z_2 \\ \hline y \end{bmatrix} = P \begin{bmatrix} \beta \\ \hline d \\ r \\ \hline u \end{bmatrix} = \left[\begin{array}{c|cc|c} 0 & W_3 & 0 & W_3 \\ \hline W_1 & W_1G_0 & W_1 & W_1G_0 \\ 0 & 0 & 0 & W_2 \\ \hline I & G_0 & 0 & G_0 \end{array}\right] \begin{bmatrix} \beta \\ \hline d \\ r \\ \hline u \end{bmatrix}. \tag{4.72}$$

Finally, one can employ LFT matrix P to synthesize controller K based on the specific design goal. For example, in H_∞ control, one can determine a satisfactory controller K by minimizing $\|T_{zw}\|_\infty$, where $w = \begin{bmatrix} d \\ r \end{bmatrix}$ and $z = \begin{bmatrix} z_1 \\ z_2 \end{bmatrix}$. More will be discussed in the rest of this book.

To sum up, this section introduced the motivations of feedback controller design. Examples were mainly designed for evaluating the presented LFT approach. Under the LFT scheme, readers can take the synthesis target as the terminator, e.g., controller K. Then, via applying the LFT manipulation, one can determine the solution of K systematically.

4.6 Inner and Co-inner Systems

This section describes inner and co-inner systems. For synthesizing a (sub)optimal control system under the LFT framework, to be elaborated in the later part of this book, properties of these systems reveal the significance of such concepts.

Definition 4.1 Let $P^*(s) = P^{\mathrm{T}}(\bar{s})$ and $P^{\sim}(s) = P^{\mathrm{T}}(-s)$. The transfer function matrix $P(s)$ is called all-pass if $P^{\sim}P = I$, for all s. An all-pass $P(s)$ is then called inner if $P(s)$ is stable and $P^*P \leq I$, $\forall\, \mathrm{Re}(s) \geq 0$. Dually, a transfer function matrix $P(s)$ is called co-all-pass if $PP^{\sim} = I$, for all s. A co-all-pass $P(s)$ is then called co-inner if $P(s)$ is stable and $PP^* \leq I$, $\forall\, \mathrm{Re}(s) \geq 0$.

Consider the $(q_1 + q_2) \times (m_1 + m_2)$ two-port system of Fig. 4.1, where $\begin{bmatrix} z \\ y \end{bmatrix} = P \begin{bmatrix} w \\ u \end{bmatrix}$ with $q_2 = m_1$ and $q_1 \geq m_2$. An inner system has the property that the 2-norm of the output signal is equal to that of the input [3, 5]. That is,

$$\left\| y\,(j\omega) \right\|_2^2 + \left\| z\,(j\omega) \right\|_2^2 = \left\| u\,(j\omega) \right\|_2^2 + \left\| w\,(j\omega) \right\|_2^2 \tag{4.73}$$

for any ω.

Example 4.6 Verify that the one-port transfer function $T(s) = \frac{s-1}{s+1}$ is inner.

From $T^{\sim}(s) = T^{\mathrm{T}}(-s)$, one has

$$T^{\sim}(s) = T^{\mathrm{T}}(-s) = \frac{-s-1}{-s+1},$$

and then

$$T^{\sim}(s)T(s) = \frac{s+1}{s-1} \times \frac{s-1}{s+1} = 1.$$

Firstly, this shows that $T(s)$ is all-pass. From $T^*(s) = T^{\mathrm{T}}(\bar{s})$, one can obtain, for $s = \sigma + j\omega$,

$$T^*T = \frac{\bar{s}-1}{\bar{s}+1} \times \frac{s-1}{s+1} = \frac{(\sigma-1)^2 + \omega^2}{(\sigma+1)^2 + \omega^2} \leq 1, \quad \forall\ \sigma \geq 0.$$

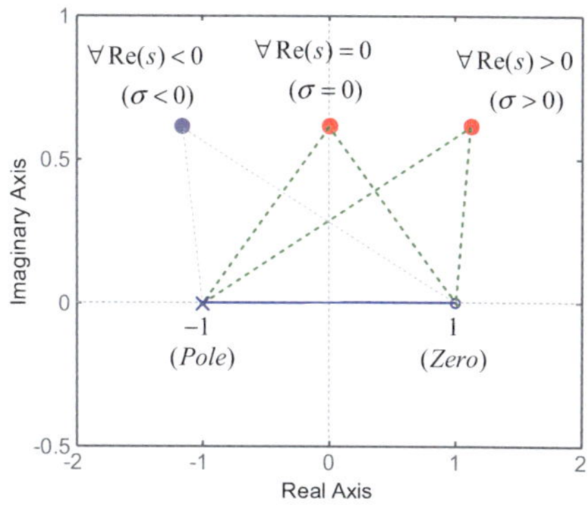

Fig. 4.33 Distances between location of s and system pole and zero

Evidently, $T^*(s)T(s) \le 1,\ \forall\, \mathrm{Re}(s) \ge 0$. By this and $T(s)$ being stable, $T(s)$ is an inner function. Furthermore, it can also be examined by the ratio of the distance between the location of s in the complex plane to the system pole and zero. Now, it can be found that

$$\begin{aligned}
&(\sigma-1)^2 < (\sigma+1)^2 \quad \text{and} \quad P^*(s)P(s) < 1, \quad \forall \mathrm{Re}(s) > 0;\\
&(\sigma-1)^2 = (\sigma+1)^2 \quad \text{and} \quad P^*(s)P(s) = 1, \quad \forall \mathrm{Re}(s) = 0;\\
&(\sigma-1)^2 > (\sigma+1)^2 \quad \text{and} \quad P^*(s)P(s) > 1; \quad \forall \mathrm{Re}(s) < 0.
\end{aligned} \tag{4.74}$$

Figure 4.33 shows the relationship.

Note that for the case of $T = \frac{s+1}{s-1}$, $T(s)$ is all-pass but is not inner. Since its pole is located in the open right-half plane,

$$T^{\sim}(s)T(s) = \left(\frac{-s+1}{-s-1}\right)\left(\frac{s+1}{s-1}\right) = 1 \tag{4.75}$$

is obtained, but $T^*(s)T(s) \ge 1,\ \forall\, \mathrm{Re}(s) \ge 0$. This means $T(s)$ is not an inner function, although $T^{\sim}(s)T(s) = 1$. ■

Example 4.7 Verify that the two-port system $P(s) = \begin{bmatrix} \frac{s-1}{s+1} & 0 \\ 0 & \frac{s-2}{s+2} \end{bmatrix}$ is inner.

From

$$P^{\sim}(s)P(s) = \begin{bmatrix} \dfrac{(s+1)}{(s-1)} & 0 \\ 0 & \dfrac{(s+2)}{(s-2)} \end{bmatrix} \begin{bmatrix} \dfrac{s-1}{s+1} & 0 \\ 0 & \dfrac{s-2}{s+2} \end{bmatrix} = I,$$

one concludes that $P(s)$ is all-pass. For $s = \sigma + j\omega$, one has

$$P^*P = \begin{bmatrix} \dfrac{\bar{s}-1}{\bar{s}+1} & 0 \\ 0 & \dfrac{\bar{s}-2}{\bar{s}+2} \end{bmatrix} \begin{bmatrix} \dfrac{s-1}{s+1} & 0 \\ 0 & \dfrac{s-2}{s+2} \end{bmatrix}$$
$$= \begin{bmatrix} \dfrac{(\sigma-1)^2+\omega^2}{(\sigma+1)^2+\omega^2} & 0 \\ 0 & \dfrac{(\sigma-2)^2+\omega^2}{(\sigma+2)^2+\omega^2} \end{bmatrix} \le I, \quad \forall\ \sigma \ge 0.$$

Since $P(s)$ is all-pass and $P^*(s)P(s) \le I$, $\forall$ Re$(s) \ge 0$ (i. e., $\sigma \ge 0$), and due to that all of its poles are in the open left-half plane, this two-port $P(s)$ is an inner system by the definition.

Example 4.8 Verify that $P(s) = \begin{bmatrix} 0 & \frac{s-1}{s+1} \\ \frac{s-2}{s+2} & 0 \end{bmatrix}$ is co-inner.

From

$$P(s)P^{\sim}(s) = \begin{bmatrix} 0 & \dfrac{s-1}{s+1} \\ \dfrac{s-2}{s+2} & 0 \end{bmatrix} \begin{bmatrix} 0 & \dfrac{s+2}{s-2} \\ \dfrac{s+1}{s-1} & 0 \end{bmatrix} = I,$$

we can gather that $P(s)$ is co-all-pass. Since one can also verify that $P(s)P^*(s) \le I$, $\forall$ Re$(s) \ge 0$ for $s = \sigma + j\omega$, and it is stable, $P(s)$ is an co-inner system.

Theorem 4.1 *Given an LFT system $P(s)$, if $P(s)$ is inner, then* $\|\mathrm{LFT_l}(P,K)\|_\infty < 1$ *for any $K \in BH_\infty$.*

Proof Since $\begin{bmatrix} z \\ y \end{bmatrix} = P \begin{bmatrix} w \\ u \end{bmatrix}$ and $P = \begin{bmatrix} P_{11} & P_{12} \\ P_{21} & P_{22} \end{bmatrix}$ is inner, one has $P^*(j\omega)P(j\omega) = I$ and $\|P_{22}\|_\infty \le 1$. The latter is true because P_{22} is part of P and $\|P\|_\infty = 1$. Then, by the small gain theory, one has $\mathrm{LFT_l}(P,K) \in RH_\infty$ for any $K \in BH_\infty$. Since P is inner, one also has

$$\|y(j\omega)\|_2{}^2 + \|z(j\omega)\|_2{}^2 = \|u(j\omega)\|_2{}^2 + \|w(j\omega)\|_2{}^2. \tag{4.76}$$

Utilizing $u = Ky$ and $K \in BH_\infty$ yields

$$\|u(j\omega)\|_2 = \|Ky(j\omega)\|_2 < \|K\|_\infty \|y(j\omega)\|_2 \le \|y(j\omega)\|_2 \tag{4.77}$$

which gives

$$\|u(j\omega)\|_2{}^2 - \|y(j\omega)\|_2{}^2 < 0. \tag{4.78}$$

Additionally, (4.76) and (4.78) together imply

$$\|z(j\omega)\|_2^{\,2} < \|w(j\omega)\|_2^{\,2}. \tag{4.79}$$

Thus, one has

$$\frac{\|z(j\omega)\|_2}{\|w(j\omega)\|_2} < 1, \quad \text{for all } \omega, \tag{4.80}$$

and further

$$\max_{w\neq 0}\frac{\|z(j\omega)\|_2}{\|w(j\omega)\|_2} < 1. \tag{4.81}$$

From $z = \mathrm{LFT_l}(P,K)w$, one concludes that $\|\mathrm{LFT_l}(P,K)\|_\infty < 1$ as

$$\|\mathrm{LFT_l}(P,K)\|_\infty = \max_{w\neq 0}\frac{\|z\|_2}{\|w\|_2}. \tag{4.82}$$

■

Example 4.9 Consider the LFT system as illustrated in Fig. 4.1, where $P(s) = \begin{bmatrix} 0 & \frac{s-1}{s+1} \\ \frac{s-2}{s+2} & 0 \end{bmatrix} \in RH_\infty$ and $K = \frac{1}{s+3} \in BH_\infty$. Verify that $\|\mathrm{LFT_l}(P,K)\|_\infty < 1$.

From (4.7), one has $\left\|\frac{1}{s+3}\right\|_\infty = \frac{1}{3}$ and $\mathrm{LFT_l}(P,K) = \frac{(s-1)(s-2)}{(s+1)(s+2)(s+3)}$.

Hence, one can find that $\|\mathrm{LFT_l}(P,K)\|_\infty = \frac{1}{3} < 1$.

As mentioned in Chap. 2, the LFT system can be transformed into a CSD one. Hence, the properties of all-pass (or co-all-pass) and inner (or co-inner) have their counterparts as J-unitary (or dual J-unitary) and J-lossless (or dual J-lossless) [5]. These properties and relations will be introduced in the next chapter.

Exercises

1. Let $w = \mathrm{LFT_l}(P,K)\begin{bmatrix} r \\ d \end{bmatrix}$ in the block diagram below. Determine P and $\mathrm{LFT_l}(P,K)$ with the given cutting point.

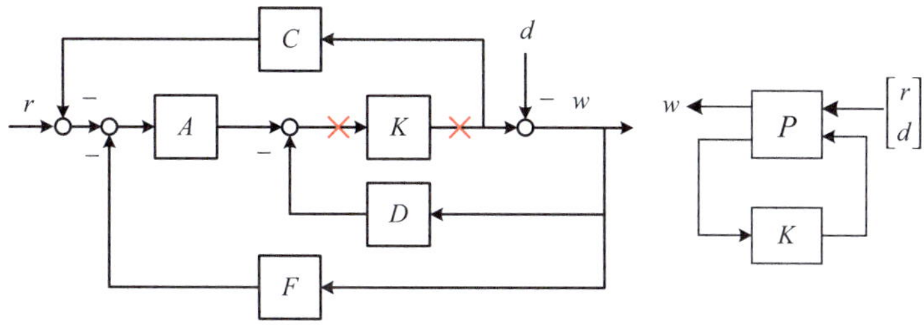

2. Consider a brushed DC motor model as given below and derive its corresponding LFT representation of $\begin{bmatrix} \omega \\ i \end{bmatrix} = P \begin{bmatrix} T_d \\ V \end{bmatrix}$, and $T_e = K_t i$.

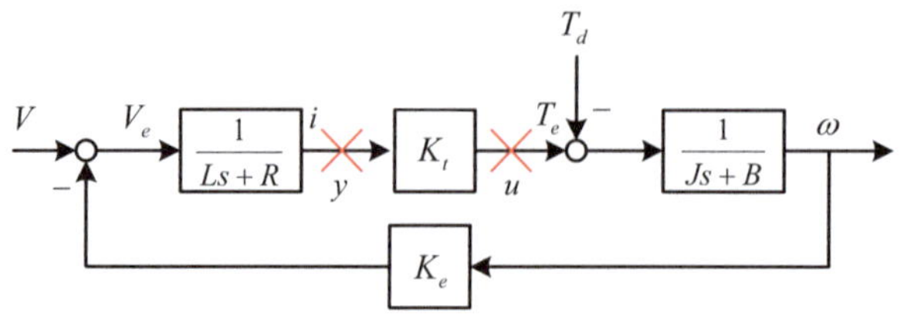

3. Determine the transfer function of the given system below from r to w. (Hint: Use the cutoffs indicated in the block diagram.)

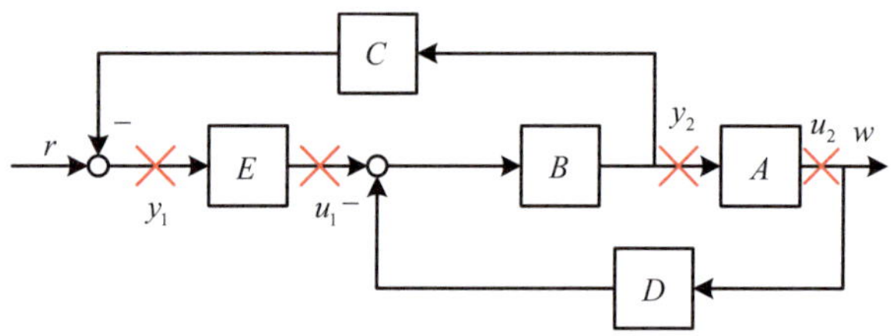

4. Find the transfer function of the following system by Mason's gain formulae as well as by LFT approach, respectively. Compare the results and discuss their relationships.

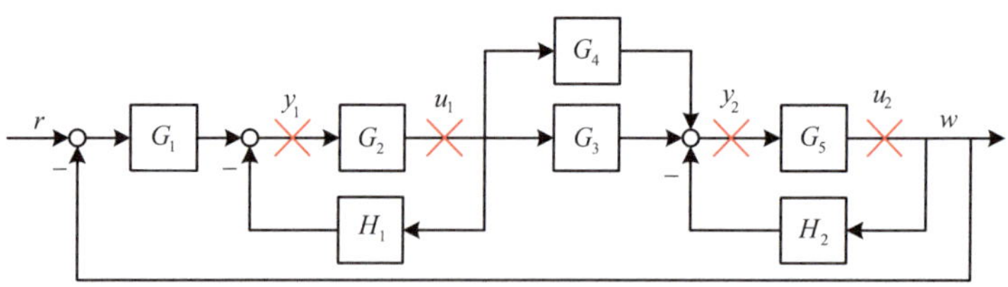

5. Consider the following block diagram. Determine the transfer function from r to w using Mason's gain formulae and LFT approach, respectively.

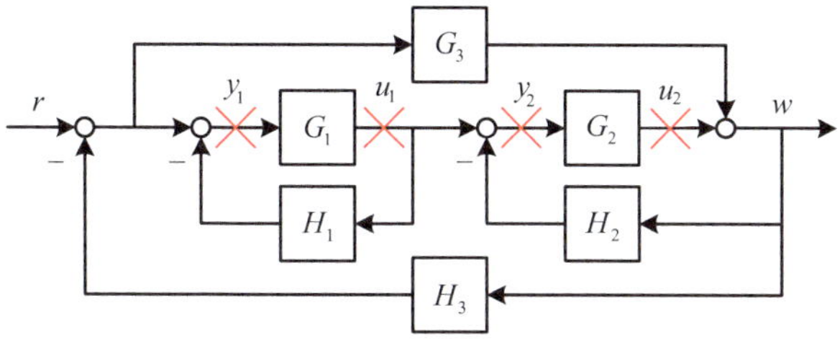

6. Consider the system block diagram below and determine its transfer function from r to w using Mason's gain formulae and LFT, respectively.

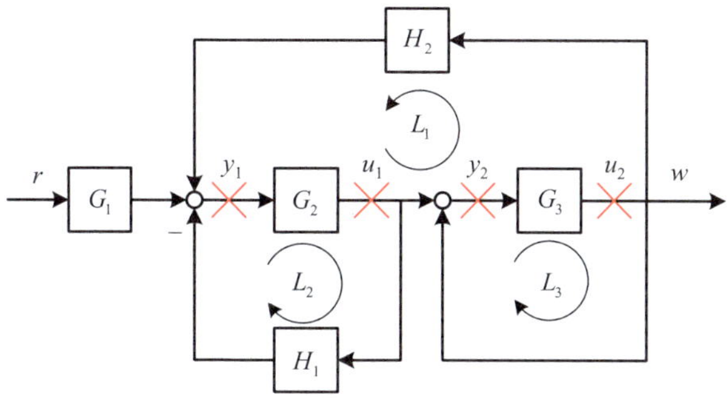

References

1. Dorf RC (1995) Modern control systems. Addison-Wesley, New York
2. Doyle JC, Francis BA, Tannenbaum AR (1992) Feedback control theory. Macmillan, New York
3. Franco S (1995) Electric circuits fundamentals. Saunders College Publishing, Orlando
4. Franklin GF, Powell JD, Emami-Naeini A (2009) Feedback control of dynamic systems, 6th edn. Addison Wesley, New York
5. Kimura H (1997) Chain-scattering approach to H_∞ control. Birkhäuser, Boston
6. Kuo BC (1976) Automatic control systems, 6th edn. Prentice Hall, Englewood Cliffs
7. Mason SJ (1953) Feedback theory-some properties of signal flow graphs. Proc IRE 41:1144–1156
8. Mason SJ (1956) Feedback theory-further properties of signal flow graphs. Proc IRE 44:920–926
9. Redheffer RM (1960) On a certain linear fractional transformation. J Math Phys 39:269–286
10. Youla DC, Jabr HA, Bongiorno JJ (1976) Modern Wiener-Hopf design of optimal controllers part II: the multivariable case. IEEE Trans Autom Control 21:319–338
11. Zhou K, Doyle JC, Glover K (1996) Robust and optimal control. Prentice Hall, Upper Saddle River

Chapter 5
Chain Scattering Descriptions

The chain scattering-matrix description (CSD) of a two-port network, first introduced in Chap. 2, originated from the conventional electrical circuit theory. In contrast to the LFT, the CSD developed in the network circuits provides a straightforward interconnection in a *cascaded* way. The CSD can transform an LFT into a two-port network connection and *vice versa*. Thus, many known results which have been developed for a two-port network can then be used in control system analysis and synthesis. Due to its benefits of describing a linear system, the CSD was later extended to the design of robust control systems [1, 2], where different structures of the CSD were investigated. In this chapter, the CSD will be formally defined and explored for transformations and the descriptions of state-space realizations. The J-lossless and dual J-lossless systems are defined in this chapter. J-lossless and dual J-lossless both play an important role in the CSD control system manipulations, analysis, and synthesis. In particular, the properties of J-lossless and dual J-lossless are essential in synthesizing H_∞ (sub)optimal controllers using the CSD approach.

5.1 CSD Definitions and Manipulations

The chain scattering representation is also called the homographic transformation [2, 3]. Figure 5.1 shows the representations using the right or left CSD matrix, respectively, as defined by terminator K being on the right or left of the interconnection system. As can be seen in the following, a two-port LFT system can be represented by coupled right or left CSD matrices. In this book, signals always flow from the lower line to the upper one. In a right CSD diagram, signals flow in the anticlockwise direction and in the clockwise direction in a left CSD diagram. For instance, in Fig. 5.1a, b, the flow follows "$w \to y \to u \to z$" but travels anticlockwise in (a) and clockwise in (b). Note that the superscript "$\sim$" symbolizes the dual description in the "left" case.

M.-C. Tsai and D.-W. Gu, *Robust and Optimal Control: A Two-port Framework Approach*, Advances in Industrial Control, DOI 10.1007/978-1-4471-6257-5_5,
© Springer-Verlag London 2014

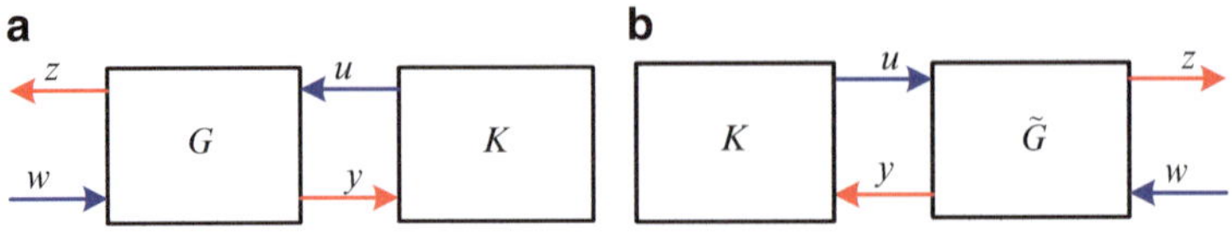

Fig. 5.1 (**a**) Right and (**b**) left CSDs

Consider the right CSD matrix $G = \begin{bmatrix} G_{11} & G_{12} \\ G_{21} & G_{22} \end{bmatrix}$ of Fig. 5.1a, where the feedback connection is given by

$$\begin{bmatrix} z \\ w \end{bmatrix} = \begin{bmatrix} G_{11} & G_{12} \\ G_{21} & G_{22} \end{bmatrix} \begin{bmatrix} u \\ y \end{bmatrix}, \quad u = Ky. \tag{5.1}$$

Note that G_{11} denotes the transfer function of $u \mapsto z$, when $y = 0$, similarly for the rest as further notated:

$$G_{11} = \frac{z}{u}\bigg|_{y=0}, \quad G_{12} = \frac{z}{y}\bigg|_{u=0}, \quad G_{21} = \frac{w}{u}\bigg|_{y=0}, \quad G_{22} = \frac{w}{y}\bigg|_{u=0}. \tag{5.2}$$

From

$$\begin{bmatrix} z \\ w \end{bmatrix} = \begin{bmatrix} G_{11}K + G_{12} \\ G_{21}K + G_{22} \end{bmatrix} y, \tag{5.3}$$

one has, when $(G_{21}K + G_{22})^{-1}$ exists,

$$z = (G_{11}K + G_{12})(G_{21}K + G_{22})^{-1}w. \tag{5.4}$$

Then, define the right CSD transformation of G and K, denoted by $\mathrm{CSD_r}(G,K)$, as

$$\mathrm{CSD_r}\left(\begin{bmatrix} G_{11} & G_{12} \\ G_{21} & G_{22} \end{bmatrix}, K\right) := (G_{12} + G_{11}K)(G_{22} + G_{21}K)^{-1}. \tag{5.5}$$

The right CSD transformation is said to be well posed if $(G_{22} + G_{21}K)$ is invertible. The whole loop transfer function of $w \mapsto z$ is given by $\mathrm{CSD_r}(G,K)$. This definition implies that G_{22} must be square.

Example 5.1 Consider the SCC plant as shown in Fig. 4.1, using MATLAB to determine its corresponding right CSD manipulations.

The following MATLAB code is an example for readers' reference.

```
clc;
clear;
disp('SCC2CSDr method')
syms P11 P12 P21 P22 K S;
syms p11 p12 p21 p22 k s;
P11=input('w => u (P11) ?');
P12=input('w=> y (P12) ?');
P21=input('r => u (P21) ?');
P22=input('r => y (P22) ?');
K=input('Please input K:');
[row col]=size(P21);
if (row~=col)
  disp('The do not exist a right CSD matrix G !')
  disp('Plz enter Ctrl c !')
  pause
else
G11=simplify(P12-P11*inv(P21)*P22);
G12=simplify(P11*inv(P21));
G21=simplify(-inv(P21)*P22);
G22=inv(P21);
G=[G11 G12;G21 G22]
end
if (G21*K+G22==0)
  disp('The transfer function is singular !')
  disp('Plz enter Ctrl c !')
  pause
else
Transfer_function_r_to_w=simplify((G11*K+G12)*inv
                             (G21*K+G22))
end
```

Similarly, the diagram shown in Fig. 5.1b represents the following two sets of equations:

$$\begin{bmatrix} u \\ y \end{bmatrix} = \begin{bmatrix} \tilde{G}_{11} & \tilde{G}_{12} \\ \tilde{G}_{21} & \tilde{G}_{22} \end{bmatrix} \begin{bmatrix} z \\ w \end{bmatrix}, u = Ky \tag{5.6}$$

where

$$\tilde{G}_{11} = \frac{u}{z}\Big|_{w=0}, \quad \tilde{G}_{12} = \frac{u}{w}\Big|_{z=0}, \quad \tilde{G}_{21} = \frac{y}{z}\Big|_{w=0}, \quad \tilde{G}_{22} = \frac{y}{w}\Big|_{z=0}. \tag{5.7}$$

From

$$\begin{bmatrix} u \\ y \end{bmatrix} = \begin{bmatrix} \tilde{G}_{11} & \tilde{G}_{12} \\ \tilde{G}_{21} & \tilde{G}_{22} \end{bmatrix} \begin{bmatrix} z \\ w \end{bmatrix} \tag{5.8}$$

and

$$\begin{bmatrix} I & -K \end{bmatrix} \begin{bmatrix} K \\ I \end{bmatrix} y = 0, \tag{5.9}$$

yield

$$0 = \begin{bmatrix} I & -K \end{bmatrix} \begin{bmatrix} \tilde{G}_{11} & \tilde{G}_{12} \\ \tilde{G}_{21} & \tilde{G}_{22} \end{bmatrix} \begin{bmatrix} z \\ w \end{bmatrix}; \tag{5.10}$$

hence, when $\left(\tilde{G}_{11} - K\tilde{G}_{21}\right)^{-1}$ exists,

$$z = -\left(\tilde{G}_{11} - K\tilde{G}_{21}\right)^{-1} \left(\tilde{G}_{12} - K\tilde{G}_{22}\right) w. \tag{5.11}$$

Then, the left CSD transformation of $\tilde{G}$ and K, denoted by $\mathrm{CSD_l}\left(\tilde{G},\ K\right)$, is defined as

$$\mathrm{CSD_l}\left(\begin{bmatrix} \tilde{G}_{11} & \tilde{G}_{12} \\ \tilde{G}_{21} & \tilde{G}_{22} \end{bmatrix}, K\right) := -\left(\tilde{G}_{11} - K\tilde{G}_{21}\right)^{-1} \left(\tilde{G}_{12} - K\tilde{G}_{22}\right). \tag{5.12}$$

The left CSD transformation is said to be well posed if $\left(\tilde{G}_{11} - K\tilde{G}_{21}\right)$ is invertible. Dually, this definition implies that $\tilde{G}_{11}$ must be square.

Example 5.2 Consider the SCC plant as shown in Fig. 4.1, using MATLAB to determine its corresponding left CSD manipulations.

The following MATLAB code is an example for readers' reference.

```
clc;
clear;
disp('SCC2CSDl method')
syms P11 P12 P21 P22 K S;
syms p11 p12 p21 p22 k s;
P11=input('w => u (P11) ?');
```

```
P12=input('w => y (P12) ?');
P21=input('r => u (P21) ?');
P22=input('r => y (P22) ?');
K=input('Please input K:');
[row col]=size(P12);
if (row~=col)
    disp('The do not exist a left CSD matrix G !')
    disp('Plz enter Ctrl c !')
    pause
else
G11=inv(P12);
G12=simplify(-inv(P12)*P11);
G21=simplify(P22*inv(P12));
G22=simplify(P21-P22*inv(P12)*P11);
G=[G11 G12;G21 G22]
end
if (G11-K*G21==0)
    disp('The transfer function is singular !')
    disp('Plz enter Ctrl c !')
    pause
else
Transfer_function_r_to_w=simplify(-inv(G11-K*G21)*
(G12-K*G22))
end
```

5.2 Cascaded Connection of Two CSD Matrices

In this section, cases of two cascaded CSD subsystems will be investigated. A comprehensive understanding of the construction and connection of CSD systems is useful for general control problems. Note that K is employed as a terminator and can also be truncated in some formulations.

Case (I) Two right CSD matrices

Figure 5.2 shows two cascaded CSD systems with the termination at the right end. The definition provides

$$\begin{bmatrix} z_1 \\ w_1 \end{bmatrix} = G_1 \begin{bmatrix} u_1 \\ y_1 \end{bmatrix} = G_1 G_2 \begin{bmatrix} u_2 \\ y_2 \end{bmatrix} = G_1 G_2 \begin{bmatrix} K \\ I \end{bmatrix} y_2 = G \begin{bmatrix} K \\ I \end{bmatrix} y_2 \tag{5.13}$$

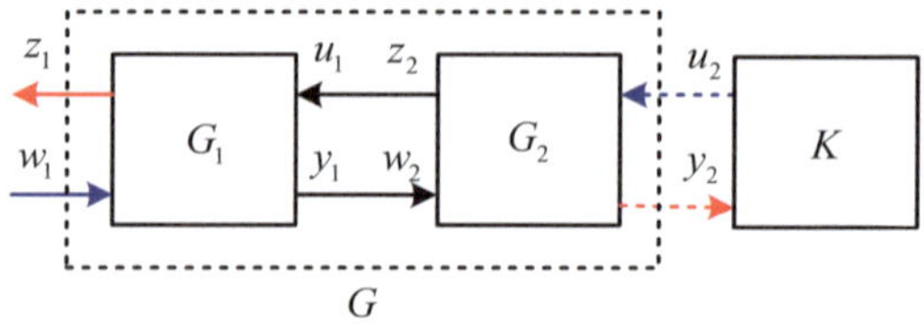

Fig. 5.2 Cascaded two CSD systems

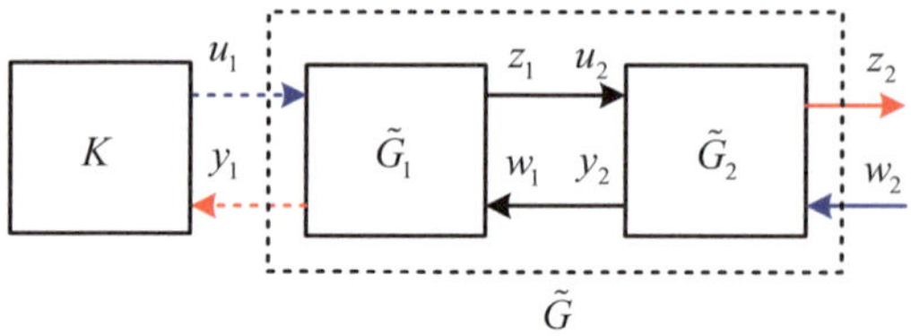

Fig. 5.3 Cascaded two CSD systems

and

$$z_1 = \mathrm{CSD_r}\left(G_1, \mathrm{CSD_r}\left(G_2, K\right)\right)\ w_1 = \mathrm{CSD_r}\left(G_1 G_2, K\right)\ w_1 = \mathrm{CSD_r}\left(G, K\right)\ w_1, \tag{5.14}$$

where the product of the two right CSD matrices follows the usual (block) matrix multiplication rule, i.e., $G = G_1 G_2$.

Property 5.1 For the cascaded connection of two right CSD matrices:

1. $\mathrm{CSD_r}(G_1, \mathrm{CSD_r}(G_2, K)) = \mathrm{CSD_r}(G_1 G_2, K)$.
2. $\mathrm{CSD_r}(G_1 G_2, K) = \mathrm{CSD_r}(I, K) = K$ if $G_1 = G_2^{-1}$.

Case (II) Two left CSD matrices
Figure 5.3 shows two cascaded CSD systems with the termination at the left end. Dually, from

$$\begin{bmatrix} u_1 \\ y_1 \end{bmatrix} = \tilde{G}_1 \begin{bmatrix} z_1 \\ w_1 \end{bmatrix} = \tilde{G}_1 \tilde{G}_2 \begin{bmatrix} z_2 \\ w_2 \end{bmatrix} = \tilde{G} \begin{bmatrix} z_2 \\ w_2 \end{bmatrix} \tag{5.15}$$

and $u_1 = K y_1$,

$$z_2 = \mathrm{CSD_l}\left(\tilde{G}, K\right)\ w_2 \tag{5.16}$$

can be obtained, where $\tilde{G} = \tilde{G}_1 \tilde{G}_2$. This can also be derived from $u_2 = z_1$ and $y_2 = w_1$ as

$$\begin{bmatrix} u_2 \\ y_2 \end{bmatrix} = \begin{bmatrix} \mathrm{CSD_l}\left(\tilde{G}_1, K\right) \\ I \end{bmatrix} w_1 = \tilde{G}_2 \begin{bmatrix} z_2 \\ w_2 \end{bmatrix} \tag{5.17}$$

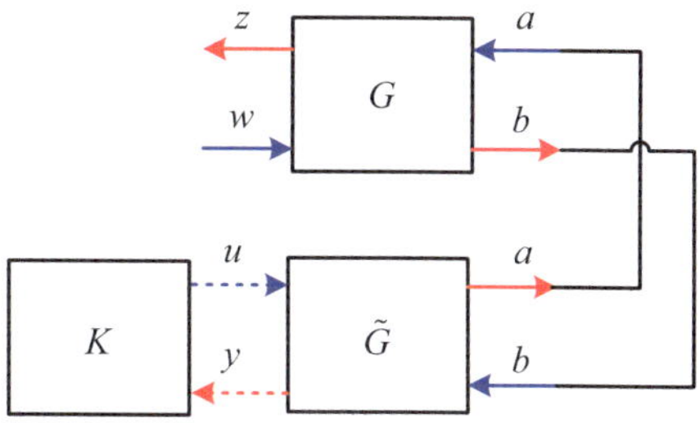

Fig. 5.4 Right CSD associated with left CSD

and then

$$0 = \begin{bmatrix} I & -\mathrm{CSD_l}\left(\tilde{G}_1, K\right) \end{bmatrix} \begin{bmatrix} \mathrm{CSD_l}\left(\tilde{G}_1, K\right) \\ I \end{bmatrix} w_1 = \begin{bmatrix} I & -\mathrm{CSD_l}\left(\tilde{G}_1, K\right) \end{bmatrix} \tilde{G}_2 \begin{bmatrix} z_2 \\ w_2 \end{bmatrix}, \tag{5.18}$$

and hence,

$$z_2 = \mathrm{CSD_l}\left(\tilde{G}_2, \mathrm{CSD_l}\left(\tilde{G}_1, K\right)\right) \ w_2 = \mathrm{CSD_l}\left(\tilde{G}, K\right) \ w_2. \tag{5.19}$$

Similarly, the product of the two left CSD matrices also follows the usual (block) matrix multiplication rule.

Property 5.2 The cascaded connection of two left CSD matrices gives the following:

1. $\mathrm{CSD_l}\left(\tilde{G}_2, \mathrm{CSD_l}\left(\tilde{G}_1, K\right)\right) = \mathrm{CSD_l}\left(\tilde{G}_1\tilde{G}_2, K\right)$.
2. $\mathrm{CSD_l}\left(\tilde{G}_1\tilde{G}_2, K\right) = \mathrm{CSD_l}\left(I, K\right) = K$ if $\tilde{G}_1 = \tilde{G}_2^{-1}$.

Case (III) Connection of right CSD associated with left CSD
Consider a feedback system of Fig. 5.4, which is represented by two coupling CSD matrices, where

$$\begin{bmatrix} z \\ w \end{bmatrix} = G \begin{bmatrix} a \\ b \end{bmatrix}, \quad \begin{bmatrix} u \\ y \end{bmatrix} = \tilde{G} \begin{bmatrix} a \\ b \end{bmatrix} \tag{5.20}$$

and the feedback termination $u = Ky$. From $a = \mathrm{CSD_l}\left(\tilde{G}, K\right) b$, one has

$$\begin{bmatrix} z \\ w \end{bmatrix} = G \begin{bmatrix} a \\ b \end{bmatrix} = G \begin{bmatrix} \mathrm{CSD_l}\left(\tilde{G}, K\right) \\ I \end{bmatrix} b, \tag{5.21}$$

and hence,

$$z = \mathrm{CSD_r}\left(G, \mathrm{CSD_l}\left(\tilde{G}, K\right)\right) w. \tag{5.22}$$

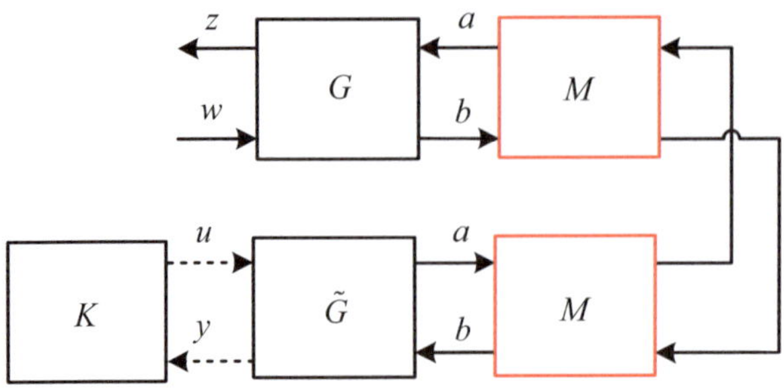

Fig. 5.5 Multiplying M at right terminal

Fig. 5.6 Left CSD associated with right CSD

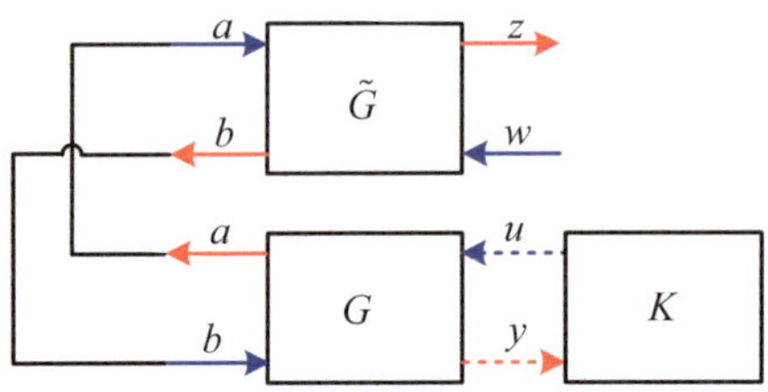

Also, from (5.20), it is clear that $\tilde{G}^{-1}\begin{bmatrix} u \\ y \end{bmatrix} = \begin{bmatrix} a \\ b \end{bmatrix}$; hence, $\begin{bmatrix} z \\ w \end{bmatrix} = G\begin{bmatrix} a \\ b \end{bmatrix} = G\tilde{G}^{-1}\begin{bmatrix} a \\ b \end{bmatrix}$ and $z = \mathrm{CSD_r}\left(G, \mathrm{CSD_l}\left(\tilde{G}, K\right)\right) w = \mathrm{CSD_r}\left(G\tilde{G}^{-1}, K\right) w$. It is trivial to show that $\mathrm{CSD_r}(I, \mathrm{CSD_l}(I,K)) = K$ and $\mathrm{CSD_r}(M, \mathrm{CSD_l}(M,K)) = K$, where M is any square and invertible transfer function matrix. Therefore, multiplying a square and invertible M at the right-hand side of both top and bottom paths, as illustrated in Fig. 5.5, will not affect the closed-loop transfer function of $w \mapsto z$.

For description ease in the connection of Fig. 5.4, the notation $\mathrm{CSD_r} \rightleftarrows \mathrm{CSD_l}$ is employed to describe such a connection.

Property 5.3 For the connection of a right CSD associated with a left CSD, one has the following:

1. $\mathrm{CSD_r}\left(G, \mathrm{CSD_l}\left(\tilde{G}, K\right)\right) = \mathrm{CSD_r}\left(GM, \mathrm{CSD_l}\left(\tilde{G}M, K\right)\right)$, where M is invertible.
2. $\mathrm{CSD_r}\left(G, \mathrm{CSD_l}\left(\tilde{G}, K\right)\right) = \mathrm{CSD_r}\left(I, \mathrm{CSD_l}\left(I, K\right)\right) = K$ if $G = \tilde{G}$ is invertible.

Case (IV) Connection of left CSD associated with right CSD

Likewise, a feedback system can also be represented by two coupling CSDs as illustrated in Fig. 5.6, where

$$\begin{bmatrix} a \\ b \end{bmatrix} = \tilde{G}\begin{bmatrix} z \\ w \end{bmatrix} = G\begin{bmatrix} u \\ y \end{bmatrix}, \tag{5.23}$$

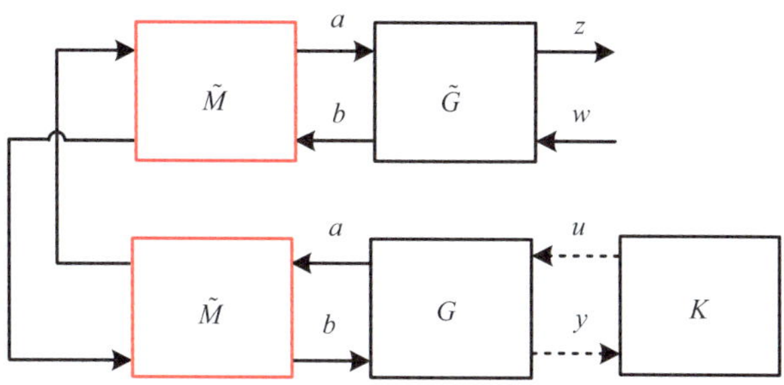

Fig. 5.7 Multiplying M at left terminal

with the feedback termination $u = Ky$. From $a = \mathrm{CSD_r}\,(G,K)b$, one has

$$\begin{bmatrix} \mathrm{CSD_r}\,(G,K) \\ I \end{bmatrix} b = \tilde{G} \begin{bmatrix} z \\ w \end{bmatrix}, \tag{5.24}$$

and therefore,

$$z = \mathrm{CSD_l}\left(\tilde{G}, \mathrm{CSD_r}\,(G,K)\right) w. \tag{5.25}$$

For easier connection description of Fig. 5.6, notation $\mathrm{CSD_l} \rightleftarrows \mathrm{CSD_r}$ is employed to sketch this connection. It is not useful to illustrate $\mathrm{CSD_l}\,(I, \mathrm{CSD_r}\,(I,K)) = K$. Additionally, one can multiply any nonsingular $\tilde{M}$ at the left terminal of both top and bottom paths, as illustrated in Fig. 5.7, which is inconsequential to the closed-loop transfer function of $w \mapsto z$.

Property 5.4 For the connection of a left CSD associated with a right CSD, one has the following:

1. $\mathrm{CSD_l}\left(\tilde{G}, \mathrm{CSD_r}\,(G,K)\right) = \mathrm{CSD_l}\left(\tilde{M}\tilde{G}, \mathrm{CSD_r}\left(\tilde{M}G, K\right)\right)$, where $\tilde{M}$ is invertible.
2. $\mathrm{CSD_l}\left(\tilde{G}, \mathrm{CSD_r}\,(G,K)\right) = \mathrm{CSD_l}\,(I, \mathrm{CSD_r}\,(I,K)) = K$ if $\tilde{G} = G$ is invertible.

5.3 Transformation from LFT to CSD Matrix

As illustrated in Fig. 5.8, for the case of $\dim(w) = \dim(y)$ and with an invertible P_{21}, the particular SCC plant P has an "equivalent" right CSD matrix G of $\begin{bmatrix} u \\ y \end{bmatrix} \mapsto \begin{bmatrix} z \\ w \end{bmatrix}$ such that $\mathrm{CSD_r}\,(G,K) = \mathrm{LFT_l}\,(P,K)$. This can be derived as follows:

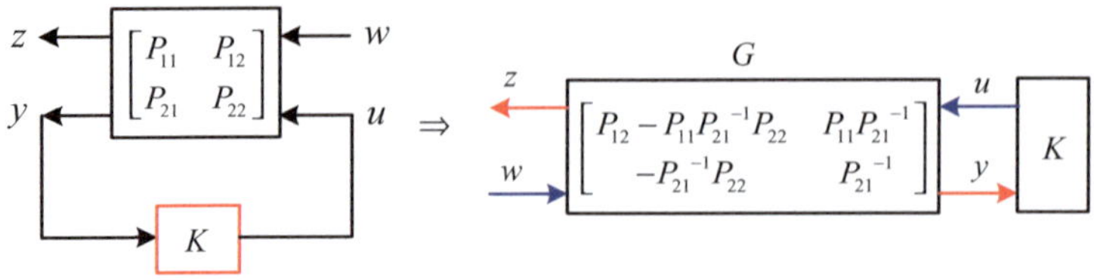

Fig. 5.8 LFT to right CSD

Let

$$\begin{bmatrix} z \\ y \\ \hline w \\ u \end{bmatrix} = \begin{bmatrix} P_{11} & P_{12} \\ P_{21} & P_{22} \\ \hline I & 0 \\ 0 & I \end{bmatrix} \begin{bmatrix} w \\ u \end{bmatrix} \Rightarrow \begin{bmatrix} z \\ w \\ \hline u \\ y \end{bmatrix} = \begin{bmatrix} P_{12} & P_{11} \\ 0 & I \\ \hline I & 0 \\ P_{22} & P_{21} \end{bmatrix} \begin{bmatrix} u \\ w \end{bmatrix} = \begin{bmatrix} G_1 \\ \hline \tilde{G}_2 \end{bmatrix} \begin{bmatrix} u \\ w \end{bmatrix} \tag{5.26}$$

where $G_1 = \begin{bmatrix} P_{12} & P_{11} \\ 0 & I \end{bmatrix}$ and $\tilde{G}_2 = \begin{bmatrix} I & 0 \\ P_{22} & P_{21} \end{bmatrix}$. It can be found that

$$\begin{bmatrix} z \\ w \end{bmatrix} = \begin{bmatrix} P_{12} & P_{11} \\ 0 & I \end{bmatrix} \begin{bmatrix} I & 0 \\ P_{22} & P_{21} \end{bmatrix}^{-1} \begin{bmatrix} u \\ y \end{bmatrix} = G \begin{bmatrix} u \\ y \end{bmatrix}, \tag{5.27}$$

where

$$G = \begin{bmatrix} P_{12} - P_{11}P_{21}{}^{-1}P_{22} & P_{11}P_{21}{}^{-1} \\ -P_{21}{}^{-1}P_{22} & P_{21}{}^{-1} \end{bmatrix}. \tag{5.28}$$

In fact, G can also be obtained directly from Property 5.3 in that

$$\begin{aligned} \mathrm{CSD_r}\left(G_1, \mathrm{CSD_l}\left(\tilde{G}_2, K\right)\right) &= \mathrm{CSD_r}\left(G_1 M, \mathrm{CSD_l}\left(\tilde{G}_2 M, K\right)\right) \\ &= \mathrm{CSD_r}\left(G_1 M, \mathrm{CSD_l}\left(I, K\right)\right) \\ &= \mathrm{CSD_r}\left(G_1 M, K\right) \\ &= \mathrm{CSD_r}\left(G, K\right) \end{aligned} \tag{5.29}$$

where $M = \begin{bmatrix} I & 0 \\ P_{22} & P_{21} \end{bmatrix}^{-1} = \tilde{G}_2^{-1}$.

Similarly, for the case of $\dim(z) = \dim(u)$ and an invertible P_{12}, there exists a left CSD matrix $\tilde{G}$ of $\begin{bmatrix} z \\ w \end{bmatrix} \mapsto \begin{bmatrix} u \\ y \end{bmatrix}$ such that $\mathrm{CSD_l}\left(\tilde{G}, K\right) = \mathrm{LFT_l}\left(P, K\right)$. One has, from (5.26), as illustrated in Fig. 5.9 in the following:

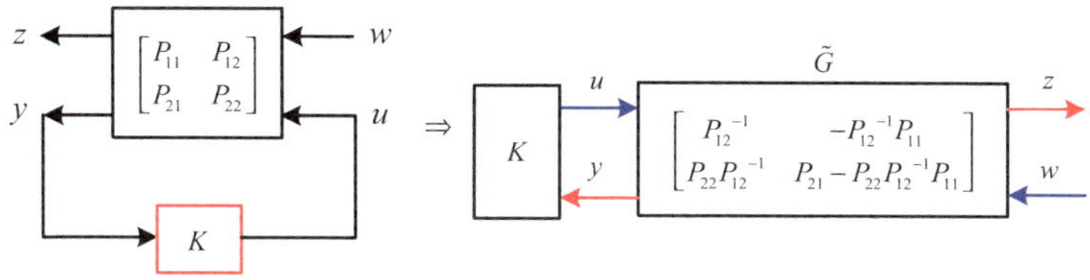

Fig. 5.9 LFT to left CSD

$$\begin{bmatrix} z' \\ y' \end{bmatrix} = \begin{bmatrix} P_{12} & 0 \\ P_{22} & -I \end{bmatrix} \begin{bmatrix} u \\ y \end{bmatrix} = \begin{bmatrix} I & -P_{11} \\ 0 & P_{21} \end{bmatrix} \begin{bmatrix} z \\ w \end{bmatrix} \Rightarrow \begin{bmatrix} z' \\ y' \end{bmatrix}$$
$$= G_2 \begin{bmatrix} u \\ y \end{bmatrix} = \tilde{G}_1 \begin{bmatrix} z \\ w \end{bmatrix}. \tag{5.30}$$

It can be seen that

$$\begin{bmatrix} u \\ y \end{bmatrix} = \begin{bmatrix} P_{12} & 0 \\ P_{22} & -I \end{bmatrix}^{-1} \begin{bmatrix} I & -P_{11} \\ 0 & P_{21} \end{bmatrix} \begin{bmatrix} z \\ w \end{bmatrix} = \tilde{G} \begin{bmatrix} z \\ w \end{bmatrix}, \tag{5.31}$$

where

$$\tilde{G} = \begin{bmatrix} P_{12}{}^{-1} & -P_{12}{}^{-1} P_{11} \\ P_{22} P_{12}{}^{-1} & P_{21} - P_{22} P_{12}{}^{-1} P_{11} \end{bmatrix}. \tag{5.32}$$

Here, $\tilde{G}$ can be also obtained directly from Property 5.4, giving

$$\begin{aligned} \mathrm{CSD_l}\left(\tilde{G}_1, \mathrm{CSD_r}(G_2, K)\right) &= \mathrm{CSD_l}\left(\tilde{M}\tilde{G}_1, \mathrm{CSD_r}\left(\tilde{M}G_2, K\right)\right) \\ &= \mathrm{CSD_r}\left(\tilde{M}\tilde{G}_1, \mathrm{CSD_r}(I, K)\right) \\ &= \mathrm{CSD_l}\left(\tilde{M}\tilde{G}_1, K\right) \\ &= \mathrm{CSD_l}\left(\tilde{G}, K\right) \end{aligned} \tag{5.33}$$

where $\tilde{M} = \begin{bmatrix} P_{12} & 0 \\ P_{22} & -I \end{bmatrix} = G_2^{-1}$. The above are summarized in the following lemma.

Lemma 5.1 *Given an LFT matrix P, there exists a right CSD matrix G such that* $\mathrm{CSD_r}(G,K) = \mathrm{LFT_l}(P,K)$ *if* P_{21} *is invertible, or there exists a left CSD matrix* $\tilde{G}$ *such that* $\mathrm{CSD_l}\left(\tilde{G}, K\right) = \mathrm{LFT_l}(P, K)$ *if* P_{12} *is invertible.*

Furthermore, if both P_{21} and P_{12} are invertible, then it can be gathered that

$$G\tilde{G} = \begin{bmatrix} P_{12} & P_{11} \\ 0 & I \end{bmatrix} \begin{bmatrix} I & 0 \\ P_{22} & P_{21} \end{bmatrix}^{-1} \begin{bmatrix} P_{12} & 0 \\ P_{22} & -I \end{bmatrix}^{-1} \begin{bmatrix} I & -P_{11} \\ 0 & P_{21} \end{bmatrix} = I. \tag{5.34}$$

Consequently,

$$T = \mathrm{LFT}_{\mathrm{l}}(P, K) = \mathrm{CSD}_{\mathrm{r}}(G, K) = \mathrm{CSD}_{\mathrm{l}}\left(\tilde{G}, K\right), \tag{5.35}$$

and conversely,

$$K = \mathrm{CSD}_{\mathrm{r}}\left(\tilde{G}, T\right) = \mathrm{CSD}_{\mathrm{l}}(G, T). \tag{5.36}$$

5.4 Transformation from LFT to Cascaded CSDs

For the transformation from LFT to CSD as described in the previous section, there is a condition that P_{12} or P_{21} should be invertible. This condition can however be relaxed, as explained in this section. For any standard control configuration (SCC) plant P of $\begin{bmatrix} w \\ u \end{bmatrix} \mapsto \begin{bmatrix} z \\ y \end{bmatrix}$ and $u = Ky$, as depicted in the left-side diagram of Fig. 5.8, it can be represented by *two* coupling CSDs as stated in the following. From

$$\begin{bmatrix} z \\ y \end{bmatrix} = \begin{bmatrix} P_{11} & P_{12} \\ P_{21} & P_{22} \end{bmatrix} \begin{bmatrix} w \\ u \end{bmatrix}, \tag{5.37}$$

we derive

$$\begin{bmatrix} z \\ y \\ \hline w \\ u \end{bmatrix} = \begin{bmatrix} P_{11} & P_{12} \\ P_{21} & P_{22} \\ \hline I & 0 \\ 0 & I \end{bmatrix} \begin{bmatrix} w \\ u \end{bmatrix} \Rightarrow \begin{bmatrix} z \\ w \\ \hline u \\ y \end{bmatrix} = \begin{bmatrix} P_{12} & P_{11} \\ 0 & I \\ \hline I & 0 \\ P_{22} & P_{21} \end{bmatrix} \begin{bmatrix} u \\ w \end{bmatrix}. \tag{5.38}$$

LFT matrix P can then be decomposed as

$$\begin{bmatrix} z \\ w \end{bmatrix} = \begin{bmatrix} P_{11} & P_{12} \\ 0 & I \end{bmatrix} \begin{bmatrix} u \\ w \end{bmatrix} \text{ and} \tag{5.39}$$

$$\begin{bmatrix} u \\ y \end{bmatrix} = \begin{bmatrix} I & 0 \\ P_{22} & P_{21} \end{bmatrix} \begin{bmatrix} u \\ w \end{bmatrix}, \tag{5.40}$$

which in turn can be represented by a right CSD associated with a left CSD, as illustrated in the right-side diagram of Fig. 5.10. Since $u = Ky$, one has, from the lower part of (5.40),

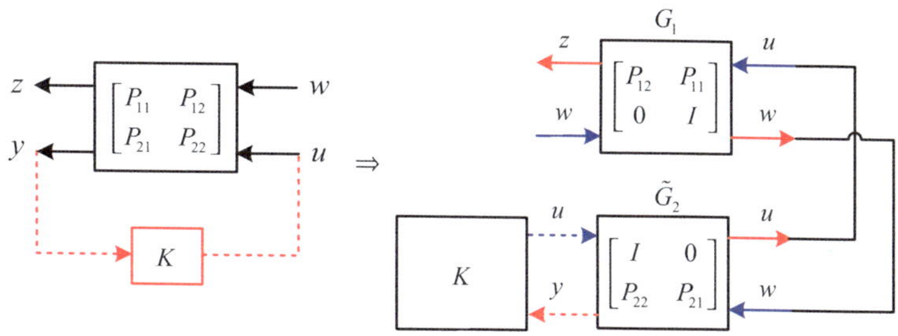

Fig. 5.10 Right CSD associated with left CSD

$$\begin{bmatrix} K \\ I \end{bmatrix} y = \begin{bmatrix} I & 0 \\ P_{22} & P_{21} \end{bmatrix} \begin{bmatrix} u \\ w \end{bmatrix}, \tag{5.41}$$

and therefore,

$$\begin{aligned} 0 &= \begin{bmatrix} I & -K \end{bmatrix} \begin{bmatrix} K \\ I \end{bmatrix} y = \begin{bmatrix} I & -K \end{bmatrix} \begin{bmatrix} I & 0 \\ P_{22} & P_{21} \end{bmatrix} \begin{bmatrix} u \\ w \end{bmatrix} \\ &= \begin{bmatrix} I - KP_{22} & -KP_{21} \end{bmatrix} \begin{bmatrix} u \\ w \end{bmatrix}. \end{aligned} \tag{5.42}$$

Consequently, this gives

$$u = (I - KP_{22})^{-1} KP_{21} w. \tag{5.43}$$

Furthermore, from (5.39), one has

$$\begin{bmatrix} z \\ w \end{bmatrix} = \begin{bmatrix} P_{12} & P_{11} \\ 0 & I \end{bmatrix} \begin{bmatrix} u \\ w \end{bmatrix} = \begin{bmatrix} P_{11} + P_{12}(I - KP_{22})^{-1} KP_{21} \\ I \end{bmatrix} w. \tag{5.44}$$

This reveals that the transfer function $w \mapsto z$ can be represented by a right CSD matrix associated with a left CSD matrix; in below, LFT_l stands for the lower linear fractional transformation as previously defined in Chap. 4:

$$\begin{aligned} z = \text{LFT}_l\left(\begin{bmatrix} P_{11} & P_{12} \\ P_{21} & P_{22} \end{bmatrix}, K \right) w = \text{CSD}_r\Bigg(\begin{bmatrix} P_{12} & P_{11} \\ 0 & I \end{bmatrix}, \\ \text{CSD}_l\left(\begin{bmatrix} I & 0 \\ P_{22} & P_{21} \end{bmatrix}, K \right) \Bigg) w. \end{aligned} \tag{5.45}$$

Example 5.3 Consider the SCC plant as given in (5.37), using MATLAB to determine its corresponding right CSD associated with a left CSD representation.

The following MATLAB code is an example for readers' reference.

```
clc;
clear;
disp('SCC_2_(CSDr associated with CSDl) method')
syms P11 P12 P21 P22 K S;
syms p11 p12 p21 p22 k s;
P11=input('z => u (P11) ?');
P12=input('z => y (P12) ?');
P21=input('w => u (P21) ?');
P22=input('w => y (P22) ?');
K=input('Please input K:');
G1_11=P12;
G1_12=P11;
G1_21=zeros(size(P12));
G1_22=eye(size(P11));
G1=[G1_11 G1_12;G1_21 G1_22 ] % upper triangle matrix
G2_11=eye(size(P22)) ;
G2_12=zeros(size(P21));
G2_21=P22;
G2_22=P21;
G2=[G2_11 G2_12;G2_21 G2_22] % lower triangle matrix
if (G2_11-K*G2_21==0)
    disp('The transfer function is singular !')
    disp('Plz enter Ctrl c !')
    pause
else
Transfer_function_w_to_u=simplify(-inv(G2_11-K*G2_21)
  *(G2_12-K*G2_22))
k=simplify(-inv(G2_11-K*G2_21)*(G2_12-K*G2_22));
Transfer_function_w_to_z=simplify((G1_11*k+G1_12)*
  inv(G1_21*k+G1_22))
end
```

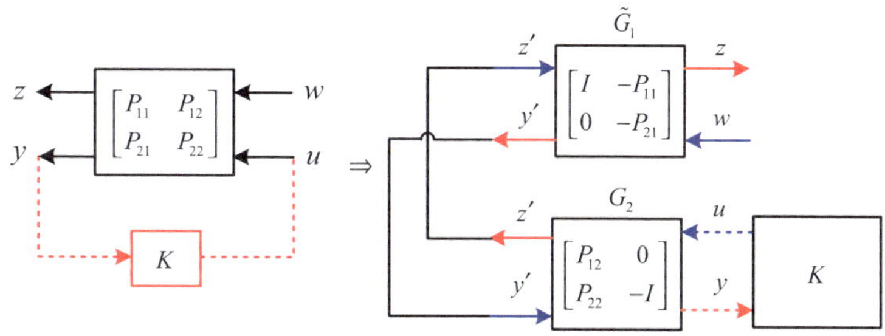

Fig. 5.11 Left CSD associated with right CSD

Likewise, (5.37) also creates

$$\begin{bmatrix} 0 \\ 0 \end{bmatrix} = \left[\begin{array}{cc|cc} I & -P_{11} & P_{12} & 0 \\ 0 & -P_{21} & P_{22} & -I \end{array}\right] \begin{bmatrix} z \\ w \\ u \\ y \end{bmatrix}. \tag{5.46}$$

Let

$$\begin{bmatrix} z' \\ y' \end{bmatrix} = \begin{bmatrix} I & -P_{11} \\ 0 & -P_{21} \end{bmatrix} \begin{bmatrix} z \\ w \end{bmatrix} = \begin{bmatrix} P_{12} & 0 \\ P_{22} & -I \end{bmatrix} \begin{bmatrix} u \\ y \end{bmatrix}. \tag{5.47}$$

Then, one can obtain the transfer function $w \mapsto z$, which is represented by a left CSD matrix associated with a right CSD matrix with the latter being terminated by K at the right end, as illustrated in Fig. 5.11, where z' and y' are the intermediate signals. From their definition in (5.47), it is clear that the dimensions of z' and y' are the same as that of z and y, respectively. It is a common usage in this book that a signal with the prime symbol means the same dimension as that of the original signal, but of course, they are different signals.

From (5.47) and $u = Ky$, one has

$$\begin{bmatrix} z' \\ y' \end{bmatrix} = \begin{bmatrix} P_{12} & 0 \\ P_{22} & -I \end{bmatrix} \begin{bmatrix} K \\ I \end{bmatrix} y = \begin{bmatrix} P_{12}K \\ P_{22}K - I \end{bmatrix} y, \tag{5.48}$$

and hence,

$$z' = \mathrm{CSD_r}\left(\begin{bmatrix} P_{12} & 0 \\ P_{22} & -I \end{bmatrix}, K \right) y' = -P_{12}K(I - P_{22}K)^{-1} y'. \tag{5.49}$$

Let

$$S = \mathrm{CSD_r}\left(\begin{bmatrix} P_{12} & 0 \\ P_{22} & -I \end{bmatrix}, K\right). \tag{5.50}$$

(5.47) offers

$$\begin{bmatrix} z' \\ y' \end{bmatrix} = \begin{bmatrix} S \\ I \end{bmatrix} y' = \begin{bmatrix} I & -P_{11} \\ 0 & -P_{21} \end{bmatrix} \begin{bmatrix} z \\ w \end{bmatrix}, \tag{5.51}$$

and thereby,

$$0 = \begin{bmatrix} I & -S \end{bmatrix} \begin{bmatrix} S \\ I \end{bmatrix} y' = \begin{bmatrix} I & -S \end{bmatrix} \begin{bmatrix} I & -P_{11} \\ 0 & -P_{21} \end{bmatrix} \begin{bmatrix} z \\ w \end{bmatrix}. \tag{5.52}$$

This shows that

$$0 = \begin{bmatrix} I & -P_{11} - \left(P_{12}K(I - P_{22}K)^{-1}\right) P_{21} \end{bmatrix} \begin{bmatrix} z \\ w \end{bmatrix} \tag{5.53}$$

and

$$z = \left[P_{11} + P_{12}K(I - P_{22}K)^{-1} P_{21}\right] w = \mathrm{LFT_l}\,(P, K)\, w. \tag{5.54}$$

One concludes that the transfer function $w \mapsto z$ can be represented by a left CSD matrix associated with a right CSD matrix as

$$z = \mathrm{LFT_l}\left(\begin{bmatrix} P_{11} & P_{12} \\ P_{21} & P_{22} \end{bmatrix}, K\right) w = \mathrm{CSD_l}\left(\begin{bmatrix} I & -P_{11} \\ 0 & -P_{21} \end{bmatrix}, \right.$$
$$\left. \mathrm{CSD_r}\left(\begin{bmatrix} P_{12} & 0 \\ P_{22} & -I \end{bmatrix}, K\right)\right) w. \tag{5.55}$$

Example 5.4 Consider the SCC plant as given in (5.37), using MATLAB to determine its corresponding left CSD associated with a right CSD representation.

The following MATLAB code is an example for readers' reference.

```
clc; clear;
disp('SCC_2_(CSDl associated with CSDr) method')
syms P11 P12 P21 P22 K S;
```

```
syms p11 p12 p21 p22 k s;
P11=input('z => u (P11) ?');
P12=input('z => y (P12) ?');
P21=input('w => u (P21) ?');
P22=input('w => y (P22) ?');
K=input('Please input K:');
G1_11=eye(size(P11));
G1_12=-P11;
G1_21=zeros(size(P21));
G1_22=-P21;
G1=[G1_11 G1_12;G1_21 G1_22 ] % upper triangle matrix
G2_11=P12;
G2_12=zeros(size(P12));
G2_21=P22;
G2_22=-eye(size(P22)) ;
G2=[G2_11 G2_12;G2_21 G2_22] % lower triangle matrix
if (G2_21*K+G2_22==0)
    disp('The transfer function is singular !')
    disp('Plz enter Ctrl c !')
    pause
else
Transfer_function_ye_to_ze=simplify((G2_11*K+G2_12)*
inv(G2_21*K+G2_22))
k=simplify((G2_11*K+G2_12)*inv(G2_21*K+G2_22));
Transfer_function_w_to_z=simplify(-inv(G1_11-k*G1_21)
 *(G1_12-k*G1_22))
end
```

5.5 Transformation from CSD to LFT matrix

Conversely, the aforementioned techniques can be employed to characterize the transformation from a given right or left CSD matrix to its corresponding LFT matrix. First, consider the transformation from a right CSD matrix G of $\begin{bmatrix} u \\ y \end{bmatrix} \mapsto \begin{bmatrix} z \\ w \end{bmatrix}$ to the LFT matrix P of $\begin{bmatrix} w \\ u \end{bmatrix} \mapsto \begin{bmatrix} z \\ y \end{bmatrix}$ in Fig. 5.12, where G_{22} is assumed invertible.

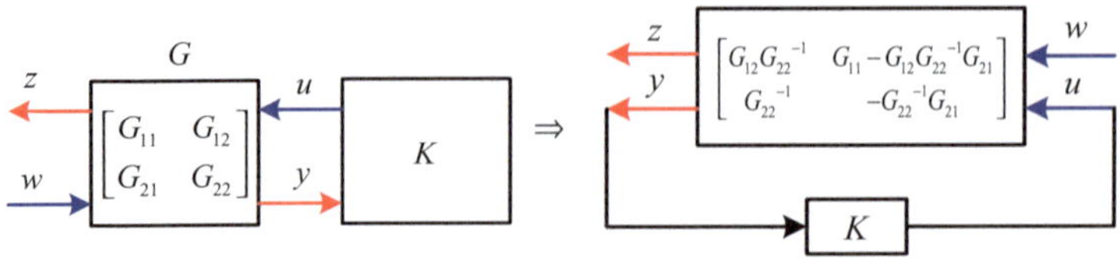

Fig. 5.12 Right CSD to LFT

From

$$\begin{bmatrix} z \\ w \end{bmatrix} = \begin{bmatrix} G_{11} & G_{12} \\ G_{21} & G_{22} \end{bmatrix} \begin{bmatrix} u \\ y \end{bmatrix}, \tag{5.56}$$

one has

$$\begin{bmatrix} z \\ w \\ \hline u \\ y \end{bmatrix} = \begin{bmatrix} G_{11} & G_{12} \\ G_{21} & G_{22} \\ \hline I & 0 \\ 0 & I \end{bmatrix} \begin{bmatrix} u \\ y \end{bmatrix} \Rightarrow \begin{bmatrix} z \\ y \\ \hline w \\ u \end{bmatrix} = \begin{bmatrix} G_{11} & G_{12} \\ 0 & I \\ \hline G_{21} & G_{22} \\ I & 0 \end{bmatrix} \begin{bmatrix} u \\ y \end{bmatrix}; \tag{5.57}$$

hence,

$$\begin{aligned} \begin{bmatrix} z \\ y \end{bmatrix} &= \begin{bmatrix} G_{11} & G_{12} \\ 0 & I \end{bmatrix} \begin{bmatrix} u \\ y \end{bmatrix} = \begin{bmatrix} G_{11} & G_{12} \\ 0 & I \end{bmatrix} \begin{bmatrix} G_{21} & G_{22} \\ I & 0 \end{bmatrix}^{-1} \begin{bmatrix} w \\ u \end{bmatrix} \\ &= \begin{bmatrix} P_{11} & P_{12} \\ P_{21} & P_{22} \end{bmatrix} \begin{bmatrix} w \\ u \end{bmatrix}, \end{aligned} \tag{5.58}$$

where

$$P = \begin{bmatrix} P_{11} & P_{12} \\ P_{21} & P_{22} \end{bmatrix} = \begin{bmatrix} G_{12}{G_{22}}^{-1} & G_{11} - G_{12}{G_{22}}^{-1}G_{21} \\ {G_{22}}^{-1} & -{G_{22}}^{-1}G_{21} \end{bmatrix}. \tag{5.59}$$

It can also be verified by direct calculations that

$$\begin{aligned} \mathrm{LFT_l}\,(P, K) &= P_{11} + P_{12}K(I - P_{22}K)^{-1}P_{21} \\ &= G_{12}{G_{22}}^{-1} + \left(G_{11} - G_{12}{G_{22}}^{-1}G_{21}\right) K\left(I + {G_{22}}^{-1}G_{21}K\right)^{-1}{G_{22}}^{-1} \\ &= (G_{12} + G_{11}K)\,(G_{22} + G_{21}K)^{-1} \\ &= \mathrm{CSD_r}\,(G, K)\,. \end{aligned} \tag{5.60}$$

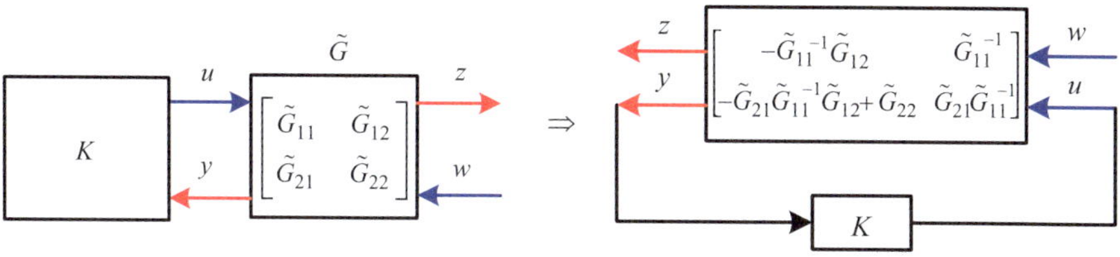

Fig. 5.13 Left CSD to LFT

Similarly, one can also derive the transformation from a left CSD matrix $\tilde{G}$ of $\begin{bmatrix} z \\ w \end{bmatrix} \mapsto \begin{bmatrix} u \\ y \end{bmatrix}$ to its LFT matrix P of $\begin{bmatrix} z \\ y \end{bmatrix} \mapsto \begin{bmatrix} w \\ u \end{bmatrix}$ as shown in Fig. 5.13, where $\tilde{G}_{11}$ is invertible.

From

$$\begin{bmatrix} u \\ y \\ \hline z \\ w \end{bmatrix} = \begin{bmatrix} \tilde{G}_{11} & \tilde{G}_{12} \\ \tilde{G}_{21} & \tilde{G}_{22} \\ \hline I & 0 \\ 0 & I \end{bmatrix} \begin{bmatrix} z \\ w \end{bmatrix} \Rightarrow \begin{bmatrix} z \\ y \\ \hline w \\ u \end{bmatrix} = \begin{bmatrix} I & 0 \\ \tilde{G}_{11} & \tilde{G}_{12} \\ \hline 0 & I \\ \tilde{G}_{21} & \tilde{G}_{22} \end{bmatrix} \begin{bmatrix} z \\ w \end{bmatrix}, \tag{5.61}$$

one has

$$\begin{bmatrix} z \\ y \end{bmatrix} = \begin{bmatrix} I & 0 \\ \tilde{G}_{21} & \tilde{G}_{22} \end{bmatrix} \begin{bmatrix} z \\ w \end{bmatrix} = \begin{bmatrix} I & 0 \\ \tilde{G}_{21} & \tilde{G}_{22} \end{bmatrix} \begin{bmatrix} 0 & I \\ \tilde{G}_{11} & \tilde{G}_{12} \end{bmatrix}^{-1}$$
$$\times \begin{bmatrix} w \\ u \end{bmatrix} = \begin{bmatrix} P_{11} & P_{12} \\ P_{21} & P_{21} \end{bmatrix} \begin{bmatrix} w \\ u \end{bmatrix}, \tag{5.62}$$

where

$$\begin{bmatrix} P_{11} & P_{12} \\ P_{21} & P_{21} \end{bmatrix} = \begin{bmatrix} -\tilde{G}_{11}^{-1}\tilde{G}_{12} & \tilde{G}_{11}^{-1} \\ -\tilde{G}_{21}\tilde{G}_{11}^{-1}\tilde{G}_{12} + \tilde{G}_{22} & \tilde{G}_{21}\tilde{G}_{11}^{-1} \end{bmatrix}. \tag{5.63}$$

This reveals that there exists a transformation from a left CSD $\tilde{G}$ to the LFT matrix P such that $\mathrm{LFT}_l(P, K) = \mathrm{CSD}_l(\tilde{G}, K)$. Similar to (5.60), one can show that

$$\begin{aligned} \mathrm{LFT}_l(P, K) &= P_{11} + P_{12}(I - KP_{22})^{-1}KP_{21} \\ &= -\tilde{G}_{11}^{-1}\tilde{G}_{12} + \tilde{G}_{11}^{-1}\left(I - K\tilde{G}_{21}\tilde{G}_{11}^{-1}\right)^{-1}K\left(-\tilde{G}_{21}\tilde{G}_{11}^{-1}\tilde{G}_{12} + \tilde{G}_{22}\right) \\ &= -\left(\tilde{G}_{11} - K\tilde{G}_{21}\right)^{-1}\left(\tilde{G}_{12} - K\tilde{G}_{22}\right) \\ &= \mathrm{CSD}_l\left(\tilde{G}, K\right). \end{aligned} \tag{5.64}$$

Fig. 5.14 Simple RC circuit

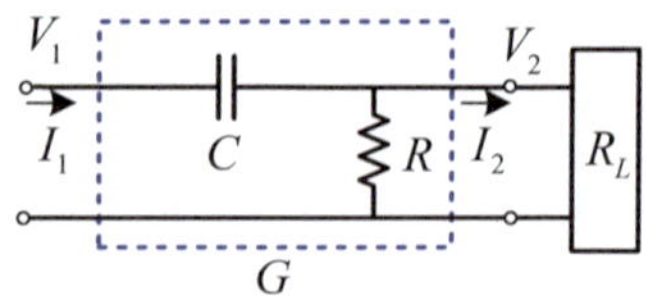

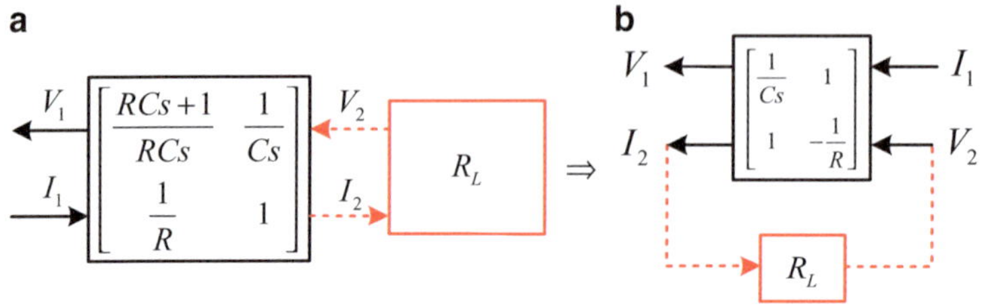

Fig. 5.15 Right CSD to LFT: (**a**) Right CSD representation and (**b**) LFT representation

Example 5.5 For an RC circuit as depicted in Fig. 5.14, determine a right CSD matrix G of $\begin{bmatrix} V_2 \\ I_2 \end{bmatrix} \mapsto \begin{bmatrix} V_1 \\ I_1 \end{bmatrix}$ and its corresponding LFT matrix P of $\begin{bmatrix} I_1 \\ V_2 \end{bmatrix} \mapsto \begin{bmatrix} V_1 \\ I_2 \end{bmatrix}$ such that $\mathrm{LFT}_1(P,\ R_L) = \mathrm{CSD}_r(G,\ R_L)$.

UtilizingKirchhoff's laws, Fig. 5.14 gives $V_2 = I_2 R_L$ and

$$\begin{cases} V_1 = \frac{1}{Cs} I_1 + V_2 \\ I_2 = I_1 - \frac{1}{R} V_2 \end{cases} \Rightarrow \begin{bmatrix} V_1 \\ I_2 \end{bmatrix} = \begin{bmatrix} P_{11} & P_{12} \\ P_{21} & P_{22} \end{bmatrix} \begin{bmatrix} I_1 \\ V_2 \end{bmatrix} = \begin{bmatrix} \frac{1}{Cs} & 1 \\ 1 & -\frac{1}{R} \end{bmatrix} \begin{bmatrix} I_1 \\ V_2 \end{bmatrix}. \tag{5.65}$$

Rearranging (5.65) yields

$$\begin{cases} V_1 = \frac{1}{Cs} I_1 + V_2 \\ I_2 = I_1 - \frac{1}{R} V_2 \end{cases} \Rightarrow \begin{cases} V_1 = \frac{RCs+1}{RCs} V_2 + \frac{1}{Cs} I_2 \\ I_1 = \frac{1}{R} V_2 + I_2 \end{cases} \Rightarrow \begin{bmatrix} V_1 \\ I_1 \end{bmatrix} = \begin{bmatrix} \frac{RCs+1}{RCs} & \frac{1}{Cs} \\ \frac{1}{R} & 1 \end{bmatrix} \begin{bmatrix} V_2 \\ I_2 \end{bmatrix}. \tag{5.66}$$

Both LFT and CSD representations are depicted in Fig. 5.15. Alternatively, the right CSD matrix G can also be found directly by the transmission parameter matrices described in Chap. 3 as

Fig. 5.16 RC circuit

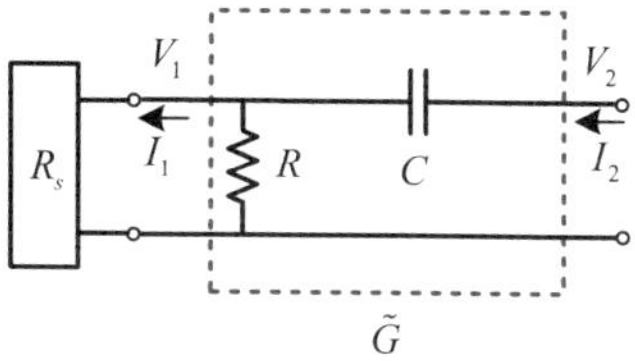

$$\begin{bmatrix} V_1 \\ I_1 \end{bmatrix} = \begin{bmatrix} 1 & \frac{1}{Cs} \\ 0 & 1 \end{bmatrix} \begin{bmatrix} 1 & 0 \\ \frac{1}{R} & 1 \end{bmatrix} \begin{bmatrix} V_2 \\ I_2 \end{bmatrix} = \begin{bmatrix} \frac{RCs+1}{RCs} & \frac{1}{Cs} \\ \frac{1}{R} & 1 \end{bmatrix} \begin{bmatrix} V_2 \\ I_2 \end{bmatrix}$$
$$= \begin{bmatrix} G_{11} & G_{12} \\ G_{21} & G_{22} \end{bmatrix} \begin{bmatrix} V_2 \\ I_2 \end{bmatrix}. \tag{5.67}$$

Furthermore, one can find LFT matrix P using (5.59) with data from (5.66) as

$$P = \begin{bmatrix} G_{12}G_{22}^{-1} & G_{11} - G_{12}G_{22}{}^{-1}G_{21} \\ G_{22}{}^{-1} & -G_{22}^{-1}G_{21} \end{bmatrix} = \begin{bmatrix} \frac{1}{Cs} & 1 \\ 1 & -\frac{1}{R} \end{bmatrix}. \tag{5.68}$$

Now consider the circuit with a load of R_L. The equivalent impedance $Z = \frac{V_1}{I_1}$ can be derived by Kirchhoff's laws from Fig. 5.14 as

$$Z = \frac{V_1}{I_1} = \frac{1}{Cs} + \frac{RR_L}{R + R_L}. \tag{5.69}$$

One can obtain the same result via the LFT or CSD approach, respectively, as

$$Z = \frac{V_1}{I_1} = \mathrm{CSD_r}\left(G, R_L\right) = \frac{1}{Cs} + \frac{RR_L}{R + R_L}, \tag{5.70}$$

$$Z = \frac{V_1}{I_1} = \mathrm{LFT_l}\left(P, R_L\right) = \frac{1}{Cs} + R_L\left(1 + \frac{R_L}{R}\right)^{-1} = \frac{1}{Cs} + \frac{RR_L}{R + R_L}. \tag{5.71}$$

Clearly, (5.69), (5.70), and (5.71) are the same. These have shown that

$$Z = \frac{V_1}{I_1} = \mathrm{LFT_l}\left(P, R_L\right) = \mathrm{CSD_r}\left(G, R_L\right). \tag{5.72}$$

Example 5.6 For the RC circuit depicted in Fig. 5.16, determine a left CSD matrix $\tilde{G}$ of $\begin{bmatrix} V_2 \\ I_2 \end{bmatrix} \mapsto \begin{bmatrix} V_1 \\ I_1 \end{bmatrix}$ and its corresponding LFT matrix P of $\begin{bmatrix} I_2 \\ V_1 \end{bmatrix} \mapsto \begin{bmatrix} V_2 \\ I_1 \end{bmatrix}$ such that $\mathrm{LFT_l}\left(P, R_s\right) = \mathrm{CSD_l}\left(\tilde{G}, R_s\right)$.

Similarly, by Kirchhoff's laws, one has $V_1 = R_s I_1$ and

$$\begin{cases} V_2 = \frac{1}{Cs} I_2 + V_1 \\ I_1 = I_2 - \frac{1}{R} V_1 \end{cases} \Rightarrow \begin{bmatrix} V_2 \\ I_1 \end{bmatrix} = \begin{bmatrix} P_{11} & P_{12} \\ P_{21} & P_{22} \end{bmatrix} \begin{bmatrix} I_2 \\ V_1 \end{bmatrix} = \begin{bmatrix} \frac{1}{Cs} & 1 \\ 1 & -\frac{1}{R} \end{bmatrix} \begin{bmatrix} I_2 \\ V_1 \end{bmatrix}. \tag{5.73}$$

Rearranging (5.73) yields

$$\begin{cases} V_1 = V_2 - \dfrac{1}{Cs} I_2 \\ I_1 = -\dfrac{1}{R} V_2 + \dfrac{RCS+1}{RCs} I_2 \end{cases} \Rightarrow \begin{bmatrix} V_1 \\ I_1 \end{bmatrix} = \begin{bmatrix} \tilde{G}_{11} & \tilde{G}_{12} \\ \tilde{G}_{21} & \tilde{G}_{22} \end{bmatrix} \begin{bmatrix} V_2 \\ I_2 \end{bmatrix}$$
$$= \begin{bmatrix} 1 & -\dfrac{1}{Cs} \\ -\dfrac{1}{R} & \dfrac{RCS+1}{RCs} \end{bmatrix} \begin{bmatrix} V_2 \\ I_2 \end{bmatrix}. \tag{5.74}$$

As in Example 5.5, the left CSD matrix $\tilde{G}$ can also be found by the transmission parameter matrices as

$$\begin{bmatrix} V_1 \\ I_1 \end{bmatrix} = \begin{bmatrix} 1 & 0 \\ -\dfrac{1}{R} & 1 \end{bmatrix} \begin{bmatrix} 1 & -\dfrac{1}{Cs} \\ 0 & 1 \end{bmatrix} \begin{bmatrix} V_2 \\ I_2 \end{bmatrix} = \begin{bmatrix} 1 & -\dfrac{1}{Cs} \\ -\dfrac{1}{R} & \dfrac{RCS+1}{RCs} \end{bmatrix} \begin{bmatrix} V_2 \\ I_2 \end{bmatrix}. \tag{5.75}$$

Similarly, the LFT matrix P of $\begin{bmatrix} I_2 \\ V_1 \end{bmatrix} \mapsto \begin{bmatrix} V_2 \\ I_1 \end{bmatrix}$ can be obtained from (5.75) such that

$$P = \begin{bmatrix} -\tilde{G}_{11}^{-1}\tilde{G}_{12} & \tilde{G}_{11}^{-1} \\ -\tilde{G}_{21}\tilde{G}_{11}^{-1}\tilde{G}_{12} + \tilde{G}_{22} & \tilde{G}_{21}\tilde{G}_{11}^{-1} \end{bmatrix} = \begin{bmatrix} \frac{1}{Cs} & 1 \\ 1 & -\frac{1}{R} \end{bmatrix}. \tag{5.76}$$

The LFT representation is depicted in Fig. 5.17. With R_s and $V_1 = R_s I_1$, the equivalent impedance $Z = \frac{V_1}{I_1}$ can be derived by Kirchhoff's laws as

$$Z = R \,\|\, R_s + \frac{1}{Cs} = \frac{RR_s}{R+R_s} + \frac{1}{Cs}. \tag{5.77}$$

One can also obtain the same result via the LFT formula or CSD approach as

$$Z = \frac{V_2}{I_2} = \mathrm{CSD}_l\left(\tilde{G}, R_s\right) = \frac{1}{Cs} + \frac{RR_s}{R+R_s} \text{ and} \tag{5.78}$$

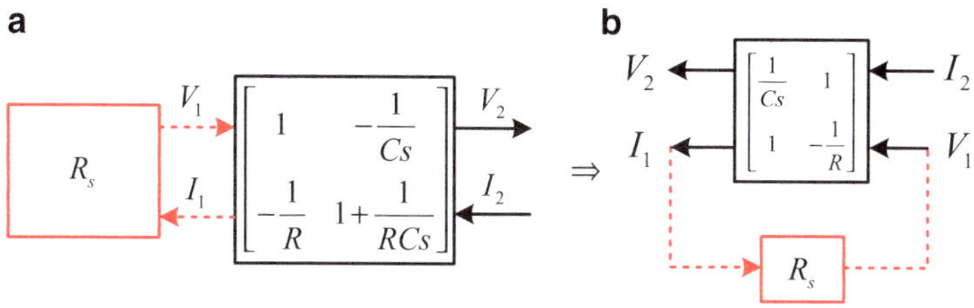

Fig. 5.17 Left CSD to LFT: (**a**) Left CSD representation and (**b**) LFT representation

$$\mathrm{LFT_l}\,(P, R_s) = \frac{1}{Cs} + R_s\left(I + \frac{R_s}{R}\right)^{-1} = \frac{1}{Cs} + \frac{RR_s}{R + R_s}. \tag{5.79}$$

Hence, $\mathrm{LFT_l}\,(P, R_s) = \mathrm{CSD_l}\left(\tilde{G}, R_s\right)$.

5.6 Applications of CSDs in State-Space Realizations

The applications of CSDs (e.g., state-space realizations, inversions, and similarity transformations) are useful and convenient for solving several control manipulations. Herein, consider a linear system $y = Pu$, where the state-space realization is $P(s) \overset{s}{=} \left[\begin{array}{c|c} A & B \\ \hline C & D \end{array}\right]$. That is,

$$\begin{bmatrix} \dot{x} \\ y \end{bmatrix} = \begin{bmatrix} A & B \\ C & D \end{bmatrix}\begin{bmatrix} x \\ u \end{bmatrix}, \tag{5.80}$$

and

$$P(s) = \mathrm{LFT_u}\left(\begin{bmatrix} A & B \\ C & D \end{bmatrix}, \frac{I}{s}\right). \tag{5.81}$$

Note that it has been assumed that all initial states are zero. The state-space realization of $P(s)$ can be represented by using a right CSD matrix associated with a left CSD matrix. From (5.80), one has

$$\begin{bmatrix} \dot{x} \\ y \\ \hline x \\ u \end{bmatrix} = \begin{bmatrix} A & B \\ C & D \\ \hline I & 0 \\ 0 & I \end{bmatrix}\begin{bmatrix} x \\ u \end{bmatrix} \Rightarrow \begin{bmatrix} y \\ u \\ \hline x \\ \dot{x} \end{bmatrix} = \begin{bmatrix} C & D \\ 0 & I \\ \hline I & 0 \\ A & B \end{bmatrix}\begin{bmatrix} x \\ u \end{bmatrix} \tag{5.82}$$

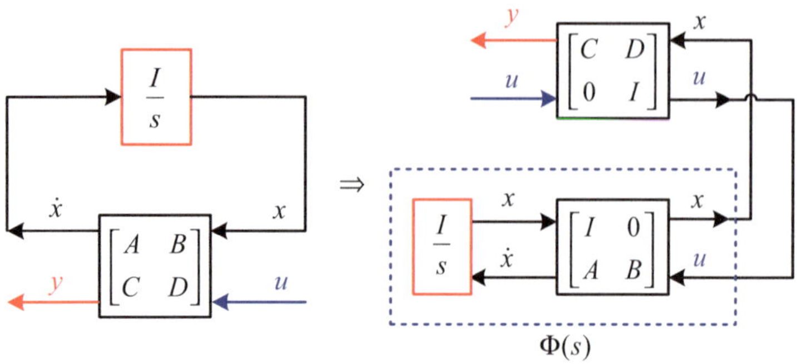

Fig. 5.18 Upper LFT to $\mathrm{CSD_r} \rightleftarrows \mathrm{CSD_l}$

as illustrated in Fig. 5.18.

Then, the transfer function $u \mapsto x$ is obtained by

$$\Phi(s) = \mathrm{CSD_l}\left(\begin{bmatrix} I & 0 \\ A & B \end{bmatrix}, \frac{I}{s}\right) = (sI - A)^{-1}B. \tag{5.83}$$

One can therefore have

$$y = \mathrm{CSD_r}\left(\begin{bmatrix} C & D \\ 0 & I \end{bmatrix}, \Phi(s)\right)u = \left(D + C(sI - A)^{-1}B\right)u. \tag{5.84}$$

This shows that, using a right CSD associated with a left CSD,

$$\begin{aligned}\mathrm{CSD_r}\left(\begin{bmatrix} C & D \\ 0 & I \end{bmatrix}, \mathrm{CSD_l}\left(\begin{bmatrix} I & 0 \\ A & B \end{bmatrix}, \frac{I}{s}\right)\right) &= \mathrm{LFT_u}\left(\begin{bmatrix} A & B \\ C & D \end{bmatrix}, \frac{I}{s}\right) \\ &= D + C(sI - A)^{-1}B. \end{aligned} \tag{5.85}$$

One can now consider the state feedback case. Let $u = Fx + Wu'$, where W is nonsingular. Then,

$$\begin{bmatrix} \dot{x} \\ y \\ u \end{bmatrix} = \left[\begin{array}{c|c} A & B \\ \hline C & D \\ 0 & I \end{array}\right]\begin{bmatrix} x \\ u \end{bmatrix} = \left[\begin{array}{c|c} A & B \\ \hline C & D \\ 0 & I \end{array}\right]\begin{bmatrix} I & 0 \\ F & W \end{bmatrix}\begin{bmatrix} x \\ u' \end{bmatrix} \Rightarrow \begin{bmatrix} \dot{x} \\ y \\ u \end{bmatrix} = \left[\begin{array}{c|c} A + BF & BW \\ \hline C + DF & DW \\ F & W \end{array}\right]\begin{bmatrix} x \\ u' \end{bmatrix}, \tag{5.86}$$

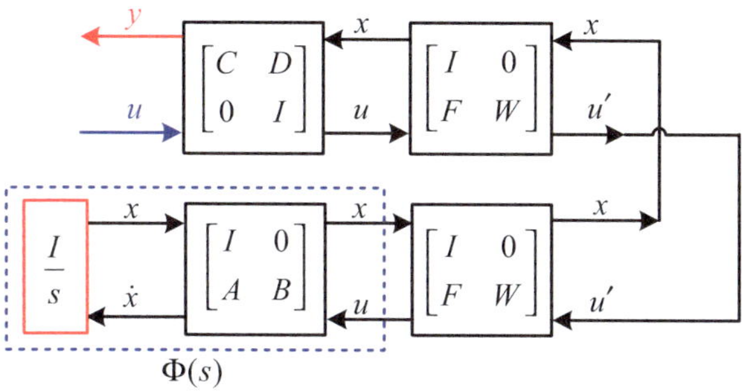

Fig. 5.19 CSD representations

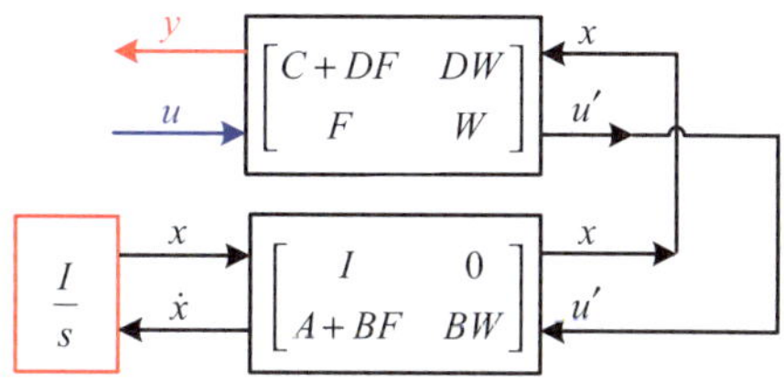

Fig. 5.20 CSD representations

and

$$\begin{bmatrix} y \\ u \\ \hline x \\ \dot{x} \end{bmatrix} = \begin{bmatrix} C & D \\ 0 & I \\ \hline I & 0 \\ A & B \end{bmatrix} \begin{bmatrix} x \\ u \end{bmatrix} = \begin{bmatrix} C & D \\ 0 & I \\ \hline I & 0 \\ A & B \end{bmatrix} \begin{bmatrix} I & 0 \\ F & W \end{bmatrix} \begin{bmatrix} x \\ u' \end{bmatrix} \Rightarrow \begin{bmatrix} y \\ u \\ \hline x \\ \dot{x} \end{bmatrix} = \begin{bmatrix} C+DF & DW \\ F & W \\ \hline I & 0 \\ A+BF & BW \end{bmatrix} \begin{bmatrix} x \\ u' \end{bmatrix}. \tag{5.87}$$

Note that by Property 5.3,

$$\mathrm{CSD_r}\left(\begin{bmatrix} I & 0 \\ F & W \end{bmatrix}, \mathrm{CSD_l}\left(\begin{bmatrix} I & 0 \\ F & W \end{bmatrix}, \Phi(s)\right)\right) = \Phi(s). \tag{5.88}$$

This concludes that the representations of Figs. 5.18 and 5.19 are equivalent, i.e., multiplying any $\begin{bmatrix} I & 0 \\ F & W \end{bmatrix}$ on the right by the top and bottom paths does not change the transfer function of $u \mapsto y$. Hence, Fig. 5.19 is equivalent to Fig. 5.20. Then, the block diagram of Fig. 5.19 gives the state-space realization of Fig. 5.21.

Furthermore, from (5.80), one has

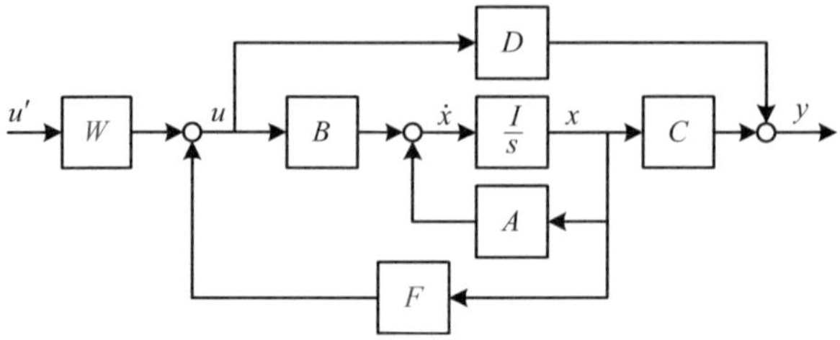

Fig. 5.21 State-space control block diagram

$$\begin{bmatrix} 0 \\ 0 \end{bmatrix} = \left[\begin{array}{cc|cc} A & B & -I & 0 \\ C & D & 0 & -I \end{array}\right] \begin{bmatrix} x \\ u \\ \dot{x} \\ y \end{bmatrix} \Rightarrow \begin{bmatrix} 0 \\ 0 \end{bmatrix} = \left[\begin{array}{cc|cc} 0 & B & A & -I \\ -I & D & C & 0 \end{array}\right] \begin{bmatrix} y \\ u \\ x \\ \dot{x} \end{bmatrix} \tag{5.89}$$

or

$$\begin{bmatrix} 0 \\ 0 \end{bmatrix} = \left[\begin{array}{cc|cc} -I & D & C & 0 \\ 0 & B & A & -I \end{array}\right] \begin{bmatrix} y \\ u \\ x \\ \dot{x} \end{bmatrix} \Rightarrow \begin{bmatrix} -I & D \\ 0 & B \end{bmatrix} \begin{bmatrix} y \\ u \end{bmatrix} = \begin{bmatrix} -C & 0 \\ -A & I \end{bmatrix} \begin{bmatrix} x \\ \dot{x} \end{bmatrix}. \tag{5.90}$$

A left CSD associated with a right CSD is then given by

Then the transfer function of $u \mapsto y$ could be given by

$$\frac{y}{u} = \mathrm{CSD_l}\left(\begin{bmatrix} -I & D \\ 0 & B \end{bmatrix}, \mathrm{CSD_r}\left(\begin{bmatrix} -C & 0 \\ -A & I \end{bmatrix}, \frac{1}{s}\right)\right) = D + C(sI - A)^{-1}B. \tag{5.91}$$

From Fig. 5.23, one can obtain Fig. 5.24. Similarly, Property 5.4 gives

$$\Omega(s) = \mathrm{CSD_l}\left(\begin{bmatrix} \tilde{W} & 0 \\ H & I \end{bmatrix}, \mathrm{CSD_r}\left(\begin{bmatrix} \tilde{W} & 0 \\ H & I \end{bmatrix}, \Omega(s)\right)\right). \tag{5.92}$$

This concludes that the representations of Figs. 5.22 and 5.24 are equivalent, i.e., multiplying any $\begin{bmatrix} \tilde{W} & 0 \\ H & I \end{bmatrix}$ on the left by both top and bottom paths does not change the transfer function of $u \mapsto y$. The block diagram of Fig. 5.23 gives the state-space realization of Fig. 5.25.

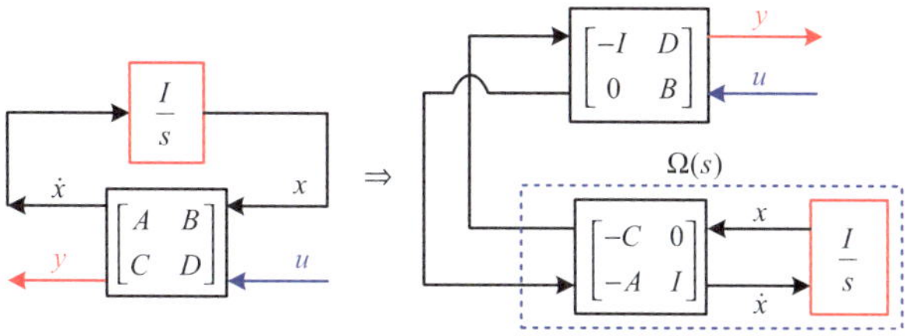

Fig. 5.22 Upper LFT to $CSD_l \rightleftarrows CSD_r$

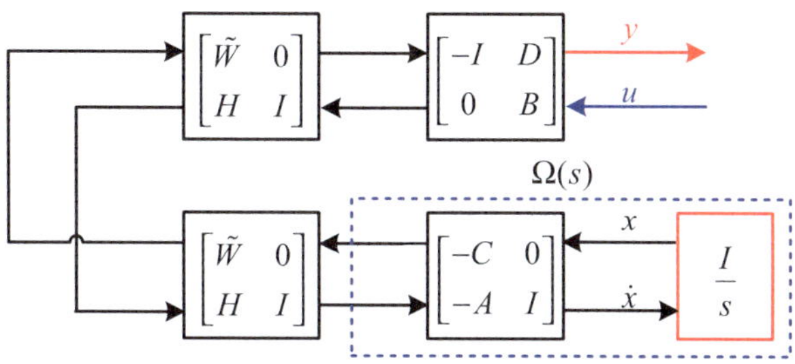

Fig. 5.23 CSD representations

Fig. 5.24 CSD representations

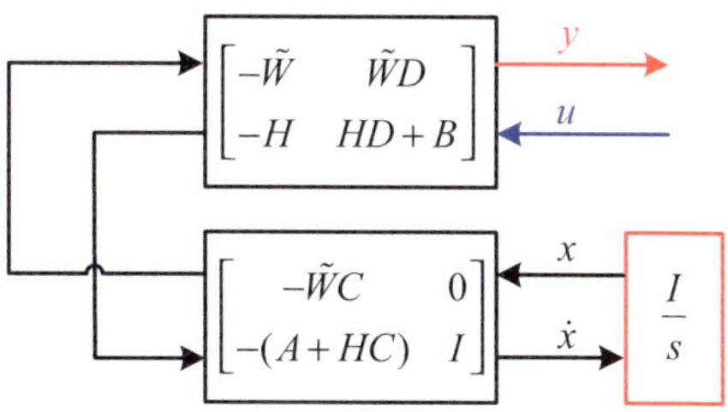

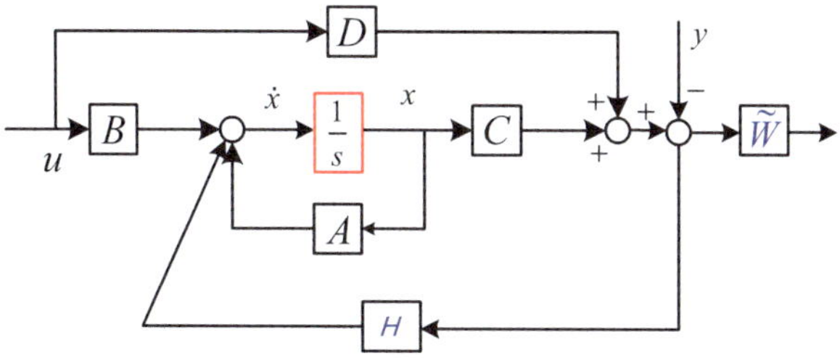

Fig. 5.25 State-space control block diagram

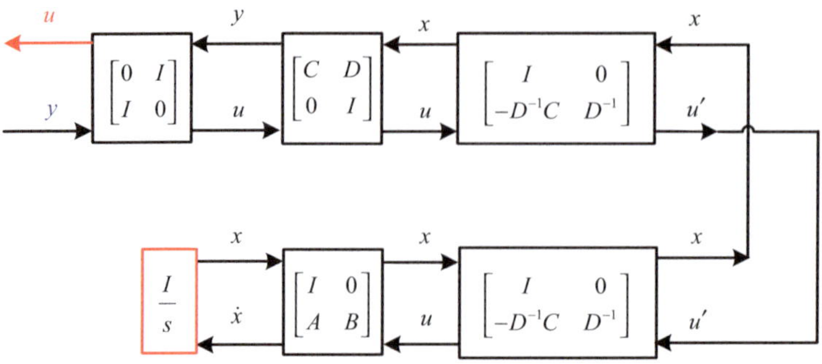

Fig. 5.26 CSD connections

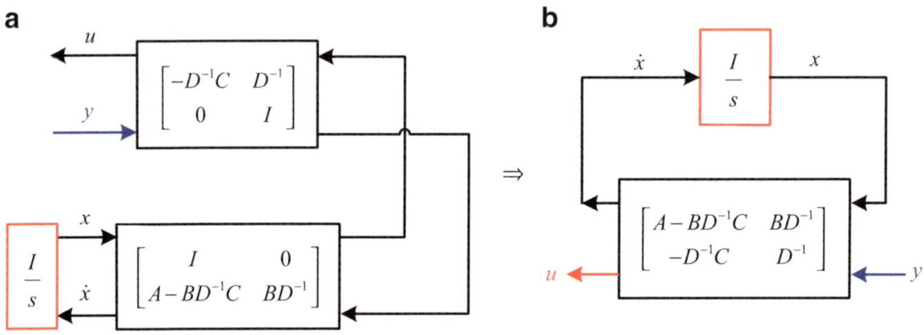

Fig. 5.27 CSD representations: (**a**) Left CSD representation and (**b**) LFT representation

Herein, $\begin{bmatrix} I & 0 \\ F & W \end{bmatrix}$ in Fig. 5.19 and $\begin{bmatrix} \tilde{W} & 0 \\ H & I \end{bmatrix}$ in Fig. 5.23 play important roles in the state-space representations of coprime factorizations, for which F and H can be chosen such that $A + BF$ and $A + HC$ are Hurwitz. Coprime factorization will be discussed in the next chapter. Apparently, as shown in Figs. 5.19 and 5.23, the coprime factorization in state-space form can be easily generated by CSD connections.

For a $P(s) \overset{s}{=} \left[\begin{array}{c|c} A & B \\ \hline C & D \end{array}\right]$ with D being nonsingular, the CSD representation is also applicable to determining the inversion $u = P(s)^{-1}y$. For this case, one can choose $F = -D^{-1}C$ and $W = D^{-1}$ in Fig. 5.19 and multiply by $\begin{bmatrix} 0 & I \\ I & 0 \end{bmatrix}$ at the input terminal $\begin{bmatrix} y \\ u \end{bmatrix}$ to obtain the CSD representation as depicted in Fig. 5.26. Further, rearranging Fig. 5.26 will yield Fig. 5.27a.

One can transform the CSDs of Fig. 5.27a back to the LFT as shown in Fig. 5.27b. Then, this gives that

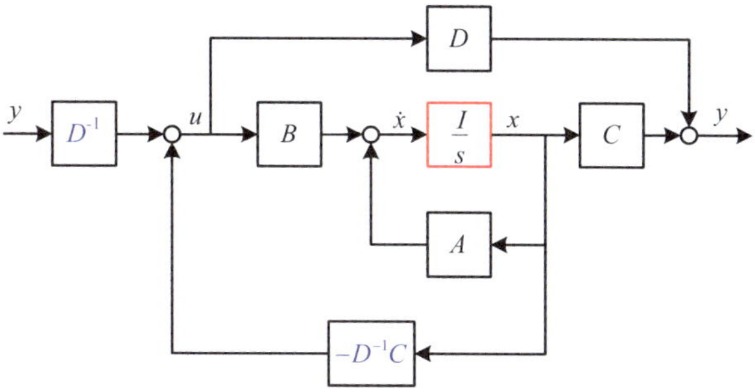

Fig. 5.28 State-space representations

$$\begin{bmatrix} \dot{x} \\ u \end{bmatrix} = \begin{bmatrix} A - BD^{-1}C & BD^{-1} \\ -D^{-1}C & D^{-1} \end{bmatrix} \begin{bmatrix} x \\ y \end{bmatrix}, \tag{5.93}$$

i.e., $u = P(s)^{-1}y$, where

$$P^{-1}(s) \stackrel{s}{=} \left[\begin{array}{c|c} A - BD^{-1}C & BD^{-1} \\ \hline -D^{-1}C & D^{-1} \end{array}\right]. \tag{5.94}$$

One can verify that the state-space representation from u to y is given by Fig. 5.28.

5.7 An Application of CSDs to Similarity Transformations

The similarity transformation of a state-space realization can be represented by the CSD approach. Let

$$x = Tx' \quad \text{and} \quad \dot{x} = T\dot{x}'. \tag{5.95}$$

In further explication, one can use "s," the Laplace transform symbol, to show the differentiation operation, i.e., $x = \frac{I}{s}T\dot{x}'$, with a bit of abuse of notations. Figure 5.18 shows LFT and CSD representations, respectively, in terms of state-space realization matrices. By inserting some CSDs, Fig. 5.18 is equivalent to Fig. 5.29. It can be found from Fig. 5.30 that the transfer function $u \mapsto x'$ is given by

$$\Psi(s) = \mathrm{CSD}_l\left(\begin{bmatrix} I & 0 \\ T^{-1}AT & T^{-1}B \end{bmatrix}, \frac{I}{s}\right) = \left(sI - T^{-1}AT\right)^{-1}T^{-1}B. \tag{5.96}$$

Therefore, this concludes that

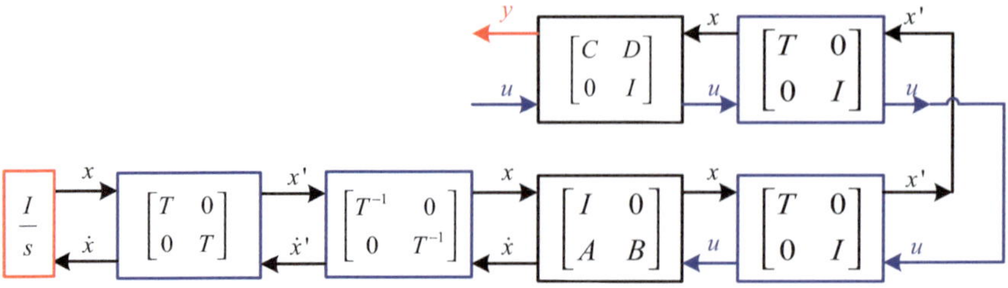

Fig. 5.29 CSD representations

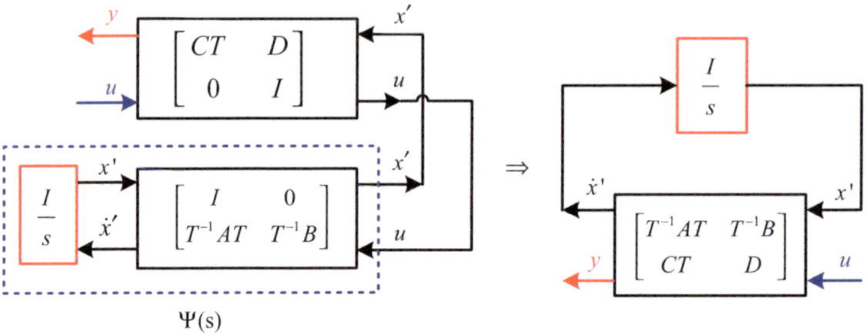

Fig. 5.30 CSD representations

$$y = \mathrm{CSD_r}\left(\begin{bmatrix} CT & D \\ 0 & I \end{bmatrix}, \Psi(s)\right) u = (D + CT\Psi(s))\, u$$
$$= \left(D + CT\left(sI - T^{-1}AT\right)^{-1} T^{-1}B\right) u = \left(D + C(sI - A)^{-1}B\right) u. \quad (5.97)$$

5.8 State-Space Formulae of CSD Matrix Transformed from LFT Matrix

Given an LFT $P(s)$ matrix with proper dynamics,

$$P(s) = \begin{bmatrix} P_{11}(s) & P_{12}(s) \\ P_{21}(s) & P_{22}(s) \end{bmatrix} \stackrel{s}{=} \left[\begin{array}{c|cc} A & B_1 & B_2 \\ \hline C_1 & D_{11} & D_{12} \\ C_2 & D_{21} & D_{22} \end{array}\right] \Rightarrow \begin{bmatrix} \dot{x} \\ \hline z \\ y \end{bmatrix} = \left[\begin{array}{c:cc} A & B_1 & B_2 \\ \hdashline C_1 & D_{11} & D_{12} \\ C_2 & D_{21} & D_{22} \end{array}\right] \begin{bmatrix} x \\ \hline w \\ u \end{bmatrix}. \quad (5.98)$$

Here the solid lines in a matrix show the matrix is a compact expression for a transfer function matrix, while the dotted line in a matrix indicates usual matrix partition for the sake of clarity. From (5.98), one can derive

$$\begin{bmatrix} \dot{x} \\ z \\ y \\ \hline x \\ w \\ u \end{bmatrix} = \begin{bmatrix} A & B_1 & B_2 \\ C_1 & D_{11} & D_{12} \\ C_2 & D_{21} & D_{22} \\ \hline I & 0 & 0 \\ 0 & I & 0 \\ 0 & 0 & I \end{bmatrix} \begin{bmatrix} x \\ w \\ u \end{bmatrix} \Rightarrow \begin{bmatrix} \dot{x} \\ z \\ w \\ \hline x \\ u \\ y \end{bmatrix} = \begin{bmatrix} A & B_1 & B_2 \\ C_1 & D_{11} & D_{12} \\ 0 & I & 0 \\ \hline I & 0 & 0 \\ 0 & 0 & I \\ C_2 & D_{21} & D_{22} \end{bmatrix} \begin{bmatrix} x \\ w \\ u \end{bmatrix}. \tag{5.99}$$

Assume that D_{21} is invertible. Then,

$$\begin{bmatrix} \dot{x} \\ z \\ w \end{bmatrix} = \begin{bmatrix} A & B_1 & B_2 \\ C_1 & D_{11} & D_{12} \\ 0 & I & 0 \end{bmatrix} \begin{bmatrix} I & 0 & 0 \\ 0 & 0 & I \\ C_2 & D_{21} & D_{22} \end{bmatrix}^{-1} \begin{bmatrix} x \\ u \\ y \end{bmatrix}. \tag{5.100}$$

One then obtains

$$\begin{bmatrix} \dot{x} \\ z \\ w \end{bmatrix} = \begin{bmatrix} A - B_1 D_{21}{}^{-1} C_2 & B_2 - B_1 D_{21}{}^{-1} D_{22} & B_1 D_{21}{}^{-1} \\ C_1 - D_{11} D_{21}{}^{-1} C_2 & D_{12} - D_{11} D_{21}{}^{-1} D_{22} & D_{11} D_{21}{}^{-1} \\ -D_{21}{}^{-1} C_2 & -D_{21}{}^{-1} D_{22} & D_{21}{}^{-1} \end{bmatrix} \begin{bmatrix} x \\ u \\ y \end{bmatrix}. \tag{5.101}$$

Therefore, for the state-space representation of $\begin{bmatrix} z \\ w \end{bmatrix} = G(s) \begin{bmatrix} u \\ y \end{bmatrix}$, one has

$$G(s) \stackrel{s}{=} \left[\begin{array}{c|cc} A - B_1 D_{21}{}^{-1} C_2 & B_2 - B_1 D_{21}{}^{-1} D_{22} & B_1 D_{21}{}^{-1} \\ \hline C_1 - D_{11} D_{21}{}^{-1} C_2 & D_{12} - D_{11} D_{21}{}^{-1} D_{22} & D_{11} D_{21}{}^{-1} \\ -D_{21}{}^{-1} C_2 & -D_{21}{}^{-1} D_{22} & D_{21}{}^{-1} \end{array} \right]. \tag{5.102}$$

Note that if P_{21} is strictly proper, then the state-space representation of G does not exist.

Dually, from

$$\begin{bmatrix} \dot{x} \\ z \\ y \\ \hline x \\ w \\ u \end{bmatrix} = \begin{bmatrix} A & B_1 & B_2 \\ C_1 & D_{11} & D_{12} \\ C_2 & D_{21} & D_{22} \\ \hline I & 0 & 0 \\ 0 & I & 0 \\ 0 & 0 & I \end{bmatrix} \begin{bmatrix} x \\ w \\ u \end{bmatrix} \Rightarrow \begin{bmatrix} \dot{x} \\ u \\ y \\ \hline x \\ z \\ w \end{bmatrix} = \begin{bmatrix} A & B_1 & B_2 \\ 0 & 0 & I \\ C_2 & D_{21} & D_{22} \\ \hline I & 0 & 0 \\ C_1 & D_{11} & D_{12} \\ 0 & I & 0 \end{bmatrix} \begin{bmatrix} x \\ w \\ u \end{bmatrix}, \tag{5.103}$$

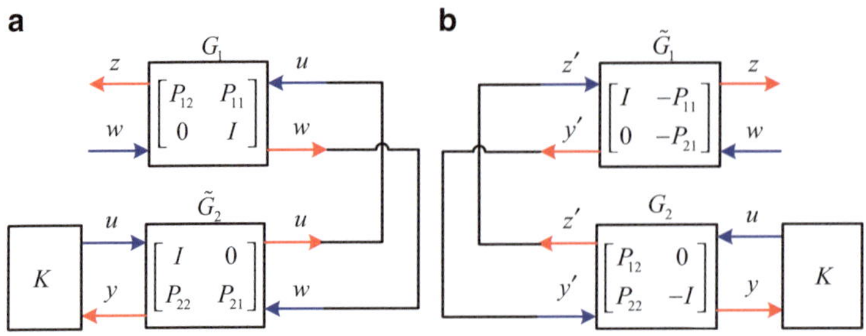

Fig. 5.31 $CSD_r \rightleftarrows CSD_l$ and $CSD_l \rightleftarrows CSD_r$: (**a**) Right CSD associated with left CSD and (**b**) Left CSD associated with right CSD

one has

$$\begin{bmatrix} \dot{x} \\ u \\ y \end{bmatrix} = \begin{bmatrix} A & B_1 & B_2 \\ 0 & 0 & I \\ C_2 & D_{21} & D_{22} \end{bmatrix} \begin{bmatrix} I & 0 & 0 \\ C_1 & D_{11} & D_{12} \\ 0 & I & 0 \end{bmatrix}^{-1} \begin{bmatrix} x \\ z \\ w \end{bmatrix}. \tag{5.104}$$

Here, D_{12} is assumed to be invertible. One then obtains

$$\begin{bmatrix} \dot{x} \\ u \\ y \end{bmatrix} = \begin{bmatrix} A - B_2 D_{12}{}^{-1} C_1 & B_2 D_{12}{}^{-1} & B_1 - B_2 D_{12}{}^{-1} D_{11} \\ -D_{12}{}^{-1} C_1 & D_{12}{}^{-1} & -D_{12}{}^{-1} D_{11} \\ C_2 - D_{22} D_{12}{}^{-1} C_1 & D_{22} D_{12}{}^{-1} & D_{21}{}^{-1} - D_{22} D_{12}{}^{-1} D_{11} \end{bmatrix} \begin{bmatrix} x \\ z \\ w \end{bmatrix}. \tag{5.105}$$

Hence, for the state-space representation of $\begin{bmatrix} u \\ y \end{bmatrix} = \tilde{G}(s) \begin{bmatrix} z \\ w \end{bmatrix}$, one has

$$\tilde{G}(s) \overset{s}{=} \left[\begin{array}{c|cc} A - B_2 D_{12}{}^{-1} C_1 & B_2 D_{12}{}^{-1} & B_1 - B_2 D_{12}{}^{-1} D_{11} \\ \hline -D_{12}{}^{-1} C_1 & D_{12}{}^{-1} & -D_{12}{}^{-1} D_{11} \\ C_2 - D_{22} D_{12}{}^{-1} C_1 & D_{22} D_{12}{}^{-1} & D_{21}{}^{-1} - D_{22} D_{12}{}^{-1} D_{11} \end{array} \right]. \tag{5.106}$$

Note that if P_{12} is strictly proper, then the state-space representation of $\tilde{G}$ does not exist.

In Sect. 5.4, it showed that a general LFT system can be transformed into a right CSD $G_1(s)$ associated with a left CSD $\tilde{G}_2(s)$ or a left CSD $\tilde{G}_1(s)$ associated with a right CSD $G_2(s)$. For the case of Fig. 5.31a where a right CSD $G_1(s)$ is associated with a left CSD $\tilde{G}_2(s)$, the corresponding state-space realization can be obtained, assuming $P(s)$ in the state-space form of (5.98), as

$$\begin{bmatrix} G_1(s) \\ \tilde{G}_2(s) \end{bmatrix} = \begin{bmatrix} P_{11}(s) & P_{12}(s) \\ I & 0 \\ \hline 0 & I \\ P_{21}(s) & P_{22}(s) \end{bmatrix} \stackrel{s}{=} \left[\begin{array}{c|cc} A & B_1 & B_2 \\ \hline C_1 & D_{11} & D_{12} \\ 0 & I & 0 \\ \hline 0 & 0 & I \\ C_2 & D_{21} & D_{22} \end{array}\right]. \tag{5.107}$$

For the case of Fig. 5.31b where a left CSD $\tilde{G}_1(s)$ is associated with a right CSD $G_2(s)$, the corresponding state-space realization can be obtained as

$$\left[\tilde{G}_1(s) \mid G_2(s)\right] = \left[\begin{array}{cc|cc} I & -P_{11}(s) & P_{12}(s) & 0 \\ 0 & -P_{21}(s) & P_{22}(s) & -I \end{array}\right] \stackrel{s}{=} \left[\begin{array}{c|cc|cc} A & 0 & -B_1 & B_2 & 0 \\ \hline C_1 & I & -D_{11} & D_{12} & 0 \\ C_2 & 0 & -D_{21} & D_{22} & -I \end{array}\right]. \tag{5.108}$$

This chapter introduces the definitions of CSD and its manipulations. These fundamentals are essential for the determination of a robust controller, which will be presented in the following chapters. The structures of $\mathrm{CSD_r} \rightleftarrows \mathrm{CSD_l}$ and $\mathrm{CSD_l} \rightleftarrows \mathrm{CSD_r}$, which circumvent the difficulty of inversion, were firstly investigated by the authors of this book and coresearchers. These unique structures offer a unified approach for the robust controller synthesis problem, which will be discussed in detail later in the book.

5.9 State-Space Formulae of LFT Matrix Transformed from CSD Matrix

In the following, the state-space formulae for the transformation between CSD and LFT will be discussed. Let

$$\begin{bmatrix} z \\ w \end{bmatrix} = \begin{bmatrix} G_{11}(s) & G_{12}(s) \\ G_{21}(s) & G_{22}(s) \end{bmatrix} \begin{bmatrix} u \\ y \end{bmatrix} \quad \text{and } u = Ky, \tag{5.109}$$

where

$$G(s) = \begin{bmatrix} G_{11}(s) & G_{12}(s) \\ G_{21}(s) & G_{22}(s) \end{bmatrix} \stackrel{s}{=} \left[\begin{array}{c|cc} A & B_1 & B_2 \\ \hline C_1 & D_{11} & D_{12} \\ C_2 & D_{21} & D_{22} \end{array}\right] \Rightarrow \begin{bmatrix} \dot{x} \\ z \\ w \end{bmatrix} = \begin{bmatrix} A & B_1 & B_2 \\ C_1 & D_{11} & D_{12} \\ C_2 & D_{21} & D_{22} \end{bmatrix} \begin{bmatrix} x \\ u \\ y \end{bmatrix}. \tag{5.110}$$

Assume that D_{22} is invertible. The problem here is to find a state-space realization of $P(s)$ such that

$$z = \mathrm{CSD_r}(G, K)\, w = \mathrm{LFT_l}(P, K)\, w. \tag{5.111}$$

From (5.110), one has

$$\begin{bmatrix} \dot{x} \\ z \\ w \\ \hline x \\ u \\ y \end{bmatrix} = \begin{bmatrix} A & B_1 & B_2 \\ C_1 & D_{11} & D_{12} \\ C_2 & D_{21} & D_{22} \\ \hdashline I & 0 & 0 \\ 0 & I & 0 \\ 0 & 0 & I \end{bmatrix} \begin{bmatrix} x \\ u \\ y \end{bmatrix}. \tag{5.112}$$

Rearrange the row by the required input-output relationship:

$$\begin{bmatrix} \dot{x} \\ z \\ y \\ \hline x \\ w \\ u \end{bmatrix} = \begin{bmatrix} A & B_1 & B_2 \\ C_1 & D_{11} & D_{12} \\ 0 & 0 & I \\ \hdashline I & 0 & 0 \\ C_2 & D_{21} & D_{22} \\ 0 & I & 0 \end{bmatrix} \begin{bmatrix} x \\ u \\ y \end{bmatrix} \tag{5.113}$$

Hence, one has

$$\begin{aligned} \begin{bmatrix} \dot{x} \\ z \\ y \end{bmatrix} &= \begin{bmatrix} A & B_1 & B_2 \\ C_1 & D_{11} & D_{12} \\ 0 & 0 & I \end{bmatrix} \begin{bmatrix} I & 0 & 0 \\ C_2 & D_{21} & D_{22} \\ 0 & I & 0 \end{bmatrix}^{-1} \begin{bmatrix} x \\ w \\ u \end{bmatrix} \\ &= \begin{bmatrix} A - B_2 D_{22}{}^{-1} C_2 & B_2 D_{22}{}^{-1} & B_1 - B_2 D_{22}{}^{-1} D_{21} \\ C_1 - D_{12} D_{22}{}^{-1} C_2 & D_{12} D_{22}{}^{-1} & D_{11} - D_{12} D_{22}{}^{-1} D_{21} \\ -D_{22}{}^{-1} C_2 & D_{22}{}^{-1} & -D_{22}{}^{-1} D_{21} \end{bmatrix} \begin{bmatrix} x \\ w \\ u \end{bmatrix}. \end{aligned} \tag{5.114}$$

Therefore,

$$P(s) \stackrel{s}{=} \left[\begin{array}{c|cc} A - B_2 D_{22}{}^{-1} C_2 & B_2 D_{22}{}^{-1} & B_1 - B_2 D_{22}{}^{-1} D_{21} \\ \hline C_1 - D_{12} D_{22}{}^{-1} C_2 & D_{12} D_{22}{}^{-1} & D_{11} - D_{12} D_{22}{}^{-1} D_{21} \\ -D_{22}{}^{-1} C_2 & D_{22}{}^{-1} & -D_{22}{}^{-1} D_{21} \end{array} \right] \tag{5.115}$$

Note that if G_{22} is strictly proper, then the state-space representation of P does not exist.

In the dual case of left CSD, let

$$\begin{bmatrix} u \\ y \end{bmatrix} = \begin{bmatrix} \tilde{G}_{11}(s) & \tilde{G}_{12}(s) \\ \tilde{G}_{21}(s) & \tilde{G}_{22}(s) \end{bmatrix} \begin{bmatrix} z \\ w \end{bmatrix} \quad \text{and} \quad u = Ky, \tag{5.116}$$

where

$$\tilde{G}(s) = \begin{bmatrix} \tilde{G}_{11}(s) & \tilde{G}_{12}(s) \\ \tilde{G}_{21}(s) & \tilde{G}_{22}(s) \end{bmatrix} \overset{s}{=} \left[\begin{array}{c|cc} A & B_1 & B_2 \\ \hline C_1 & D_{11} & D_{12} \\ C_2 & D_{21} & D_{22} \end{array}\right] \Rightarrow \begin{bmatrix} \dot{x} \\ u \\ y \end{bmatrix} = \begin{bmatrix} A & B_1 & B_2 \\ C_1 & D_{11} & D_{12} \\ C_2 & D_{21} & D_{22} \end{bmatrix} \begin{bmatrix} x \\ z \\ w \end{bmatrix}. \tag{5.117}$$

If D_{11} is invertible, the problem is to find a state-space realization of $P(s)$ such that

$$z = \mathrm{CSD_l}\left(\tilde{G}, K\right) w = \mathrm{LFT_l}\left(P, K\right) w. \tag{5.118}$$

From (5.117), one has

$$\begin{bmatrix} \dot{x} \\ u \\ y \\ \hline x \\ z \\ w \end{bmatrix} = \left[\begin{array}{ccc} A & B_1 & B_2 \\ C_1 & D_{11} & D_{12} \\ C_2 & D_{21} & D_{22} \\ \hdashline I & 0 & 0 \\ 0 & I & 0 \\ 0 & 0 & I \end{array}\right] \begin{bmatrix} x \\ z \\ w \end{bmatrix} \tag{5.119}$$

and hence,

$$\begin{bmatrix} \dot{x} \\ z \\ y \\ \hline x \\ w \\ u \end{bmatrix} = \left[\begin{array}{ccc} A & B_1 & B_2 \\ 0 & I & 0 \\ C_2 & D_{21} & D_{22} \\ \hdashline I & 0 & 0 \\ 0 & 0 & I \\ C_1 & D_{11} & D_{12} \end{array}\right] \begin{bmatrix} x \\ z \\ w \end{bmatrix}. \tag{5.120}$$

This gives

$$\begin{bmatrix} \dot{x} \\ z \\ y \end{bmatrix} = \begin{bmatrix} A & B_1 & B_2 \\ 0 & I & 0 \\ C_2 & D_{21} & D_{22} \end{bmatrix} \begin{bmatrix} I & 0 & 0 \\ 0 & 0 & I \\ C_1 & D_{11} & D_{12} \end{bmatrix}^{-1} \begin{bmatrix} x \\ w \\ u \end{bmatrix}. \tag{5.121}$$

One then obtains

$$\begin{bmatrix} \dot{x} \\ z \\ y \end{bmatrix} = \begin{bmatrix} A - B_1 D_{11}^{-1} C_1 & B_2 - B_1 D_{11}^{-1} D_{12} & B_1 D_{11}^{-1} \\ -D_{11}^{-1} C_1 & -D_{11}^{-1} D_{12} & D_{11}^{-1} \\ C_2 - D_{21} D_{11}^{-1} C_1 & D_{22} - D_{21} D_{11}^{-1} D_{12} & D_{21} D_{11}^{-1} \end{bmatrix} \begin{bmatrix} x \\ w \\ u \end{bmatrix}$$

$$= P(s) \begin{bmatrix} x \\ w \\ u \end{bmatrix}, \tag{5.122}$$

and therefore,

$$P(s) \overset{s}{=} \left[\begin{array}{c|cc} A - B_1 D_{11}^{-1} C_1 & B_2 - B_1 D_{11}^{-1} D_{12} & B_1 D_{11}^{-1} \\ \hline -D_{11}^{-1} C_1 & -D_{11}^{-1} D_{12} & D_{11}^{-1} \\ C_2 - D_{21} D_{11}^{-1} C_1 & D_{22} - D_{21} D_{11}^{-1} D_{12} & D_{21} D_{11}^{-1} \end{array} \right]. \tag{5.123}$$

Note that if $\tilde{G}_{11}$ is strictly proper, then the state-space representation of P does not exist.

5.10 Star Connection

In this section, the relations between LFT (SCC) and CSD will be further investigated. The CSD originates from two-port networks and has benefits for cascading multisystems. Unlike the CSDs, the interconnections of two LFT matrices, namely, the *star product* [3], look much more complicated in their representation. Figure 4.29 in Chap. 4 shows the star product of two LFT matrices, where

$$z = \mathrm{LFT}_l \left(\begin{bmatrix} P_{11} & P_{12} \\ P_{21} & P_{22} \end{bmatrix}, \mathrm{LFT}_l \left(\begin{bmatrix} L_{11} & L_{12} \\ L_{21} & L_{22} \end{bmatrix}, \Phi \right) \right) w. \tag{5.124}$$

Next, how to transform the star product into its equivalent CSDs will be expounded. Figure 4.29 can be converted into CSDs with the termination Φ connected at the right port, as shown in Fig. 5.32. When L_{12} is invertible, one can insert $\begin{bmatrix} L_{12} & 0 \\ L_{22} & -I \end{bmatrix}^{-1}$ in the configuration as in Fig. 5.33, which does not change the input/output relation. In Fig. 5.33, the LFT matrix P is first expressed as a right CSD followed by a left CSD, which in turn is connected to two left CSDs and another two right CSDs. By rearranging the middle part of the CSDs, one can obtain

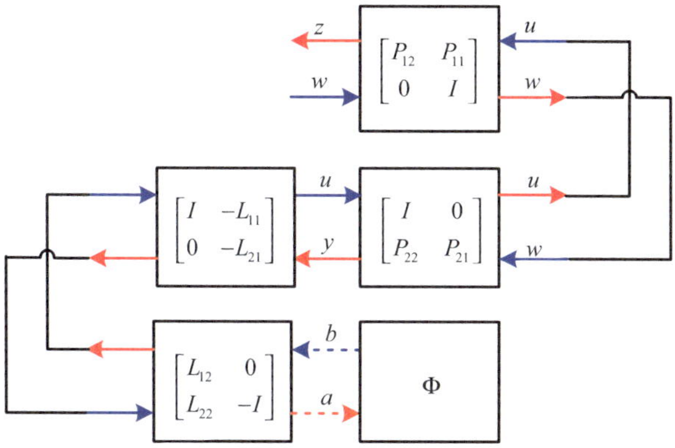

Fig. 5.32 CSD formulation of star product

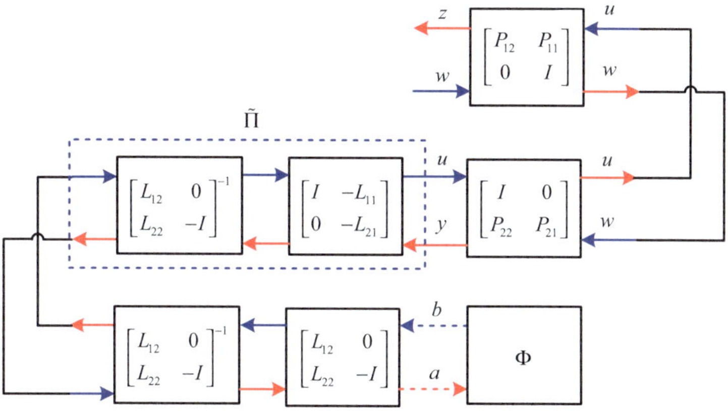

Fig. 5.33 CSD formulation of star product

$$z = \mathrm{CSD_r}\left(\begin{bmatrix} P_{12} & P_{11} \\ 0 & I \end{bmatrix}, \mathrm{CSD_l}\left(\begin{bmatrix} I & 0 \\ P_{22} & P_{21} \end{bmatrix}, \mathrm{CSD_l}\left(\begin{bmatrix} I & -L_{11} \\ 0 & -L_{21} \end{bmatrix},\right.\right.\right.$$
$$\left.\left.\left.\mathrm{CSD_l}\left(\begin{bmatrix} L_{12} & 0 \\ L_{22} & -I \end{bmatrix}^{-1}, \Phi\right)\right)\right)\right) w.$$

Controller K could be rewritten as

$$K = \mathrm{CSD_l}\left(\tilde{\Pi}, \Phi\right), \tag{5.125}$$

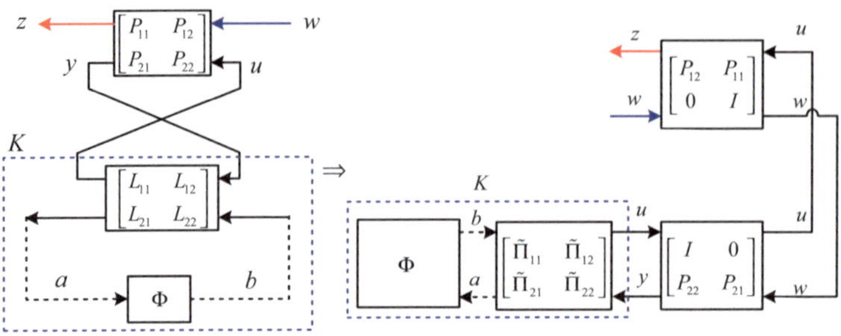

Fig. 5.34 Equivalent representations

where

$$\tilde{\Pi} = \begin{bmatrix} \tilde{\Pi}_{11} & \tilde{\Pi}_{12} \\ \tilde{\Pi}_{21} & \tilde{\Pi}_{22} \end{bmatrix} = \begin{bmatrix} {L_{12}}^{-1} & -{L_{12}}^{-1}L_{11} \\ L_{22}{L_{12}}^{-1} & L_{21} - L_{22}{L_{12}}^{-1}L_{11} \end{bmatrix}. \tag{5.126}$$

As can be seen in Fig. 5.34, the star product of LFT can be formulated into $\text{CSD}_r \rightleftarrows \text{CSD}_l$.

For the dual case that if L_{21} is invertible, one can convert Fig. 4.29 into CSDs with the termination Φ connected at the left port as shown in Fig. 5.35. Then, by rearranging the middle CSDs, one can acquire Fig. 5.36. Similarly, controller K could be rewritten as

$$K = \text{CSD}_r\left(\Pi, \Phi\right), \tag{5.127}$$

where

$$\Pi = \begin{bmatrix} \Pi_{11} & \Pi_{12} \\ \Pi_{21} & \Pi_{22} \end{bmatrix} = \begin{bmatrix} L_{12} - L_{11}{L_{21}}^{-1}L_{22} & L_{11}{L_{21}}^{-1} \\ -{L_{21}}^{-1}L_{22} & {L_{21}}^{-1} \end{bmatrix}. \tag{5.128}$$

As can be seen in Fig. 5.36, the star product of LFT can be formulated into $\text{CSD}_r \rightleftarrows \text{CSD}_l$. Evidently, Figs. 5.34 and 5.36 are definitely equivalent.

5.11 J-Lossless and Dual J-Lossless Systems

The properties of lossless two-port networks in the viewpoint of power wave propagation were discussed in Chap. 3. The lossless (inner) systems in the described LFT system were defined in Chap. 4. In this section, the J-lossless and dual J-lossless systems will be further investigated for general control problems. They

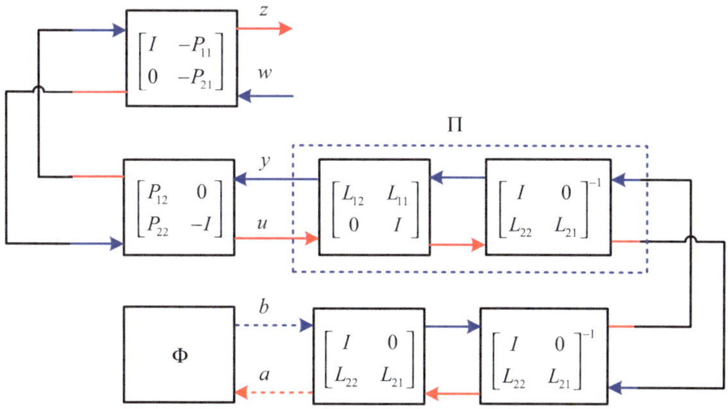

Fig. 5.35 Equivalent CSD formulations

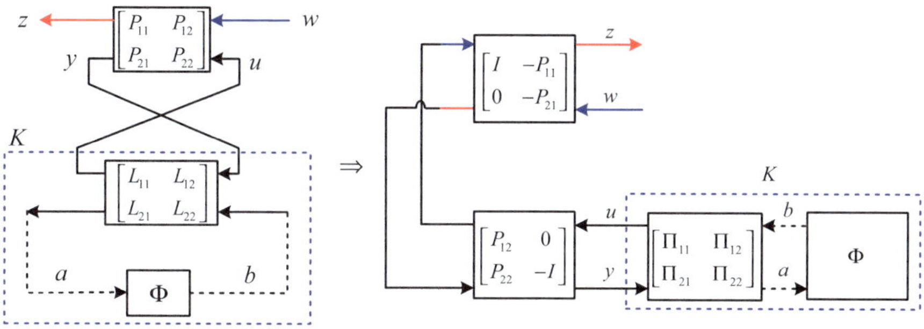

Fig. 5.36 Equivalent CSD formulations

represent the energy balance of the two-port networks between the left and right ports. Consider the system as illustrated in Fig. 5.1; this means

$$\|z(jw)\|_2{}^2 - \|w(jw)\|_2{}^2 = \|u(jw)\|_2{}^2 - \|y(jw)\|_2{}^2. \tag{5.129}$$

Definition 5.1 Let $J_{n,k} = \begin{bmatrix} I_n & 0 \\ 0 & -I_k \end{bmatrix}$. A $(n_1 + n_2) \times (k_1 + k_2)$ right CSD matrix $G(s)$ is called J-unitary if $G^{\sim} J_{n_1,n_2} G = J_{k_1,k_2}$, where $n_2 = k_2$. A J-unitary G is then called J-lossless if $G^* J_{n_1,n_2} G \leq J_{k_1,k_2}$ and $\forall\, \mathrm{Re}(s) \geq 0$. A $(n_1 + n_2) \times (k_1 + k_2)$ left CSD matrix $\tilde{G}(s)$ is called dual J-unitary if $\tilde{G} J_{k_1,k_2} \tilde{G}^{\sim} = J_{n_1,n_2}$, where $n_1 = k_1$. A dual J-unitary G is then called dual J-lossless if $\tilde{G}(s) J_{k_1,k_2} \tilde{G}^*(s) \geq J_{n_1,n_2}$ and $\forall\, \mathrm{Re}(s) \geq 0$.

Example 5.7 Verify that $G(s) = \begin{bmatrix} \frac{s-1}{s+1} & 0 \\ 0 & \frac{s+1}{s-1} \end{bmatrix}$ is J-unitary and J-lossless.

From

$$G^{\sim}(s)JG(s) = \begin{bmatrix} \dfrac{-s-1}{-s+1} & 0 \\ 0 & \dfrac{-s+1}{-s-1} \end{bmatrix} \begin{bmatrix} 1 & 0 \\ 0 & -1 \end{bmatrix} \begin{bmatrix} \dfrac{s-1}{s+1} & 0 \\ 0 & \dfrac{s+1}{s-1} \end{bmatrix}$$

$$= \begin{bmatrix} 1 & 0 \\ 0 & -1 \end{bmatrix} = J, \tag{5.130}$$

it concludes that $G(s)$ is a J-unitary system. From (5.130) and $s = \sigma + j\omega$, $\forall\, \sigma \geq 0$,

$$G^{*}(s)JG(s) = \begin{bmatrix} \dfrac{\bar{s}-1}{\bar{s}+1} & 0 \\ 0 & \dfrac{\bar{s}-1}{\bar{s}+1} \end{bmatrix} \begin{bmatrix} \dfrac{s-1}{s+1} & 0 \\ 0 & -\dfrac{s+1}{s-1} \end{bmatrix}$$

$$= \begin{bmatrix} \dfrac{(\sigma-1)^2+\omega^2}{(\sigma+1)^2+\omega^2} & 0 \\ 0 & -\dfrac{(\sigma+1)^2+\omega^2}{(\sigma-1)^2+\omega^2} \end{bmatrix} \leq \begin{bmatrix} 1 & 0 \\ 0 & -1 \end{bmatrix}.$$

It concludes that $G(s)$ is a J-lossless system. It should be noted that an inner (lossless) LFT matrix $P(s)$ of $\begin{bmatrix} w \\ u \end{bmatrix} \mapsto \begin{bmatrix} z \\ y \end{bmatrix}$ must be stable; however, a J-lossless CSD matrix $G(s)$ of $\begin{bmatrix} u \\ y \end{bmatrix} \mapsto \begin{bmatrix} z \\ w \end{bmatrix}$ is not always stable. It can be verified that the J-lossless matrix $G(s)$ is not a stable system; however, its LFT matrix $P(s)$ is lossless (inner) and stable as

$$P(s) = \begin{bmatrix} \dfrac{s-1}{s+1} & 0 \\ 0 & 1 \end{bmatrix} \begin{bmatrix} 0 & \dfrac{s+1}{s-1} \\ 1 & 0 \end{bmatrix}^{-1}$$

$$= \begin{bmatrix} \dfrac{s-1}{s+1} & 0 \\ 0 & 1 \end{bmatrix} \begin{bmatrix} 0 & 1 \\ \dfrac{s-1}{s+1} & 0 \end{bmatrix} = \begin{bmatrix} 0 & \dfrac{s-1}{s+1} \\ \dfrac{s-1}{s+1} & 0 \end{bmatrix}.$$

As mentioned in Sect. 4.6, the J-lossless and dual J-lossless properties are counterparts of inner and co-inner. Hence, the relationship is maintained during the transformation between LFT and CSD. These properties are discussed in the following.

Lemma 5.2 *A right CSD matrix G(s) is J-lossless if and only if its corresponding LFT matrix P(s) which makes* $\mathrm{CSD_r}(G,\ K) = \mathrm{LFT_l}(P,\ K)$ *is inner.*

Proof Let $G(s) = \begin{bmatrix} G_{11}(s) & G_{12}(s) \\ G_{21}(s) & G_{22}(s) \end{bmatrix}$ be a right CSD matrix of $\begin{bmatrix} u \\ y \end{bmatrix} \mapsto \begin{bmatrix} z \\ w \end{bmatrix}$ with $\dim(y) = \dim(w)$. By (5.58), one can obtain

$$P = \begin{bmatrix} G_{11} & G_{12} \\ 0 & I \end{bmatrix} \begin{bmatrix} G_{21} & G_{22} \\ I & 0 \end{bmatrix}^{-1} =: G_{1*} G_{2*}^{-1}, \tag{5.131}$$

which makes $\mathrm{CSD_r}(G,\ K) = \mathrm{LFT_l}(P,\ K)$. Then, one has

$$\begin{aligned} G^{\sim} J_{z,w} G &= \begin{bmatrix} G_{11}^{\sim} & G_{21}^{\sim} \\ G_{12}^{\sim} & G_{22}^{\sim} \end{bmatrix} \begin{bmatrix} G_{11} & G_{12} \\ -G_{21} & -G_{22} \end{bmatrix} \\ &= \begin{bmatrix} G_{11}^{\sim} G_{11} - G_{21}^{\sim} G_{21} & G_{11}^{\sim} G_{12} - G_{21}^{\sim} G_{22} \\ G_{12}^{\sim} G_{11} - G_{22}^{\sim} G_{21} & G_{12}^{\sim} G_{12} - G_{22}^{\sim} G_{22} \end{bmatrix}, \end{aligned}$$

and on the other hand,

$$\begin{aligned} G_{2*}^{\sim} G_{2*} - G_{1*}^{\sim} G_{1*} &= \begin{bmatrix} G_{21}^{\sim} & I \\ G_{22}^{\sim} & 0 \end{bmatrix} \begin{bmatrix} G_{21} & G_{22} \\ I & 0 \end{bmatrix} - \begin{bmatrix} G_{11}^{\sim} & 0 \\ G_{12}^{\sim} & I \end{bmatrix} \begin{bmatrix} G_{11} & G_{12} \\ 0 & I \end{bmatrix} \\ &= \begin{bmatrix} G_{21}^{\sim} G_{21} - G_{11}^{\sim} G_{11} + I & G_{21}^{\sim} G_{22} - G_{11}^{\sim} G_{12} \\ G_{22}^{\sim} G_{21} - G_{12}^{\sim} G_{11} & G_{22}^{\sim} G_{22} - G_{12}^{\sim} G_{12} - I \end{bmatrix} \\ &= J_{u,y} - G^{\sim} J_{z,w} G. \end{aligned}$$

This implies that $G^{\sim} J_{z,w} G = J_{u,y}$ if and only if $G_{2*}^{\sim} G_{2*} = G_{1*}^{\sim} G_{1*}$ and, further,

$$P^{\sim} P = \left(G_{1*} G_{2*}^{-1}\right)^{\sim} G_{1*} G_{2*}^{-1} = G_{2*}^{-1^{\sim}} G_{1*}{}^{\sim} G_{1*} G_{2*}^{-1} = G_{2*}^{-1^{\sim}} G_{2*}^{\sim} G_{2*} G_{2*}^{-1} = I. \tag{5.132}$$

This concludes that $P^{\sim} P = I$ if and only if $G^{\sim} J_{z,w} G = J_{u,y}$. That is, $G(s)$ is J-unitary if and only if $P(s)$ is all pass. Furthermore, when $G(s)$ is J-lossless, one has $G_{2*}^{*}(s) G_{2*}(s) \geq G_{1*}^{*}(s) G_{1*}(s)$ and $\forall\, \mathrm{Re}(s) \geq 0$, which implies that

$$\begin{aligned} P^{*}(s) P(s) &= \left(G_{1*}(s) G_{2*}^{-1}(s)\right)^{*} G_{1*}(s) G_{2*}^{-1}(s) \\ &= G_{2*}^{-1^{*}}(s) G_{1*}{}^{*}(s) G_{1*}(s) G_{2*}^{-1}(s) \leq I, \forall \mathrm{Re}(s) \geq 0. \end{aligned} \tag{5.133}$$

Hence, it concludes that $G(s)$ is J-lossless if and only if $P(s)$ is inner (or lossless). ■

Lemma 5.3 *A left CSD matrix $\tilde{G}(s)$ is dual J-lossless if and only if its corresponding LFT matrix P(s) which makes* $\mathrm{CSD_l}\left(\tilde{G}, K\right) = \mathrm{LFT_l}(P, K)$ *is co-inner.*

Proof Let $\tilde{G}(s) = \begin{bmatrix} \tilde{G}_{11}(s) & \tilde{G}_{12}(s) \\ \tilde{G}_{21}(s) & \tilde{G}_{22}(s) \end{bmatrix}$ be a left CSD matrix of $\begin{bmatrix} z \\ w \end{bmatrix} \mapsto \begin{bmatrix} u \\ y \end{bmatrix}$ with $\dim(y) = \dim(w)$. From (5.62), one has

$$P = \begin{bmatrix} I & 0 \\ \tilde{G}_{21} & \tilde{G}_{22} \end{bmatrix} \begin{bmatrix} 0 & I \\ \tilde{G}_{11} & \tilde{G}_{12} \end{bmatrix}^{-1} =: \tilde{G}_{*2}\tilde{G}_{*1}^{-1}. \tag{5.134}$$

Now, from

$$\begin{aligned} \tilde{G}_{*2}\tilde{G}_{*2}^{\sim} - \tilde{G}_{*1}\tilde{G}_{*1}^{\sim} &= \begin{bmatrix} \tilde{G}_{12} & -I \\ \tilde{G}_{22} & 0 \end{bmatrix} \begin{bmatrix} \tilde{G}_{12}^{\sim} & \tilde{G}_{22}^{\sim} \\ -I & 0 \end{bmatrix} - \begin{bmatrix} -\tilde{G}_{11} & 0 \\ -\tilde{G}_{21} & I \end{bmatrix} \begin{bmatrix} -\tilde{G}_{11}^{\sim} & -G_{21}^{\sim} \\ 0 & I \end{bmatrix} \\ &= \begin{bmatrix} \tilde{G}_{12}\tilde{G}_{12}^{\sim} - \tilde{G}_{11}\tilde{G}_{11}^{\sim} + I & \tilde{G}_{12}\tilde{G}_{22}^{\sim} - \tilde{G}_{11}\tilde{G}_{21}^{\sim} \\ \tilde{G}_{22}\tilde{G}_{12}^{\sim} - \tilde{G}_{21}\tilde{G}_{11}^{\sim} & \tilde{G}_{22}\tilde{G}_{22}^{\sim} - \tilde{G}_{21}\tilde{G}_{21}^{\sim} - I \end{bmatrix} \\ &= J_{u,y} - \tilde{G} J_{z,w} \tilde{G}^{\sim}, \end{aligned} \tag{5.135}$$

one can obtain that $J_{u,y} - \tilde{G} J_{z,w} \tilde{G}^{\sim} = 0$ if and only if $PP^{\sim} = I$. Furthermore, $\tilde{G}(s) J_{z,w} \tilde{G}^{*}(s) \geq J_{u,y}$, and $\forall$ Re $s \geq 0$ implies that

$$\begin{aligned} P(s)P^{*}(s) &= \tilde{G}_{*1}{}^{-}1(s)\tilde{G}_{*2}(s)\left(\tilde{G}_{*1}^{-1}(s)\tilde{G}_{*2}(s)\right)^{*} \\ &= \tilde{G}_{*1}^{-1}(s)\tilde{G}_{*2}(s)\tilde{G}_{*2}^{*}(s)\tilde{G}_{*1}^{-1*}(s) \leq I, \quad \forall \mathrm{Re}s \geq 0. \end{aligned} \tag{5.136}$$

Hence, it concludes that $\tilde{G}(s)$ is dual J-lossless if and only if $P(s)$ is co-inner. ■

As illustrated in Fig. 5.1, the CSD matrix (G or $\tilde{G}$) has relations to its terminator K, especially when the CSD matrix satisfies J-lossless (or dual J-lossless). These properties are introduced in the following lemmas.

Lemma 5.4 *If a right CSD matrix G(s) is J-lossless, then* $\mathrm{CSD_r}(G,\Phi) \in BH_\infty$ *for* $\forall\, \Phi \in BH_\infty$.

Proof Let $u = \Phi y$, where

$$\|u(jw)\|_2 = \|\Phi y(jw)\|_2 \leq \|\Phi\|_\infty \|y(jw)\|_2 < \|y(jw)\|_2. \tag{5.137}$$

From $z = \mathrm{CSD_r}(G,\Phi)w$ and (5.129),

$$\|z(jw)\|_2{}^2 - \|w(jw)\|_2{}^2 = \|u(jw)\|_2{}^2 - \|y(jw)\|_2{}^2, \tag{5.138}$$

and because

$$\|u(jw)\|_2^{\,2} - \|y(jw)\|_2^{\,2} < 0, \tag{5.139}$$

one has $\|z(jw)\|_2^{\,2} - \|w(jw)\|_2^{\,2} < 0$. Hence, $\mathrm{CSD_r}(G,\Phi) \in BH_\infty$, $\forall\, \Phi \in BH_\infty$. ■

Example 5.8 Given a J-lossless $G(s) = \begin{bmatrix} \frac{s-1}{s+1} & 0 \\ 0 & \frac{s+1}{s-1} \end{bmatrix}$ and $\Phi = \frac{1}{s+2} \in BH_\infty$, verify that $\mathrm{CSD_r}(G,\ \Phi) \in BH_\infty$.

From (5.5), one has

$$\mathrm{CSD_r}\,(G,\Phi) = \frac{(s-1)^2}{(s+1)^2\,(s+2)} \in BH_\infty.$$

Lemma 5.5 *Dually, if a left CSD system $\tilde{G}(s)$ is dual J-lossless, then* $\mathrm{CSD_l}\left(\tilde{G},\Phi\right) \in BH_\infty$ *for* $\forall\, \Phi \in BH_\infty$.

Until now, we have introduced the inner and co-inner LFT systems in Chap. 4 and J-lossless and dual J-lossless CSD systems in this chapter. In the next chapter, the coprime factorization to LFT and CSD systems will be discussed. These factorization approaches offer a straightforward way to describe a control system using the RH_∞ functions.

Exercises

1. Let P be an SCC plant shown below, where $P(s) = \begin{bmatrix} \frac{1}{s+1} & \frac{1}{s(s+1)} \\ 1 & \frac{1}{s} \end{bmatrix}$, $K = -3$.

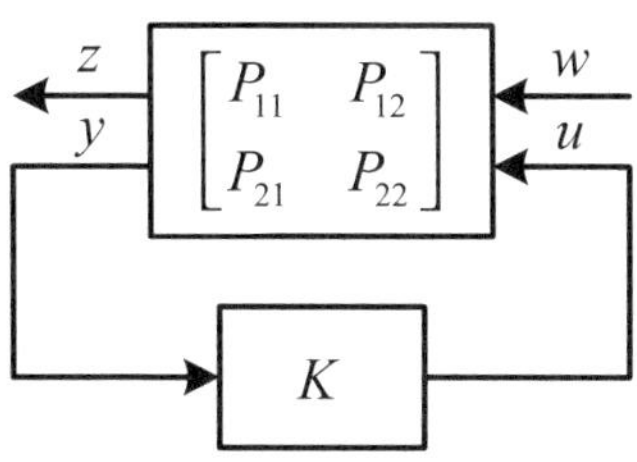

   (a) Find the transfer matrix of $\mathrm{LFT_l}(P,K)$.
   (b) Transform the SCC plant P into a right CSD matrix G, and calculate the transfer function of $\mathrm{CSD_r}(G,K)$.
   (c) Let P be represented by a right CSD matrix P_{1*}, associated with a left CSD matrix P_{2*} as in the following figure. Find P_{1*} and P_{2*}, and determine $\mathrm{CSD_r}(P_{1*}, \mathrm{CSD_l}(P_{2*},K))$.

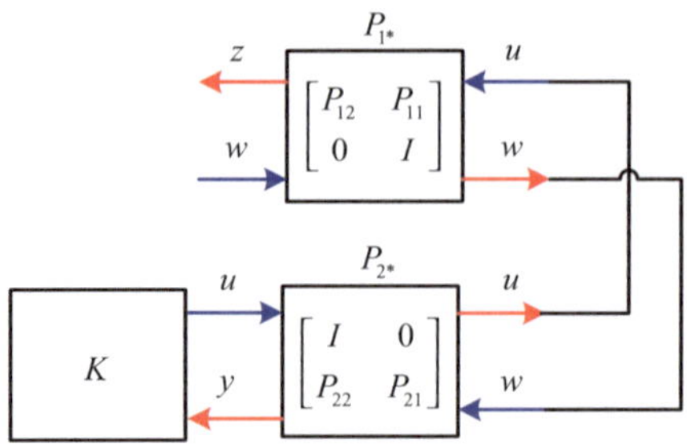

2. Determine the interconnected matrix P in the SCC plant for the following system, where $w = d, z = \begin{bmatrix} v \\ u_f \end{bmatrix}$ and transform it into a CSD form.

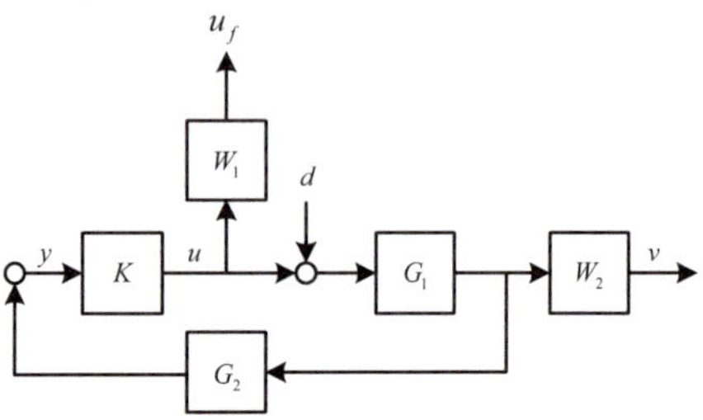

3. Find the transfer function $\frac{V_2}{V_1}$ of the network circuit given below, using LFT and CSD approaches, respectively, where $R_1 = 9$, $R_2 = 1$, and $C = 1/9$.

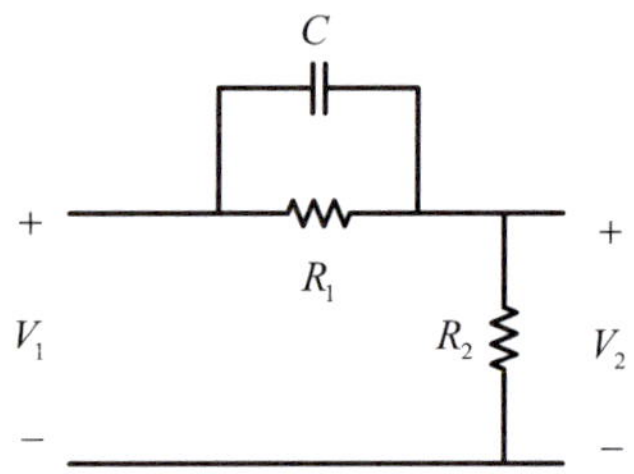

4. Use the LFT method to obtain the transfer function of $w \mapsto z$ for the block diagram shown below and then derive its left CSD form.

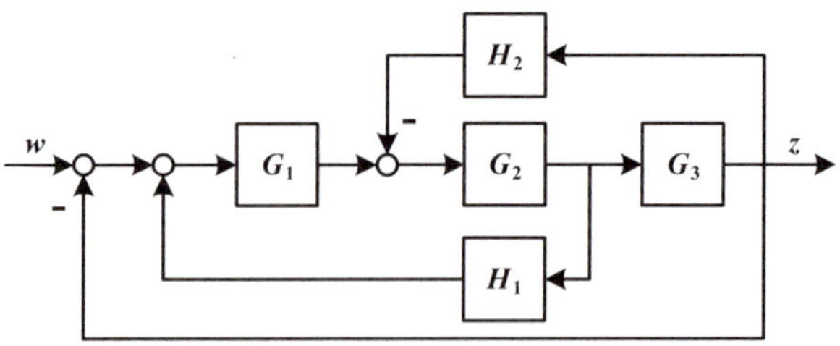

5. Use the techniques presented in this chapter to find the transfer function of the following:

   (a) $\mathrm{LFT_l}(P,K)$
   (b) $\mathrm{CSD_r}(G,K)$
   (c) $\mathrm{CSD_l}\left(\tilde{G}, K\right)$
   (d) $\mathrm{CSD_r}\left(G_1, \mathrm{CSD_l}\left(\tilde{G}_2, K\right)\right)$
   (e) $\mathrm{CSD_l}\left(\tilde{G}_1, \mathrm{CSD_r}\left(G_2, K\right)\right)$, where $P(s) = \begin{bmatrix} \frac{1}{s+1} & \frac{1}{s(s+1)} \\ 1 & \frac{1}{s} \end{bmatrix}$, and stabilizing K.

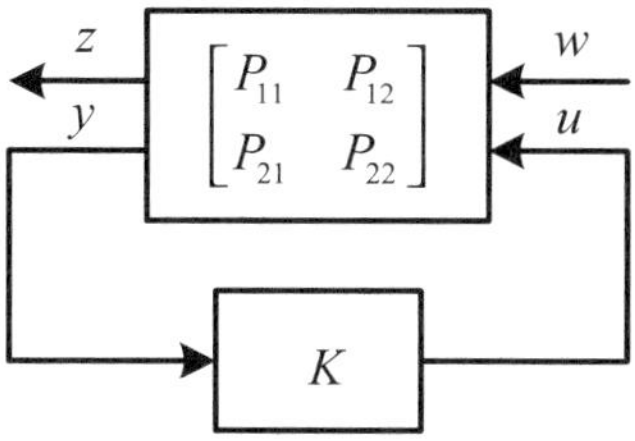

References

1. Green M (1992) H_∞ controller synthesis by J-lossless coprime factorization. SIAM J Control Optim 30:522–547
2. Kimura H (1997) Chain-scattering approach to H_∞ control. Birkhäuser, Boston
3. Redheffer RM (1960) On a certain linear fractional transformation. J Math Phys 39:269–286

Chapter 6
Coprime Factorizations

Coprime factorization originates from algebra studied by French mathematician E. Bezout [1]. In recent years, it has been used to describe dynamic systems [6]. Coprime factorization can be applied in controller synthesis for a given dynamic system with uncertainties [7, 8]. The factorizations can be further employed to construct the set of all stabilizing controllers for the system and to represent a simple parameterization of all stabilized closed-loop transfer functions. In addition, the normalized coprime factorization which will be introduced in this chapter has a strong link to the H_∞ loop-shaping problem [4]. It is also relevant to the spectral factorizations and internal stability.

6.1 Coprimeness and Coprime Factorization

One can start with the simplest case of real numbers. Consider a real rational number $r = \frac{n}{d}$, where d and n are two integers. If the greatest common divisor (g.c.d.) of the pair of integers (d, n) is 1, then d and n are called coprime, and $r = \frac{n}{d}$ is called the coprime factorization of r over the integers. It is well known that if a pair of integers (d, n) is coprime, there exists a row vector of two integers $\begin{bmatrix} \tilde{x} & -\tilde{y} \end{bmatrix}$ such that

$$\begin{bmatrix} \tilde{x} & -\tilde{y} \end{bmatrix} \begin{bmatrix} d \\ n \end{bmatrix} = 1, \tag{6.1}$$

i.e., $\tilde{x}d - \tilde{y}n = 1$, where $\begin{bmatrix} \tilde{x} & -\tilde{y} \end{bmatrix}$ is called the left inverse of $\begin{bmatrix} d \\ n \end{bmatrix}$. For the example $r = \frac{3}{2}$, it can be found that

$$\begin{bmatrix} -1 & 1 \end{bmatrix} \begin{bmatrix} 2 \\ 3 \end{bmatrix} = 1.$$

M.-C. Tsai and D.-W. Gu, *Robust and Optimal Control: A Two-port Framework Approach*, Advances in Industrial Control, DOI 10.1007/978-1-4471-6257-5_6,
© Springer-Verlag London 2014

Clearly, 3 and 2 are coprime, and $1.5 = \frac{3}{2}$ is the coprime factorization of the real rational number 1.5. In fact, the factorization $\frac{6}{4}$ is also equal to 1.5. However, 6 and 4 are not coprime, since it is evident that the g.c.d. of (4,6) is 2, and the vector $\begin{bmatrix} 4 \\ 6 \end{bmatrix}$ does not have a left inverse which only has integer elements. This reveals that $1.5 = \frac{6}{4}$ is not a coprime factorization.

Proceed forward and consider the coprimeness over a ring of polynomials with real coefficients. Two polynomials are called coprime if they do not share common zeros. Let $d(s)$ be a polynomial with real coefficients denoted as

$$d(s) = d_n s^n + d_{n-1} s^{n-1} + \cdots + d_1 s + d_0. \tag{6.2}$$

The polynomial $d(s)$ is said to be of *degree n* if $d_n \neq 0$. If the leading coefficient d_n is equal to 1, the polynomial is called *monic*. Consider two polynomials $d(s)$ and $n(s)$ with $d(s) \neq 0$. Then, it can be proved [2] that the pair of polynomials $(d(s), n(s))$ are coprime over the polynomial ring if and only if there exist two polynomials $(x(s), y(s))$ such that

$$xd - yn = 1. \tag{6.3}$$

Consider a simple, illustrative example $F(s) = \frac{n(s)}{d(s)}$ with $d(s) = s + 1$ and $n(s) = s + 2$. Trivially, this is a coprime factorization over the polynomial ring since one can find that these two polynomials do not share common zero and $\begin{bmatrix} s+1 & -s \end{bmatrix} \begin{bmatrix} s+1 \\ s+2 \end{bmatrix} = 1$. For an alternative factorization $F(s) = \frac{n(s)}{d(s)}$ with $d(s) = s^2 + 4s + 3$ and $n(s) = s^2 + 5s + 6$, apparently, the pair of $(n(s), d(s))$ is not coprime since $s = -3$ is a common zero. It can also be seen that this factorization is *reducible* as

$$F(s) = \frac{s^2 + 5s + 6}{s^2 + 4s + 3} = \frac{(s+2)\cancel{(s+3)}}{(s+1)\cancel{(s+3)}}. \tag{6.4}$$

If two polynomials $d(s)$ and $n(s)$ are coprime over the polynomial ring, the rational function of $\frac{n(s)}{d(s)}$ should be *irreducible* over the polynomial ring. Note that the coprime factorization is not unique up to a unit (i.e., the real number) in the polynomial ring. For instance, one can check that

$$F(s) = \frac{s+2}{s+1} = \frac{2s+4}{2s+2}, \tag{6.5}$$

where $[(2s + 2),(2s + 4)]$ is also coprime, since

$$\begin{bmatrix} s+\frac{3}{2} & -\left(s+\frac{1}{2}\right) \end{bmatrix} \begin{bmatrix} 2s+2 \\ 2s+4 \end{bmatrix} = 1. \tag{6.6}$$

The (coprime) factorization of a rational function need be expanded when the stability issue of a transfer function is to be considered in control system analysis and synthesis. A transfer function is a rational, fractional function with real coefficients. This problem can naturally be solved if the coprime factors belong to the ring of *stable* rational functions. That is, one can consider the coprime factorization of a real rational function $T(s)$ over stable proper rational functions (i.e., the Hardy space RH_∞). It is defined that two stable rational functions $M(s)$ and $N(s)$ are coprime if there exist two stable rational functions $\left(\tilde{X}(s), \tilde{Y}(s)\right)$ such that

$$\begin{bmatrix} \tilde{X}(s) & -\tilde{Y}(s) \end{bmatrix} \begin{bmatrix} M(s) \\ N(s) \end{bmatrix} = I. \tag{6.7}$$

For instance, consider an unstable transfer function $T(s) = \frac{s-2}{s-3}$. One can find a coprime factorization of $T(s)$ over stable rational functions as $T(s) = \frac{N(s)}{M(s)}$, where $M(s) = \frac{s-3}{s+1}$ and $N(s) = \frac{s-2}{s+1}$. It can be seen that $M(s)$ and $N(s)$ are both stable and they are coprime over the stable transfer functions, since there exists a left inverse in the following:

$$\begin{bmatrix} \frac{-4s-1}{s+1} & \frac{5s+1}{s+1} \end{bmatrix} \begin{bmatrix} \frac{s-3}{s+1} \\ \frac{s-2}{s+1} \end{bmatrix} = 1. \tag{6.8}$$

Note that the coprime factorization is not unique but just up to a unit in RH_∞ (i.e., outer (bistable) rational function). In the above example, one can easily find another coprime factorization such as $M(s) = \frac{(s-3)(s+6)}{(s+2)(s+5)}$ and $N(s) = \frac{(s-2)(s+6)}{(s+2)(s+5)}$ which are coprime factors of $T(s)$, since $T(s) = \frac{N(s)}{M(s)} = \frac{s-2}{s-3}$. It can be verified that $M(s)$ and $N(s)$ are coprime to each other, because there is a left inverse as

$$\begin{bmatrix} \frac{(-4s-1)(s+5)}{(s+1)(s+6)} & \frac{(5s+1)(s+5)}{(s+1)(s+6)} \end{bmatrix} \begin{bmatrix} \frac{(s-3)(s+6)}{(s+1)(s+5)} \\ \frac{(s-2)(s+6)}{(s+1)(s+5)} \end{bmatrix} = 1. \tag{6.9}$$

Until now, one has discussed the coprimeness of SISO systems over integers, polynomials, and stable rational functions, respectively. In the following, one will expand the coprimeness over stable rational transfer function matrices which will be introduced for general MIMO cases in the development of control system analysis and synthesis.

6.2 Coprime Factorization over RH_∞

Given a transfer function matrix $T(s)$, a basic problem is to find four transfer function matrices $N(s)$, $M(s)$, $\tilde{M}(s)$, and $\tilde{N}(s)$ in RH_∞ such that

$$T(s) = N(s)M^{-1}(s) = \tilde{M}^{-1}(s)\tilde{N}(s), \tag{6.10}$$

where the pair of $\{M(s), N(s)\}$ is the right coprime factors and $\{\tilde{M}(s), \tilde{N}(s)\}$ the left coprime factors. Such coprime factors actually always exist. For instance, for a transfer function matrix $T(s) = \begin{bmatrix} \frac{s+2}{(s+3)^2} & 0 \\ \frac{1}{s+1} & \frac{1}{s+1} \end{bmatrix}$, there is a right coprime factorization of $T(s) = N(s)M(s)^{-1}$, where

$$M(s) = \begin{bmatrix} \dfrac{s+3}{s+2} & 0 \\ -\dfrac{s+3}{s+2} & 1 \end{bmatrix} \in RH_\infty, \quad N(s) = \begin{bmatrix} \dfrac{1}{s+3} & 0 \\ 0 & \dfrac{1}{s+1} \end{bmatrix} \in RH_\infty, \tag{6.11}$$

since one can find a stable $\begin{bmatrix} \tilde{X} & -\tilde{Y} \end{bmatrix}$ such that $\begin{bmatrix} \tilde{X} & -\tilde{Y} \end{bmatrix}\begin{bmatrix} M \\ N \end{bmatrix} = I$, with

$$\tilde{X}(s) = \begin{bmatrix} \dfrac{s+2}{s+3} & 0 \\ \dfrac{s+2}{s+3} & \dfrac{s+2}{s+3} \end{bmatrix} \in RH_\infty \text{ and } -\tilde{Y}(s) = \begin{bmatrix} 0 & 0 \\ 0 & \dfrac{s+1}{s+3} \end{bmatrix} \in RH_\infty. \tag{6.12}$$

Moreover, there is a left coprime factorization of $T(s) = \tilde{M}(s)^{-1}\tilde{N}(s)$, where

$$\tilde{M}(s) = \begin{bmatrix} \dfrac{s+3}{s+2} & 0 \\ -\dfrac{(s+3)^2}{(s+1)(s+2)} & 1 \end{bmatrix} \in RH_\infty,$$

$$\tilde{N}(s) = \begin{bmatrix} \dfrac{1}{s+3} & 0 \\ 0 & \dfrac{1}{s+1} \end{bmatrix} \in RH_\infty, \tag{6.13}$$

since one can verify a stable $\begin{bmatrix} X \\ Y \end{bmatrix}$ such that $\begin{bmatrix} \tilde{M} & \tilde{N} \end{bmatrix}\begin{bmatrix} X \\ -Y \end{bmatrix} = I$ where

$$X(s) = \begin{bmatrix} 1 & 0 \\ \dfrac{(s+3)^2}{(s+1)(s+2)} & \dfrac{s+3}{s+2} \end{bmatrix} \in RH_\infty \text{ and}$$

$$-Y(s) = \begin{bmatrix} -\dfrac{s+3}{s+2} & 0 \\ 0 & -\dfrac{s+1}{s+2} \end{bmatrix} \in RH_\infty. \tag{6.14}$$

Definition 6.1 Two matrices $M(s)$ and $N(s)$ in RH_∞ are right coprime over RH_∞ if they have the same number of columns and if there exist matrices $\tilde{X}(s)$ and $\tilde{Y}(s)$ in RH_∞ such that

$$\tilde{X}(s)M(s) - \tilde{Y}(s)N(s) = I. \tag{6.15}$$

Similarly, two matrices $\tilde{M}(s)$ and $\tilde{N}(s)$ in RH_∞ are left coprime over RH_∞ if they have the same number of rows and if there exist matrices $X(s)$ and $Y(s)$ in RH_∞ such that

$$\tilde{M}(s)X(s) - \tilde{N}(s)Y(s) = I. \tag{6.16}$$

This is equivalent to stating that the stacked matrix $\begin{bmatrix} M \\ N \end{bmatrix}$ is left invertible in RH_∞, and $\begin{bmatrix} \tilde{M} & \tilde{N} \end{bmatrix}$ is right invertible in RH_∞.

For readers who are more familiar with the block diagram manipulations, the following may help to understand the relationships. Assume that $T(s) = N(s)M^{-1}(s) = \tilde{M}^{-1}(s)\tilde{N}(s)$ are the right and left coprime factorizations, respectively [3]. Let $y = Tu$ and $u = Mu'$, where u' is a virtual signal. Then, from

$$\begin{bmatrix} u \\ y \end{bmatrix} = \begin{bmatrix} I \\ T \end{bmatrix} u = \begin{bmatrix} M \\ TM \end{bmatrix} u', \tag{6.17}$$

one has

$$\begin{bmatrix} u \\ y \end{bmatrix} = \begin{bmatrix} M \\ N \end{bmatrix} u'. \tag{6.18}$$

Figure 6.1 gives a representation of $y = Tu = NM^{-1}u$.

Dually, from

$$0 = \begin{bmatrix} T & -I \end{bmatrix} \begin{bmatrix} u \\ y \end{bmatrix} \tag{6.19}$$

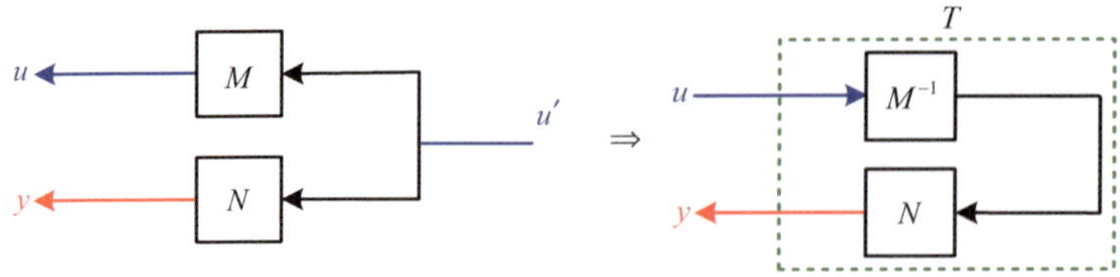

Fig. 6.1 Right coprime factorization

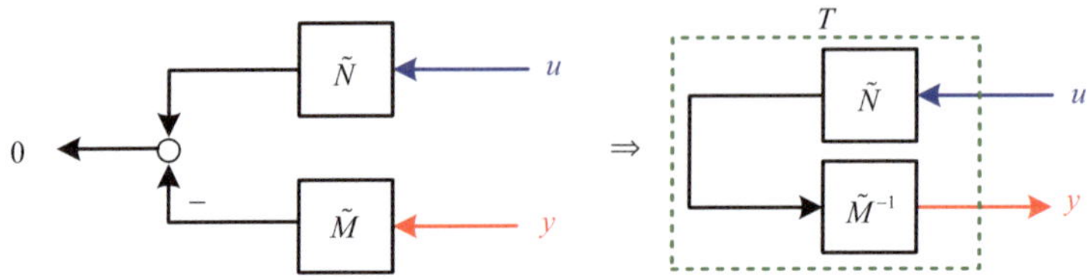

Fig. 6.2 Left coprime factorization

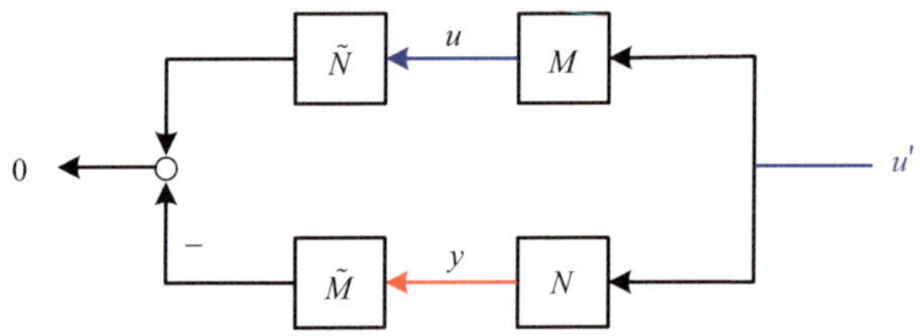

Fig. 6.3 Double coprime factors

and multiplying by $\tilde{M}$, one obtains

$$0 = \begin{bmatrix} \tilde{M}P & -\tilde{M} \end{bmatrix} \begin{bmatrix} u \\ y \end{bmatrix} = \begin{bmatrix} \tilde{N} & \tilde{M} \end{bmatrix} \begin{bmatrix} u \\ -y \end{bmatrix}. \tag{6.20}$$

Figure 6.2 shows another representation of $y = Tu = \tilde{M}^{-1}\tilde{N}u$.

Combining (6.18) and (6.20) will yield

$$\begin{bmatrix} \tilde{N}(s) & -\tilde{M}(s) \end{bmatrix} \begin{bmatrix} M(s) \\ N(s) \end{bmatrix} u' = 0, \tag{6.21}$$

which holds for any u' which implies $\begin{bmatrix} \tilde{N}(s) & -\tilde{M}(s) \end{bmatrix} \begin{bmatrix} M(s) \\ N(s) \end{bmatrix} = 0$. The control block diagram of Fig. 6.3 gives a cascaded representation.

The following lemma shows the coprimeness of the constructed state-space realizations using this approach. In the proof of the lemma, by multiplying $\begin{bmatrix} I & 0 \\ F & W \end{bmatrix}$

and $\begin{bmatrix} I & H \\ 0 & \tilde{W} \end{bmatrix}$, respectively, state-space realizations are derived, similar to Figs. 5.21 and 5.25 in Sect. 5.6.

Lemma 6.1 *Let* $T(s) \overset{s}{=} \left[\begin{array}{c|c} A & B \\ \hline C & D \end{array}\right]$ *be a proper real-rational transfer function matrix with (A, B) stabilizable and (C, A) detectable. A double coprime factorization of* $T(s) = N(s)M^{-1}(s) = \tilde{M}^{-1}(s)\tilde{N}(s)$ *in the state-space form is given by*

$$\begin{bmatrix} M(s) \\ N(s) \end{bmatrix} \overset{s}{=} \left[\begin{array}{c|c} A+BF & BW \\ \hline F & W \\ C+DF & DW \end{array}\right] \in RH_\infty \text{ and} \tag{6.22}$$

$$\begin{bmatrix} \tilde{N}(s) & \tilde{M}(s) \end{bmatrix} \overset{s}{=} \left[\begin{array}{c|cc} A+HC & B+HD & H \\ \hline \tilde{W}C & \tilde{W}D & \tilde{W} \end{array}\right] \in RH_\infty, \tag{6.23}$$

where W and $\tilde{W}$ *are any square constant matrices which are nonsingular and F and H are chosen such that both* $A + BF$ *and* $A + HC$ *are Hurwitz.*

Proof Let

$$\begin{bmatrix} \dot{x} \\ y \end{bmatrix} = \begin{bmatrix} A & B \\ C & D \end{bmatrix} \begin{bmatrix} x \\ u \end{bmatrix}, \tag{6.24}$$

which is not required to be in minimal realization. Since (A, B) is stabilizable, a gain matrix F can be introduced such that $A + BF$ is Hurwitz. Then, let $\begin{bmatrix} x \\ u \end{bmatrix} = \begin{bmatrix} I & 0 \\ F & W \end{bmatrix} \begin{bmatrix} x \\ u' \end{bmatrix}$, where W is invertible. Hence, (6.24) can be expanded to include the signal u that leads to

$$\begin{bmatrix} \dot{x} \\ y \\ u \end{bmatrix} = \begin{bmatrix} A & B \\ C & D \\ 0 & I \end{bmatrix} \begin{bmatrix} x \\ u \end{bmatrix} = \begin{bmatrix} A & B \\ C & D \\ 0 & I \end{bmatrix} \begin{bmatrix} I & 0 \\ F & W \end{bmatrix} \begin{bmatrix} x \\ u' \end{bmatrix}. \tag{6.25}$$

Then,

$$\begin{bmatrix} \dot{x} \\ y \\ u \end{bmatrix} = \begin{bmatrix} A+BF & BW \\ C+DF & DW \\ F & W \end{bmatrix} \begin{bmatrix} x \\ u' \end{bmatrix} \Rightarrow \begin{bmatrix} \dot{x} \\ u \\ y \end{bmatrix} = \begin{bmatrix} A+BF & BW \\ F & W \\ C+DF & DW \end{bmatrix} \begin{bmatrix} x \\ u' \end{bmatrix}. \tag{6.26}$$

By (6.18), one can conclude that

$$\begin{bmatrix} M(s) \\ N(s) \end{bmatrix} \overset{s}{=} \left[\begin{array}{c|c} A+BF & BW \\ \hline F & W \\ C+DF & DW \end{array}\right] \in RH_{\infty}, \tag{6.27}$$

where $T(s) = N(s)M^{-1}(s)$.

Dually, the left coprime factorization gives

$$\begin{bmatrix} \dot{x} \\ y_e \end{bmatrix} = \begin{bmatrix} A & B & 0 \\ C & D & I \end{bmatrix} \begin{bmatrix} x \\ u \\ -y \end{bmatrix}. \tag{6.28}$$

Let $\begin{bmatrix} \dot{\hat{x}} \\ y' \end{bmatrix} = \begin{bmatrix} I & H \\ 0 & \tilde{W} \end{bmatrix} \begin{bmatrix} \dot{x} \\ y_e \end{bmatrix}$. Dual to (6.25), this is equivalent to multiplying $\begin{bmatrix} I & H \\ 0 & \tilde{W} \end{bmatrix}$ by the left of (6.28). Then, the left factorization can yield

$$\begin{bmatrix} \dot{\hat{x}} \\ y' \end{bmatrix} = \begin{bmatrix} I & H \\ 0 & \tilde{W} \end{bmatrix} \begin{bmatrix} A & B & 0 \\ C & D & I \end{bmatrix} \begin{bmatrix} x \\ u \\ -y \end{bmatrix} = \begin{bmatrix} A+HC & B+HD & H \\ \tilde{W}C & \tilde{W}D & \tilde{W} \end{bmatrix} \begin{bmatrix} x \\ u \\ -y \end{bmatrix}, \tag{6.29}$$

where $\tilde{W}$ is any nonsingular matrix. Since (C, A) is detectable, the injected gain matrix H is chosen such that $A + HC$ is Hurwitz. This structure is in fact a closed-loop state observer where u and y are the inputs. Then, by comparing (6.20) and (6.29), one can obtain

$$\begin{bmatrix} \tilde{N}(s) & \tilde{M}(s) \end{bmatrix} \overset{s}{=} \left[\begin{array}{c|cc} A+HC & B+HD & H \\ \hline \tilde{W}C & \tilde{W}D & \tilde{W} \end{array}\right] \in RH_{\infty}, \tag{6.30}$$

where $T(s) = \tilde{M}^{-1}(s)\tilde{N}(s)$.

To ensure that the state-space realization (6.22) is a pair of right coprime factors $\begin{bmatrix} M \\ N \end{bmatrix}$, one now needs Definition 6.1 to construct a left inverse such that $\tilde{X}M - \tilde{Y}N = I$. From $\begin{bmatrix} x \\ u' \end{bmatrix} = \begin{bmatrix} I & 0 \\ F & W \end{bmatrix}^{-1} \begin{bmatrix} x \\ u \end{bmatrix} = \begin{bmatrix} I & 0 \\ -W^{-1}F & W^{-1} \end{bmatrix} \begin{bmatrix} x \\ u \end{bmatrix}$, one has $u' = -W^{-1}Fx + W^{-1}u$. Then, as illustrated in Fig. 6.4, a state-space realization $\begin{bmatrix} \tilde{X} & \tilde{Y} \end{bmatrix} \in RH_{\infty}$ can be constructed based on $\begin{bmatrix} \tilde{N}(s) & \tilde{M}(s) \end{bmatrix}$ in (6.29) as

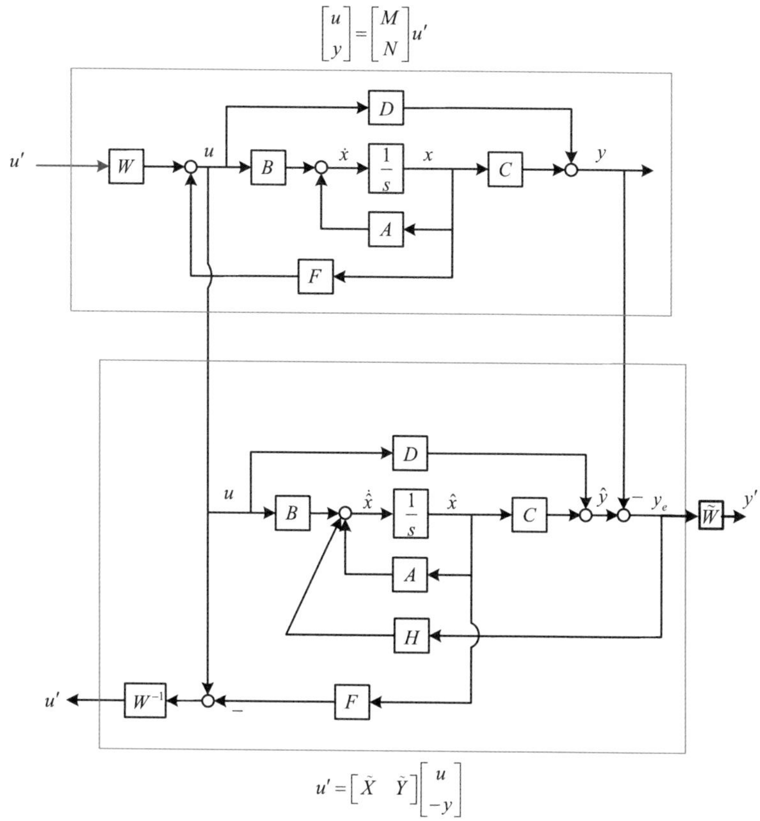

Fig. 6.4 State-space realization $u' \mapsto u'$

$$\begin{bmatrix} \dot{\hat{x}} \\ u' \end{bmatrix} = \begin{bmatrix} A+HC & B+HD & H \\ -W^{-1}F & W^{-1} & 0 \end{bmatrix} \begin{bmatrix} \hat{x} \\ u \\ -y \end{bmatrix} \Rightarrow \begin{bmatrix} \tilde{X}(s) & \tilde{Y}(s) \end{bmatrix} \stackrel{s}{=} \left[\begin{array}{c|cc} A+HC & B+HD & H \\ \hline -W^{-1}F & W^{-1} & 0 \end{array}\right]. \tag{6.31}$$

For $\hat{x} = x$ and applying the state-space algebra of (2.72) will yield

$$\begin{aligned} &\begin{bmatrix} I & 0 & 0 & 0 \\ 0 & A+HC & B+HD & H \\ 0 & -W^{-1}F & W^{-1} & 0 \\ 0 & \tilde{W}C & \tilde{W}D & \tilde{W} \end{bmatrix} \times \begin{bmatrix} A+BF & 0 & BW \\ 0 & I & 0 \\ F & 0 & W \\ -(C+DF) & 0 & -DW \end{bmatrix} \\ &\Rightarrow \left[\begin{array}{cc|c} A+BF & 0 & BW \\ BF-HC & A+HC & BW \\ \hline W^{-1}F & -W^{-1}F & I \\ -\tilde{W}C & \tilde{W}C & 0 \end{array}\right]. \end{aligned} \tag{6.32}$$

This gives

$$\begin{bmatrix} \tilde{X}(s) & \tilde{Y}(s) \\ \tilde{N}(s) & \tilde{M}(s) \end{bmatrix} \begin{bmatrix} M(s) \\ -N(s) \end{bmatrix} \overset{s}{=} \left[\begin{array}{cc|c} A+BF & 0 & BW \\ BF-HC & A+HC & BW \\ \hline W^{-1}F & -W^{-1}F & I \\ -\tilde{W}C & \tilde{W}C & 0 \end{array} \right] \tag{6.33}$$

Then $\tilde{X}M - \tilde{Y}N = I$ can be determined by a similarity transformation $T = \begin{bmatrix} I & 0 \\ I & I \end{bmatrix}$ as

$$\left[\begin{array}{cc|c} A+BF & 0 & BW \\ BF-HC & A+HC & BW \\ \hline W^{-1}F & -W^{-1}F & I \\ -\tilde{W}C & \tilde{W}C & 0 \end{array} \right] \overset{s}{=} \left[\begin{array}{cc|c} A+BF & 0 & BW \\ 0 & A+HC & 0 \\ \hline 0 & -W^{-1}F & I \\ 0 & \tilde{W}C & 0 \end{array} \right]. \tag{6.34}$$

This concludes that $\begin{bmatrix} M \\ N \end{bmatrix}$ given by (6.22) is left invertible in RH_∞, and then $T = NM^{-1}$ is a right coprime factorization.

Likewise, to ensure that the state-space realization (6.23) is of coprime factors, one now needs to construct a right inverse $\begin{bmatrix} Y \\ X \end{bmatrix} \in RH_\infty$ in the state-space form such that $\tilde{M}X - \tilde{N}Y = I$. From $\begin{bmatrix} \dot{x} \\ y_e \end{bmatrix} = \begin{bmatrix} I & H \\ 0 & \tilde{W} \end{bmatrix}^{-1} \begin{bmatrix} \dot{\hat{x}} \\ y' \end{bmatrix}$ and based on (6.26), one has, from Fig. 6.5,

$$\begin{bmatrix} \dot{x} \\ u \\ y_e \end{bmatrix} = \begin{bmatrix} A+BF & BW & -H\tilde{W}^{-1} \\ F & W & 0 \\ C+DF & DW & \tilde{W}^{-1} \end{bmatrix} \begin{bmatrix} x \\ u' \\ y' \end{bmatrix} \Rightarrow \begin{bmatrix} Y(s) \\ X(s) \end{bmatrix} \overset{s}{=} \left[\begin{array}{c|c} A+BF & -H\tilde{W}^{-1} \\ \hline F & 0 \\ C+DF & \tilde{W}^{-1} \end{array} \right]. \tag{6.35}$$

For $\hat{x} = x$, applying the state-space algebra of (2.72) will yield

$$\begin{bmatrix} \tilde{N}(s) & \tilde{M}(s) \end{bmatrix} \begin{bmatrix} M(s) & -Y(s) \\ -N(s) & X(s) \end{bmatrix} \overset{s}{=} \left[\begin{array}{cc|cc} A+BF & 0 & BW & H\tilde{W}^{-1} \\ BF-HC & A+HC & BW & H\tilde{W}^{-1} \\ \hline -\tilde{W}C & \tilde{W}C & 0 & I \end{array} \right], \tag{6.36}$$

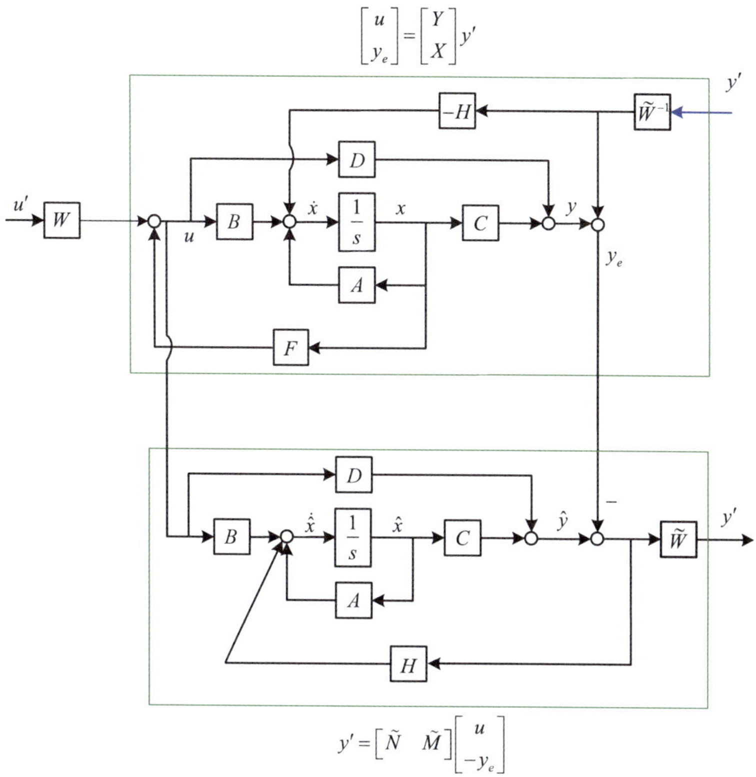

Fig. 6.5 State-space realization $y' \mapsto y'$

where

$$\left[\begin{array}{cccc} A+BF & 0 & BW & H\tilde{W}^{-1} \\ BF-HC & A+HC & BW & H\tilde{W}^{-1} \\ -\tilde{W}C & \tilde{W}C & 0 & I \end{array}\right]$$
$$= \begin{bmatrix} I & 0 & 0 & 0 \\ 0 & A+HC & B+HD & H \\ 0 & \tilde{W}C & \tilde{W}D & \tilde{W} \end{bmatrix} \begin{bmatrix} A+BF & 0 & BW & H\tilde{W}^{-1} \\ 0 & I & 0 & 0 \\ F & 0 & W & 0 \\ -(C+DF) & 0 & -DW & \tilde{W}^{-1} \end{bmatrix}. \tag{6.37}$$

Then $-\tilde{N}Y + \tilde{M}X = I$ can be determined by a similarity transformation $T = \begin{bmatrix} I & 0 \\ I & I \end{bmatrix}$ as

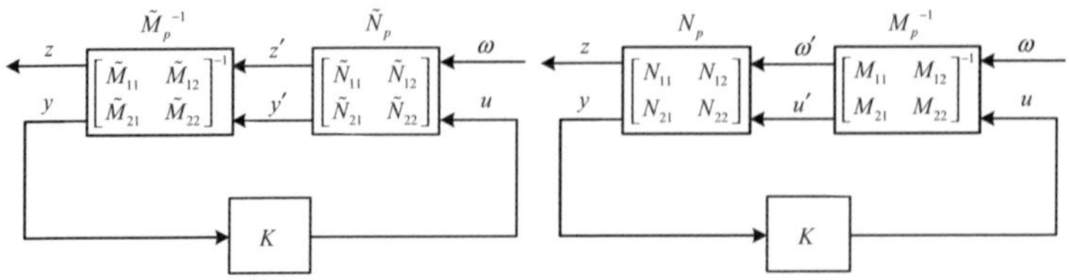

Fig. 6.6 SCC in right or left coprime forms

$$\overset{I}{\curvearrowleft}\;(-I)\left[\begin{array}{cc|cc} A+BF & 0 & BW & H\tilde{W}^{-1} \\ BF-HC & A+HC & BW & H\tilde{W}^{-1} \\ \hline -\tilde{W}C & \tilde{W}C & 0 & I \end{array}\right] \overset{s}{=} \left[\begin{array}{cc|cc} A+BF & 0 & BW & H\tilde{W}^{-1} \\ 0 & A+HC & 0 & 0 \\ \hline 0 & \tilde{W}C & 0 & I \end{array}\right]. \tag{6.38}$$

This concludes that $\left[\tilde{N}(s) \quad \tilde{M}(s)\right]$ given by (6.23) is right invertible in RH_∞, and then $T = \tilde{M}^{-1}\tilde{N}$ is a left coprime factorization. ■

Furthermore, it can be verified from (6.31) and (6.35) that

$$0 = \left[\tilde{X}(s) \quad \tilde{Y}(s)\right]\begin{bmatrix} -Y(s) \\ X(s) \end{bmatrix} \overset{s}{=} \left[\begin{array}{cc|c} A+BF & 0 & H\tilde{W}^{-1} \\ BF-HC & A+HC & H\tilde{W}^{-1} \\ \hline W^{-1}F & -W^{-1}F & 0 \end{array}\right]. \tag{6.39}$$

Hence, (6.33) and (6.36) will form the Bezout identity as summarized in the following lemma.

Lemma 6.2 *For any proper real-rational matrix $T(s)$, there always exists a double (left and right) coprime factorization given by* (6.10), *where $N(s)$, $M(s)$, $\tilde{N}(s)$, and $\tilde{M}(s)$ are in RH_∞, respectively. For the double coprime factorization, there exist RH_∞ transfer matrices $X(s)$, $Y(s)$, $\tilde{X}(s)$, and $\tilde{Y}(s)$ satisfying the Bezout identity*

$$\begin{bmatrix} \tilde{X}(s) & \tilde{Y}(s) \\ \tilde{N}(s) & \tilde{M}(s) \end{bmatrix}\begin{bmatrix} M(s) & -Y(s) \\ -N(s) & X(s) \end{bmatrix} = \begin{bmatrix} I & 0 \\ 0 & I \end{bmatrix}. \tag{6.40}$$

As stated in Chap. 5, the coprime factorization can also arise in the two-port representation. Referring to the chain scattering approach proposed by Tsai [5], the coprime factorization of $P = N_p M_p^{-1}$ (or $P = \tilde{M}_p^{-1}\tilde{N}_p$) as illustrated in Fig. 6.6 is utilized.

As shown in Fig. 6.6, let

$$N_p = \begin{bmatrix} N_{11} & N_{12} \\ N_{21} & N_{22} \end{bmatrix} \in RH_\infty, \quad M_p = \begin{bmatrix} M_{11} & M_{12} \\ M_{21} & M_{22} \end{bmatrix} \in RH_\infty, \tag{6.41}$$

and

$$\tilde{N}_p = \begin{bmatrix} \tilde{N}_{11} & \tilde{N}_{12} \\ \tilde{N}_{21} & \tilde{N}_{22} \end{bmatrix} \in RH_\infty, \quad \tilde{M}_p = \begin{bmatrix} \tilde{M}_{11} & \tilde{M}_{12} \\ \tilde{M}_{21} & \tilde{M}_{22} \end{bmatrix} \in RH_\infty. \tag{6.42}$$

Recall from the SCC of Fig. 5.8. Let

$$\begin{bmatrix} z \\ y \end{bmatrix} = \begin{bmatrix} P_{11} & P_{21} \\ P_{12} & P_{22} \end{bmatrix} \begin{bmatrix} w \\ u \end{bmatrix} = N_p M_p^{-1} \begin{bmatrix} w \\ u \end{bmatrix}, \tag{6.43}$$

where

$$\begin{bmatrix} z \\ y \end{bmatrix} = N_p \begin{bmatrix} w' \\ u' \end{bmatrix} = \begin{bmatrix} N_{11} & N_{12} \\ N_{21} & N_{22} \end{bmatrix} \begin{bmatrix} w' \\ u' \end{bmatrix}, \tag{6.44}$$

and

$$\begin{bmatrix} w \\ u \end{bmatrix} = M_p \begin{bmatrix} w' \\ u' \end{bmatrix} = \begin{bmatrix} M_{11} & M_{12} \\ M_{21} & M_{22} \end{bmatrix} \begin{bmatrix} w' \\ u' \end{bmatrix}. \tag{6.45}$$

Similarly, the left coprime factorization, as shown in Fig. 6.6, can be found in the dual way. From the SCC of Fig. 5.8, one has

$$\begin{bmatrix} z \\ y \end{bmatrix} = \begin{bmatrix} P_{11} & P_{21} \\ P_{12} & P_{22} \end{bmatrix} \begin{bmatrix} w \\ u \end{bmatrix} = \tilde{M}_p^{-1} \tilde{N}_p \begin{bmatrix} w \\ u \end{bmatrix}, \tag{6.46}$$

where

$$\begin{bmatrix} z' \\ y' \end{bmatrix} = \tilde{N}_p \begin{bmatrix} w \\ u \end{bmatrix} = \begin{bmatrix} \tilde{N}_{11} & \tilde{N}_{12} \\ \tilde{N}_{21} & \tilde{N}_{22} \end{bmatrix} \begin{bmatrix} w \\ u \end{bmatrix}, \tag{6.47}$$

and

$$\begin{bmatrix} z' \\ y' \end{bmatrix} = \tilde{M}_p \begin{bmatrix} z \\ y \end{bmatrix} = \begin{bmatrix} \tilde{M}_{11} & \tilde{M}_{12} \\ \tilde{M}_{21} & \tilde{M}_{22} \end{bmatrix} \begin{bmatrix} z \\ y \end{bmatrix}. \tag{6.48}$$

Lemma 6.3 *Let* $P(s) \overset{s}{=} \left[\begin{array}{c|cc} A & B_1 & B_2 \\ \hline C_1 & D_{11} & D_{12} \\ C_2 & D_{21} & 0 \end{array}\right]$ *be a proper real-rational matrix with* (A,B_2) *stabilizable and* (C_2,A) *detectable. A double coprime factorization of* $P(s) = N_p(s)M_p^{-1}(s) = \tilde{M}_p^{-1}(s)\tilde{N}_p(s)$ *in the state-space form is given by*

$$\begin{bmatrix} M_p(s) \\ N_p(s) \end{bmatrix} = \left[\begin{array}{cc} M_{p11}(s) & M_{p12}(s) \\ M_{p21}(s) & M_{p22}(s) \\ \hdashline N_{p11}(s) & N_{p12}(s) \\ N_{p21}(s) & N_{p22}(s) \end{array}\right] \overset{s}{=} \left[\begin{array}{c|cc} A+B_1F_w+B_2F_u & B_1W_{ww} & B_1W_{wu}+B_2W_{uu} \\ \hline F_w & W_{ww} & W_{wu} \\ F_u & 0 & W_{uu} \\ \hdashline C_1+D_{11}F_w+D_{12}F_u & D_{11}W_{ww} & D_{11}W_{wu}+D_{12}W_{uu} \\ C_2+D_{21}F_w & D_{21}W_{ww} & D_{21}W_{wu} \end{array}\right], \tag{6.49}$$

and

$$\begin{bmatrix} \tilde{N}_p(s) & \tilde{M}_p(s) \end{bmatrix} = \left[\begin{array}{cc:cc} \tilde{N}_{p11}(s) & \tilde{N}_{p12}(s) & \tilde{M}_{p11}(s) & \tilde{M}_{p12}(s) \\ \tilde{N}_{p21}(s) & \tilde{N}_{p22}(s) & \tilde{M}_{p21}(s) & \tilde{M}_{p22}(s) \end{array}\right]$$

$$\overset{s}{=} \left[\begin{array}{c|cc:cc} A+H_zC_1+H_yC_2 & B_1+H_zD_{11}+H_yD_{21} & B_2+H_zD_{12} & H_z & H_y \\ \hline \tilde{W}_{zz}C_1 & \tilde{W}_{zz}D_{11} & \tilde{W}_{zz}D_{12} & \tilde{W}_{zz} & 0 \\ \tilde{W}_{yz}C_1+\tilde{W}_{yy}C_2 & \tilde{W}_{yz}D_{11}+\tilde{W}_{yy}D_{21} & \tilde{W}_{yz}D_{12} & \tilde{W}_{yz} & \tilde{W}_{yy} \end{array}\right] \tag{6.50}$$

where W_{uu}, W_{ww}, $\tilde{W}_{zz}$, *and* $\tilde{W}_{yy}$ *are nonsingular and* $F = \begin{bmatrix} F_u \\ F_w \end{bmatrix}$ *and* $H = [H_z \; H_y]$ *are chosen such that both* $A + B_1F_w + B_2F_u$ *and* $A + H_zC_1 + H_yC_2$ *are Hurwitz.*

Proof Let

$$\begin{bmatrix} \dot{x} \\ z \\ y \\ w \\ u \end{bmatrix} = \begin{bmatrix} A & B_1 & B_2 \\ C_1 & D_{11} & D_{12} \\ C_2 & D_{21} & 0 \\ 0 & I & 0 \\ 0 & 0 & I \end{bmatrix} \begin{bmatrix} x \\ w \\ u \end{bmatrix}. \tag{6.51}$$

Additionally, let $\begin{bmatrix} x \\ w \\ u \end{bmatrix} = \begin{bmatrix} I & 0 & 0 \\ F_w & W_{ww} & W_{wu} \\ F_u & 0 & W_{uu} \end{bmatrix} \begin{bmatrix} x \\ w' \\ u' \end{bmatrix}$, where $F = \begin{bmatrix} F_u \\ F_w \end{bmatrix}$ is chosen such that $A + B_1F_w + B_2F_u$ is Hurwitz and W_{uu} *and* W_{ww} are nonsingular.

This is equivalent to multiplying $\begin{bmatrix} I & 0 & 0 \\ F_w & W_{ww} & W_{wu} \\ F_u & 0 & W_{uu} \end{bmatrix}$ in the right-hand side of the above formulation, yielding

$$\begin{bmatrix} \dot{x} \\ z \\ y \\ w \\ u \end{bmatrix} = \begin{bmatrix} A & B_1 & B_2 \\ C_1 & D_{11} & D_{12} \\ C_2 & D_{21} & 0 \\ 0 & I & 0 \\ 0 & 0 & I \end{bmatrix} \begin{bmatrix} I & 0 & 0 \\ F_w & W_{ww} & W_{wu} \\ F_u & 0 & W_{uu} \end{bmatrix} \begin{bmatrix} x \\ w' \\ u' \end{bmatrix}$$

$$= \begin{bmatrix} A + B_1 F_w + B_2 F_u & B_1 W_{ww} & B_1 W_{wu} + B_2 W_{uu} \\ C_1 + D_{11} F_w + D_{12} F_u & D_{11} W_{ww} & D_{11} W_{wu} + D_{12} W_{uu} \\ C_2 + D_{21} F_w & D_{21} W_{ww} & D_{21} W_{wu} \\ F_w & W_{ww} & W_{wu} \\ F_u & 0 & W_{uu} \end{bmatrix} \begin{bmatrix} x \\ w' \\ u' \end{bmatrix}. \tag{6.52}$$

From (6.44) and (6.45), one concludes that

$$\begin{bmatrix} N_p(s) \\ \hdashline M_p(s) \end{bmatrix} \overset{s}{=} \left[\begin{array}{c|cc} A + B_1 F_w + B_2 F_u & B_1 W_{ww} & B_1 W_{wu} + B_2 W_{uu} \\ \hline C_1 + D_{11} F_w + D_{12} F_u & D_{11} W_{ww} & D_{11} W_{wu} + D_{12} W_{uu} \\ C_2 + D_{21} F_w & D_{21} W_{ww} & D_{21} W_{wu} \\ \hdashline F_w & W_{ww} & W_{wu} \\ F_u & 0 & W_{uu} \end{array}\right] \tag{6.53}$$

$$\Rightarrow \begin{bmatrix} M_p(s) \\ \hdashline N_p(s) \end{bmatrix} \overset{s}{=} \left[\begin{array}{c|cc} A + B_1 F_w + B_2 F_u & B_1 W_{ww} & B_1 W_{wu} + B_2 W_{uu} \\ \hline F_w & W_{ww} & W_{wu} \\ F_u & 0 & W_{uu} \\ \hdashline C_1 + D_{11} F_w + D_{12} F_u & D_{11} W_{ww} & D_{11} W_{wu} + D_{12} W_{uu} \\ C_2 + D_{21} F_w & D_{21} W_{ww} & D_{21} W_{wu} \end{array}\right].$$

Furthermore, from

$$\begin{bmatrix} \dot{x} \\ z_e \\ y_e \end{bmatrix} = \begin{bmatrix} A & B_1 & B_2 & 0 & 0 \\ C_1 & D_{11} & D_{12} & I & 0 \\ C_2 & D_{21} & 0 & 0 & I \end{bmatrix} \begin{bmatrix} x \\ w \\ u \\ -z \\ -y \end{bmatrix}, \tag{6.54}$$

let $\begin{bmatrix} \dot{\hat{x}} \\ z' \\ y' \end{bmatrix} = \begin{bmatrix} I & H_z & H_y \\ 0 & \tilde{W}_{zz} & 0 \\ 0 & \tilde{W}_{yz} & \tilde{W}_{yy} \end{bmatrix} \begin{bmatrix} \dot{x} \\ z_e \\ y_e \end{bmatrix}$, where $H = [H_z \ H_y]$ is chosen such that $A + H_z C_1 + H_y C_2$ is Hurwitz and $\tilde{W}_{zz}$ and $\tilde{W}_{yy}$ are nonsingular. This is equivalent to multiplying $\begin{bmatrix} I & H_z & H_y \\ 0 & \tilde{W}_{zz} & 0 \\ 0 & \tilde{W}_{yz} & \tilde{W}_{yy} \end{bmatrix}$ by the left-hand side of the above formulation, and then one has

$$\begin{bmatrix} \dot{\hat{x}} \\ z' \\ y' \end{bmatrix} = \begin{bmatrix} I & H_z & H_y \\ 0 & \tilde{W}_{zz} & 0 \\ 0 & \tilde{W}_{yz} & \tilde{W}_{yy} \end{bmatrix} \begin{bmatrix} A & B_1 & B_2 & 0 & 0 \\ C_1 & D_{11} & D_{12} & I & 0 \\ C_2 & D_{21} & 0 & 0 & I \end{bmatrix} \begin{bmatrix} x \\ w \\ u \\ -z \\ -y \end{bmatrix}$$

$$= \begin{bmatrix} A + H_z C_1 + H_y C_2 & B_1 + H_z D_{11} + H_y D_{21} & B_2 + H_z D_{12} & H_z & H_y \\ \tilde{W}_{zz} C_1 & \tilde{W}_{zz} D_{11} & \tilde{W}_{zz} D_{12} & \tilde{W}_{zz} & 0 \\ \tilde{W}_{yz} C_1 + \tilde{W}_{yy} C_2 & \tilde{W}_{yz} D_{11} + \tilde{W}_{yy} D_{21} & \tilde{W}_{yz} D_{12} & \tilde{W}_{yz} & \tilde{W}_{yy} \end{bmatrix} \times \begin{bmatrix} x \\ w \\ u \\ -z \\ -y \end{bmatrix}. \tag{6.55}$$

From (6.47) and (6.48), one concludes

$$\left[\tilde{N}_p(s) \mid \tilde{M}_p(s)\right] \stackrel{s}{=} \left[\begin{array}{c|cc|cc} A + H_z C_1 + H_y C_2 & B_1 + H_z D_{11} + H_y D_{21} & B_2 + H_z D_{12} & H_z & H_y \\ \hline \tilde{W}_{zz} C_1 & \tilde{W}_{zz} D_{11} & \tilde{W}_{zz} D_{12} & \tilde{W}_{zz} & 0 \\ \tilde{W}_{yz} C_1 + \tilde{W}_{yy} C_2 & \tilde{W}_{yz} D_{11} + \tilde{W}_{yy} D_{21} & \tilde{W}_{yz} D_{12} & \tilde{W}_{yz} & \tilde{W}_{yy} \end{array}\right]. \tag{6.56}$$

To ensure that (6.53) is a pair of right coprime factors, one needs Definition 6.1 to construct a left inverse based on (6.56) in RH_∞, i.e., $\tilde{X} M_p - \tilde{Y} N_p = I$. From

$$\begin{bmatrix} x \\ w' \\ u' \end{bmatrix} = \begin{bmatrix} I & 0 & 0 \\ F_w & W_{ww} & W_{wu} \\ F_u & 0 & W_{uu} \end{bmatrix}^{-1} \begin{bmatrix} x \\ w \\ u \end{bmatrix}$$

$$= \begin{bmatrix} I & 0 & 0 \\ -W_{ww}^{-1} F_w + W_{ww}^{-1} W_{wu} W_{uu}^{-1} F_u & W_{ww}^{-1} & -W_{ww}^{-1} W_{wu} W_{uu}^{-1} \\ -W_{uu}^{-1} F_u & 0 & W_{uu}^{-1} \end{bmatrix} \begin{bmatrix} x \\ w \\ u \end{bmatrix}, \tag{6.57}$$

one has

$$\left[\tilde{X}(s) \mid \tilde{Y}(s)\right] = \left[\begin{array}{cc|cc} \tilde{X}_{11}(s) & \tilde{X}_{12}(s) & \tilde{Y}_{11}(s) & \tilde{Y}_{12}(s) \\ \tilde{X}_{21}(s) & \tilde{X}_{22}(s) & \tilde{Y}_{21}(s) & \tilde{Y}_{22}(s) \end{array}\right]$$

$$\stackrel{s}{=} \left[\begin{array}{c|cc|cc} A + H_z C_1 + H_y C_2 & B_1 + H_z D_{11} + H_y D_{21} & B_2 + H_z D_{12} & H_z & H_y \\ \hline -W_{ww}^{-1} F_w + W_{ww}^{-1} W_{wu} W_{uu}^{-1} F_u & W_{ww}^{-1} & -W_{ww}^{-1} W_{wu} W_{uu}^{-1} & 0 & 0 \\ -W_{uu}^{-1} F_u & 0 & W_{uu}^{-1} & 0 & 0 \end{array}\right], \tag{6.58}$$

Then, one has from (6.53) and (6.58)

$$\left[\tilde{X}(s) \quad \tilde{Y}(s)\right] \begin{bmatrix} M_p(s) \\ -N_p(s) \end{bmatrix}$$

$$\stackrel{s}{=} \left[\begin{array}{cc|cc} A + B_1 F_w + B_2 F_u & 0 & B_1 W_{ww} & B_1 W_{wu} + B_2 W_{uu} \\ B_1 F_w + B_2 F_u - H_z C_1 - H_y C_2 & A + H_z C_1 + H_y C_2 & B_1 W_{ww} & B_1 W_{wu} + B_2 W_{uu} \\ \hline W_{ww}^{-1} F_w - W_{ww}^{-1} W_{wu} W_{uu}^{-1} F_u & -W_{ww}^{-1} F_w + W_{ww}^{-1} W_{wu} W_{uu}^{-1} F_u & I & 0 \\ W_{uu}^{-1} F_u & -W_{uu}^{-1} F_u & 0 & I \end{array}\right]. \tag{6.59}$$

The similarity transformation $T = \begin{bmatrix} I & 0 \\ I & I \end{bmatrix}$ will yield

$$\begin{array}{c} I \\ -I \end{array} \left[\begin{array}{cc|cc} A + B_1 F_w + B_2 F_u & 0 & B_1 W_{ww} & B_1 W_{wu} + B_2 W_{uu} \\ B_1 F_w + B_2 F_u - H_z C_1 - H_y C_2 & A + H_z C_1 + H_y C_2 & B_1 W_{ww} & B_1 W_{wu} + B_2 W_{uu} \\ \hline W_{ww}^{-1} F_w - W_{ww}^{-1} W_{wu} W_{uu}^{-1} F_u & -W_{ww}^{-1} F_w + W_{ww}^{-1} W_{wu} W_{uu}^{-1} F_u & I & 0 \\ W_{uu}^{-1} F_u & -W_{uu}^{-1} F_u & 0 & I \end{array}\right]$$

$$\stackrel{s}{=} \left[\begin{array}{cc|cc} A + B_1 F_w + B_2 F_u & 0 & B_1 W_{ww} & B_1 W_{wu} + B_2 W_{uu} \\ 0 & A + H_z C_1 + H_y C_2 & 0 & 0 \\ \hline 0 & -W_{ww}^{-1} F_w + W_{ww}^{-1} W_{wu} W_{uu}^{-1} F_u & I & 0 \\ 0 & -W_{uu}^{-1} F_u & 0 & I \end{array}\right]. \tag{6.60}$$

This concludes that (6.49) is left invertible in RH_∞ such that $P(s) = N_p(s) M_p^{-1}(s)$ is a coprime factorization.

Analogously, to ensure (6.50) consists of coprime factors, one needs to construct, based on (6.52), a right inverse in RH_∞. From

$$\begin{bmatrix} \dot{\hat{x}} \\ z' \\ y' \end{bmatrix} = \begin{bmatrix} I & H_z & H_y \\ 0 & \tilde{W}_{zz} & 0 \\ 0 & \tilde{W}_{yz} & \tilde{W}_{yy} \end{bmatrix}^{-1} \begin{bmatrix} \dot{\hat{x}} \\ z_e \\ y_e \end{bmatrix}$$

$$= \begin{bmatrix} I & -H_z \tilde{W}_{zz}^{-1} + H_y \tilde{W}_{yy}^{-1} \tilde{W}_{yz} \tilde{W}_{zz}^{-1} & -H_y \tilde{W}_{yy}^{-1} \\ 0 & \tilde{W}_{zz}^{-1} & 0 \\ 0 & -\tilde{W}_{yy}^{-1} \tilde{W}_{yz} \tilde{W}_{zz}^{-1} & \tilde{W}_{yy}^{-1} \end{bmatrix} \begin{bmatrix} \dot{\hat{x}} \\ z_e \\ y_e \end{bmatrix}, \tag{6.61}$$

let

$$\begin{bmatrix} Y(s) \\ \hdashline X(s) \end{bmatrix} = \begin{bmatrix} Y_{11}(s) & Y_{12}(s) \\ Y_{21}(s) & Y_{22}(s) \\ \hdashline X_{11}(s) & X_{12}(s) \\ X_{21}(s) & X_{22}(s) \end{bmatrix}$$

$$\stackrel{s}{=} \left[\begin{array}{c|cc} A+B_2F_u+B_1F_w & -H_z\tilde{W}_{zz}^{-1}+H_y\tilde{W}_{yy}^{-1}\tilde{W}_{yz}\tilde{W}_{zz}^{-1} & -H_y\tilde{W}_{yy}^{-1} \\ \hline F_w & 0 & 0 \\ F_u & 0 & 0 \\ \hdashline C_1+D_{12}F_u+D_{11}F_w & \tilde{W}_{zz}^{-1} & 0 \\ C_2+D_{21}F_w & -\tilde{W}_{yy}^{-1}\tilde{W}_{yz}\tilde{W}_{zz}^{-1} & \tilde{W}_{yy}^{-1} \end{array}\right]. \tag{6.62}$$

Then one has, from (6.56) and (6.62),

$$\begin{bmatrix} \tilde{N}_p(s) & \tilde{M}_p(s) \end{bmatrix} \begin{bmatrix} -Y(s) \\ X(s) \end{bmatrix}$$

$$\stackrel{s}{=} \left[\begin{array}{cc|cc} A+B_2F_u+B_1F_w & 0 & -H_z\tilde{W}_{zz}^{-1}+H_y\tilde{W}_{yy}^{-1}\tilde{W}_{yz}\tilde{W}_{zz}^{-1} & -H_y\tilde{W}_{yy}^{-1} \\ -B_1F_w-B_2F_u+H_zC_1+H_yC_2 & A+H_zC_1+H_yC_2 & H_z\tilde{W}_{zz}^{-1}-H_y\tilde{W}_{yy}^{-1}\tilde{W}_{yz}\tilde{W}_{zz}^{-1} & H_y\tilde{W}_{yy}^{-1} \\ \hline \tilde{W}_{zz}C_1 & \tilde{W}_{zz}C_1 & I & 0 \\ \tilde{W}_{yz}C_1+\tilde{W}_{yy}C_2 & \tilde{W}_{yz}C_1+\tilde{W}_{yy}C_2 & 0 & I \end{array}\right]. \tag{6.63}$$

By the state similarity transformation $T = \begin{bmatrix} I & 0 \\ I & I \end{bmatrix}$, it yields $-\tilde{N}_pY + \tilde{M}_pX = I$, since

$$\overset{-I}{\curvearrowleft} \quad I\circlearrowleft \left[\begin{array}{cc|cc} A+B_2F_u+B_1F_w & 0 & -H_z\tilde{W}_{zz}^{-1}+H_y\tilde{W}_{yy}^{-1}\tilde{W}_{yz}\tilde{W}_{zz}^{-1} & -H_y\tilde{W}_{yy}^{-1} \\ \begin{pmatrix} -B_1F_w-B_2F_u \\ +H_zC_1+H_yC_2 \end{pmatrix} & A+H_zC_1+H_yC_2 & H_z\tilde{W}_{zz}^{-1}-H_y\tilde{W}_{yy}^{-1}\tilde{W}_{yz}\tilde{W}_{zz}^{-1} & H_y\tilde{W}_{yy}^{-1} \\ \hline \tilde{W}_{zz}C_1 & \tilde{W}_{zz}C_1 & I & 0 \\ \tilde{W}_{yz}C_1+\tilde{W}_{yy}C_2 & \tilde{W}_{yz}C_1+\tilde{W}_{yy}C_2 & 0 & I \end{array}\right] \tag{6.64}$$

$$\stackrel{s}{=} \left[\begin{array}{cc|cc} A+B_2F_u+B_1F_w & 0 & -H_z\tilde{W}_{zz}^{-1}+H_y\tilde{W}_{yy}^{-1}\tilde{W}_{yz}\tilde{W}_{zz}^{-1} & -H_y\tilde{W}_{yy}^{-1} \\ 0 & A+H_zC_1+H_yC_2 & 0 & 0 \\ \hline 0 & \tilde{W}_{zz}C_1 & I & 0 \\ 0 & \tilde{W}_{yz}C_1+\tilde{W}_{yy}C_2 & 0 & I \end{array}\right].$$

This concludes that (6.50) is right invertible in RH_∞ such that $P(s) = \tilde{M}_p^{-1}(s)\tilde{N}_p(s)$ is a coprime factorization. ■

Equivalently, the Bezout identity as given in Lemma 6.2 can also be checked from

$$\begin{bmatrix} \tilde{X}(s) & \tilde{Y}(s) \\ \tilde{N}_p(s) & \tilde{M}_p(s) \end{bmatrix} \begin{bmatrix} M_p(s) & -Y(s) \\ -N_p(s) & X(s) \end{bmatrix} = \begin{bmatrix} I & 0 \\ 0 & I \end{bmatrix}, \tag{6.65}$$

where

$$\begin{bmatrix} \tilde{X}(s) & \tilde{Y}(s) \\ \tilde{N}_p(s) & \tilde{M}_p(s) \end{bmatrix} \stackrel{s}{=} \left[\begin{array}{c|cc:cc} A+H_zC_1+H_yC_2 & B_1+H_zD_{11}+H_yD_{21} & B_2+H_zD_{12} & H_z & H_y \\ \hline -W_{ww}^{-1}F_w+W_{ww}^{-1}W_{wu}W_{uu}^{-1}F_u & W_{ww}^{-1} & -W_{ww}^{-1}W_{wu}W_{uu}^{-1} & 0 & 0 \\ -W_{uu}^{-1}F_u & 0 & W_{uu}^{-1} & 0 & 0 \\ \hdashline \tilde{W}_{zz}C_1 & \tilde{W}_{zz}D_{11} & \tilde{W}_{zz}D_{12} & \tilde{W}_{zz} & 0 \\ \tilde{W}_{yz}C_1+\tilde{W}_{yy}C_2 & \tilde{W}_{yz}D_{11}+\tilde{W}_{yy}D_{21} & \tilde{W}_{yz}D_{12} & \tilde{W}_{yz} & \tilde{W}_{yy} \end{array}\right],$$

and

$$\begin{bmatrix} M_p(s) & -Y(s) \\ -N_p(s) & X(s) \end{bmatrix} \stackrel{s}{=} \left[\begin{array}{c|cc:cc} A+B_1F_w+B_2F_u & B_1W_{ww} & B_1W_{wu}+B_2W_{uu} & \begin{pmatrix} H_z\tilde{W}_{zz}^{-1} \\ -H_y\tilde{W}_{yy}^{-1}\tilde{W}_{yz}\tilde{W}_{zz}^{-1} \end{pmatrix} & H_y\tilde{W}_{yy}^{-1} \\ \hline F_w & W_{ww} & W_{wu} & 0 & 0 \\ F_u & 0 & W_{uu} & 0 & 0 \\ \hdashline -\begin{pmatrix} C_1+D_{11}F_w \\ +D_{12}F_u \end{pmatrix} & -D_{11}W_{ww} & -D_{11}W_{wu}-D_{12}W_{uu} & \tilde{W}_{zz}^{-1} & 0 \\ -(C_2+D_{21}F_w) & -D_{21}W_{ww} & -D_{21}W_{wu} & -\tilde{W}_{yy}^{-1}\tilde{W}_{yz}\tilde{W}_{zz}^{-1} & \tilde{W}_{yy}^{-1} \end{array}\right].$$

Consequently, the Bezout identity is also held in the two-port transfer matrix.

Further on, the coprime factorizations for the configurations of $\mathrm{CSD_r} \rightleftarrows \mathrm{CSD_l}$ and $\mathrm{CSD_l} \rightleftarrows \mathrm{CSD_r}$ are discussed in the following. The two-port SCC plant can be transformed into the description $\mathrm{CSD_r} \rightleftarrows \mathrm{CSD_l}$, as illustrated in Fig. 6.7. Note that multiplication by M_* does not change the overall transfer function from w to z.

Then, a right coprime factorization of $\begin{bmatrix} P_{12} & P_{11} \\ 0 & I \\ \hdashline I & 0 \\ P_{22} & P_{21} \end{bmatrix} = \begin{bmatrix} G_1 \\ \tilde{G}_2 \end{bmatrix} M_*^{-1}$ can be found by applying Lemma 6.3 such that

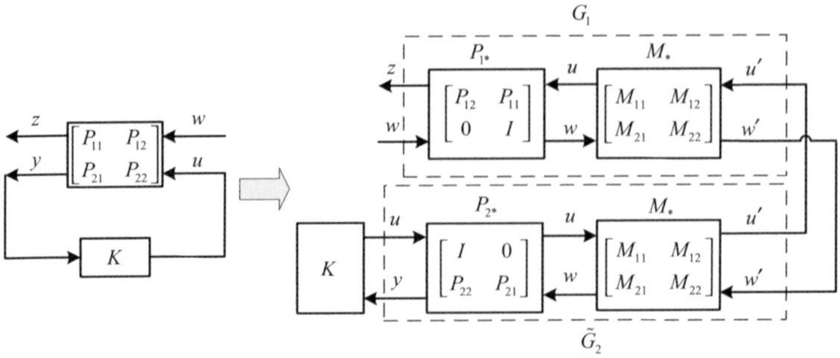

Fig. 6.7 Multiplying by M^* at right terminals

$$\left[\begin{array}{c|cc} A & B_2 & B_1 \\ \hline C_1 & D_{12} & D_{11} \\ 0 & 0 & I \\ \hline 0 & I & 0 \\ C_2 & 0 & D_{21} \\ \hline 0 & I & 0 \\ 0 & 0 & I \end{array}\right] \times \left[\begin{array}{c|cc} I & 0 & 0 \\ \hline F_u & W_{uu} & 0 \\ F_w & W_{wu} & W_{ww} \end{array}\right] \Rightarrow$$

$$\begin{bmatrix} G_1(s) \\ \tilde{G}_2(s) \\ \hline M_*(s) \end{bmatrix} \overset{s}{=} \left[\begin{array}{c|cc} A+B_1F_w+B_2F_u & B_1W_{wu}+B_2W_{uu} & B_1W_{ww} \\ \hline C_1+D_{11}F_w+D_{12}F_u & D_{11}W_{wu}+D_{12}W_{uu} & D_{11}W_{ww} \\ F_w & W_{wu} & W_{ww} \\ F_u & W_{uu} & 0 \\ C_2+D_{21}F_w & D_{21}W_{wu} & D_{21}W_{ww} \\ \hline F_u & W_{uu} & 0 \\ F_w & W_{wu} & W_{ww} \end{array}\right] \in RH_\infty. \tag{6.66}$$

Then, one has

$$G_1(s) = \begin{bmatrix} G_{11}(s) & G_{12}(s) \\ G_{12}(s) & G_{22}(s) \end{bmatrix} \overset{s}{=} \left[\begin{array}{c|cc} A+B_1F_w+B_2F_u & B_1W_{wu}+B_2W_{uu} & B_1W_{ww} \\ \hline C_1+D_{11}F_w+D_{12}F_u & D_{11}W_{wu}+D_{12}W_{uu} & D_{11}W_{ww} \\ F_w & W_{wu} & W_{ww} \end{array}\right] \in RH_\infty, \tag{6.67}$$

$$\tilde{G}_2(s) = \begin{bmatrix} \tilde{G}_{11}(s) & \tilde{G}_{12}(s) \\ \tilde{G}_{21}(s) & \tilde{G}_{22}(s) \end{bmatrix} \overset{s}{=} \left[\begin{array}{c|cc} A+B_1F_w+B_2F_u & B_1W_{wu}+B_2W_{uu} & B_1W_{ww} \\ \hline F_u & W_{uu} & 0 \\ C_2+D_{21}F_w & D_{21}W_{wu} & D_{21}W_{ww} \end{array}\right] \in RH_\infty, \tag{6.68}$$

and

$$M_*(s)=\begin{bmatrix} M_{11}(s) & M_{12}(s) \\ M_{21}(s) & M_{22}(s)\end{bmatrix} \stackrel{s}{=} \left[\begin{array}{c|cc} A+B_1F_w+B_2F_u & B_1W_{wu}+B_2W_{uu} & B_1W_{ww} \\ \hline F_u & W_{uu} & 0 \\ F_w & W_{wu} & W_{ww} \end{array}\right] \in RH_\infty, \tag{6.69}$$

where W_{uu} and W_{ww} are nonsingular and $F = \begin{bmatrix} F_u \\ F_w \end{bmatrix}$ is chosen such that $A + B_1F_w + B_2F_u$ is Hurwitz. The stable left inverse $\left[\tilde{Y}_* \mid \tilde{X}_*\right]$ that satisfies $\left[\tilde{Y}_* \mid \tilde{X}_*\right]\begin{bmatrix} -G_1 \\ -\tilde{G}_2 \\ \hline M_* \end{bmatrix} = I$ is given by

$$\begin{aligned} \left[\tilde{Y}_*(s) \mid \tilde{X}_*(s)\right] &= \left[\begin{array}{cccc|cc} \tilde{Y}_{21}(s) & 0 & 0 & \tilde{Y}_{22}(s) & \tilde{X}_{22}(s) & \tilde{X}_{21}(s) \\ \tilde{Y}_{11}(s) & 0 & 0 & \tilde{Y}_{12}(s) & \tilde{X}_{12}(s) & \tilde{X}_{11}(s) \end{array}\right] \\ &\stackrel{s}{=} \left[\begin{array}{c|cccc|cc} A+H_zC_1+H_yC_2 & H_z & 0 & 0 & H_y & B_2+H_zD_{12} & B_1+H_zD_{11}+H_yD_{21} \\ \hline -W_{uu}^{-1}F_u & 0 & 0 & 0 & 0 & W_{uu}^{-1} & 0 \\ -W_{ww}^{-1}F_w+W_{ww}^{-1}W_{wu}W_{uu}^{-1}F_u & 0 & 0 & 0 & 0 & -W_{ww}^{-1}W_{wu}W_{uu}^{-1} & W_{ww}^{-1} \end{array}\right]. \end{aligned} \tag{6.70}$$

It should be emphasized that from (6.49) and (6.66), one can assume $\begin{bmatrix} G_1(s) \\ \hline \tilde{G}_2(s) \\ \hline M_*(s) \end{bmatrix} = \begin{bmatrix} N_{p12}(s) & N_{p11}(s) \\ M_{p12}(s) & M_{p11}(s) \\ \hline M_{p22}(s) & M_{p21}(s) \\ N_{p22}(s) & N_{p21}(s) \\ \hline M_{p22}(s) & M_{p21}(s) \\ M_{p12}(s) & M_{p11}(s) \end{bmatrix}$. This concludes that $\begin{bmatrix} G_1 \\ \tilde{G}_2 \\ \hline M_* \end{bmatrix}$ is left invertible in RH_∞ such that $\begin{bmatrix} G_1 \\ \tilde{G}_2 \end{bmatrix} M_*^{-1}$ is a coprime factorization. Additionally, by Property 5.3, one has

$$\begin{aligned} \text{LFT}_\text{l}(P, K) &= \text{CSD}_\text{r}(P_{1*}, \text{CSD}_\text{l}(P_{2*}, K)) \\ &= \text{CSD}_\text{r}(P_{1*}M_*, \text{CSD}_\text{l}(P_{2*}M_*, K)) \\ &= \text{CSD}_\text{r}\left(G_1, \text{CSD}_\text{l}\left(\tilde{G}_2, K\right)\right). \end{aligned} \tag{6.71}$$

On the other hand, the two-port SCC plant can be transformed into the description of $\text{CSD}_\text{l} \rightleftarrows \text{CSD}_\text{r}$, as illustrated in Fig. 6.8. As mentioned before, multiplication by $\tilde{M}_*$ does not change the overall transfer function from w to z.

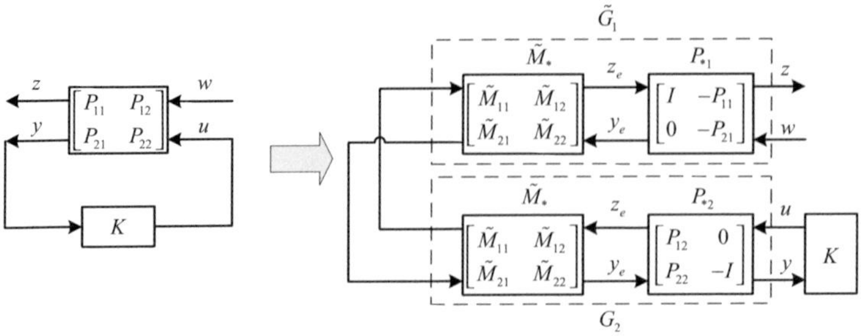

Fig. 6.8 Multiplying by $\tilde{M}_*$ at left terminal

Then, a left coprime factorization $\left[\begin{array}{cc|cc} I & -P_{11} & P_{12} & 0 \\ 0 & -P_{21} & P_{22} & -I \end{array}\right] = \tilde{M}_*^{-1}\left[\tilde{G}_1 \mid G_2\right]$ can be found by applying Lemma 6.3 such that

$$
\left[\begin{array}{c|cc} I & H_z & H_y \\ \hline 0 & \tilde{W}_{zz} & \tilde{W}_{yz} \\ 0 & 0 & \tilde{W}_{yy} \end{array}\right] \times \left[\begin{array}{c|ccccc|cc} A & 0 & -B_1 & B_2 & 0 & 0 & 0 \\ \hline C_1 & I & -D_{11} & D_{12} & 0 & I & 0 \\ C_2 & 0 & -D_{21} & 0 & -I & 0 & I \end{array}\right] \Rightarrow
$$

$$
\left[\tilde{G}_1(s) \quad G_2(s) \mid \tilde{M}_*(s)\right] \overset{s}{=} \tag{6.72}
$$

$$
\left[\begin{array}{c|ccccc|cc} A+H_zC_1+H_yC_2 & H_z & -\left(B_1+H_zD_{11}+H_yD_{21}\right) & B_2+H_zD_{12} & -H_y & H_z & H_y \\ \hline \tilde{W}_{zz}C_1 & \tilde{W}_{zz} & -\tilde{W}_{zz}D_{11} & \tilde{W}_{zz}D_{12} & 0 & \tilde{W}_{zz} & 0 \\ \tilde{W}_{yz}C_1+\tilde{W}_{yy}C_2 & \tilde{W}_{yz} & -\left(\tilde{W}_{yz}D_{11}+\tilde{W}_{yy}D_{21}\right) & \tilde{W}_{yz}D_{12} & -\tilde{W}_{yy} & \tilde{W}_{yz} & \tilde{W}_{yy} \end{array}\right].
$$

Then, one has

$$
\tilde{G}_1(s) = \begin{bmatrix} \tilde{G}_{11}(s) & \tilde{G}_{12}(s) \\ \tilde{G}_{12}(s) & \tilde{G}_{22}(s) \end{bmatrix} \overset{s}{=} \left[\begin{array}{c|cc} A+H_zC_1+H_yC_2 & H_z & -\left(B_1+H_zD_{11}+H_yD_{21}\right) \\ \hline \tilde{W}_{zz}C_1 & \tilde{W}_{zz} & -\tilde{W}_{zz}D_{11} \\ \tilde{W}_{yz}C_1+\tilde{W}_{yy}C_2 & \tilde{W}_{yz} & -\left(\tilde{W}_{yz}D_{11}+\tilde{W}_{yy}D_{21}\right) \end{array}\right] \in RH_\infty, \tag{6.73}
$$

$$
G_2(s) = \begin{bmatrix} G_{11}(s) & G_{12}(s) \\ G_{21}(s) & G_{22}(s) \end{bmatrix} \overset{s}{=} \left[\begin{array}{c|cc} A+H_zC_1+H_yC_2 & B_2+H_zD_{12} & -H_y \\ \hline \tilde{W}_{zz}C_1 & \tilde{W}_{zz}D_{12} & 0 \\ \tilde{W}_{yz}C_1+\tilde{W}_{yy}C_2 & \tilde{W}_{yz}D_{12} & -\tilde{W}_{yy} \end{array}\right] \in RH_\infty, \text{ and} \tag{6.74}
$$

$$
\tilde{M}_*(s) = \begin{bmatrix} \tilde{M}_{11}(s) & \tilde{M}_{12}(s) \\ \tilde{M}_{21}(s) & \tilde{M}_{22}(s) \end{bmatrix} \overset{s}{=} \left[\begin{array}{c|cc} A+H_zC_1+H_yC_2 & H_z & H_y \\ \hline \tilde{W}_{zz}C_1 & \tilde{W}_{zz} & 0 \\ \tilde{W}_{yz}C_1+\tilde{W}_{yy}C_2 & \tilde{W}_{yz} & \tilde{W}_{yy} \end{array}\right] \in RH_\infty, \tag{6.75}
$$

where $\tilde{W}_{zz}$ and $\tilde{W}_{yy}$ are nonsingular and $H = \begin{bmatrix} H_z & H_y \end{bmatrix}$ is chosen such that $A + H_z C_1 + H_y C_2$ is Hurwitz. The stable left inverse $\begin{bmatrix} -Y_* \\ \hline X_* \end{bmatrix}$ that satisfies $\begin{bmatrix} -\tilde{G}_1 & G_2 & | & \tilde{M}_* \end{bmatrix} \begin{bmatrix} -Y_* \\ \hline X_* \end{bmatrix} = I$ is given by

$$
\begin{bmatrix} Y_*(s) \\ \hline X_*(s) \end{bmatrix} = \begin{bmatrix} 0 & 0 \\ Y_{11}(s) & Y_{12}(s) \\ Y_{21}(s) & Y_{22}(s) \\ 0 & 0 \\ \hline X_{11}(s) & X_{12}(s) \\ X_{21}(s) & X_{22}(s) \end{bmatrix}
$$

$$
\stackrel{s}{=} \left[\begin{array}{c|cc} A + B_2 F_u + B_1 F_w & -H_z \tilde{W}_{zz}^{-1} + H_y \tilde{W}_{yy}^{-1} \tilde{W}_{yz} \tilde{W}_{zz}^{-1} & -H_y \tilde{W}_{yy}^{-1} \\ \hline 0 & 0 & 0 \\ F_w & 0 & 0 \\ F_u & 0 & 0 \\ 0 & 0 & 0 \\ \hline C_1 + D_{12} F_u + D_{11} F_w & \tilde{W}_{zz}^{-1} & 0 \\ C_2 + D_{21} F_w & -\tilde{W}_{yy}^{-1} \tilde{W}_{yz} \tilde{W}_{zz}^{-1} & \tilde{W}_{yy}^{-1} \end{array}\right]. \tag{6.76}
$$

Dually, from (6.50) and (6.72), one has

$$
\begin{bmatrix} \tilde{G}_1(s) & G_2(s) & | & \tilde{M}_*(s) \end{bmatrix} = \left[\begin{array}{cccc|cc} \tilde{M}_{p11}(s) & -\tilde{N}_{p11}(s) & \tilde{N}_{p12}(s) & -\tilde{M}_{p12}(s) & \tilde{M}_{p11}(s) & \tilde{M}_{p12}(s) \\ \tilde{M}_{p21}(s) & -\tilde{N}_{p21}(s) & \tilde{N}_{p22}(s) & -\tilde{M}_{p22}(s) & \tilde{M}_{p21}(s) & \tilde{M}_{p22}(s) \end{array}\right].
$$

This concludes that $\begin{bmatrix} \tilde{G}_1 & G_2 & | & \tilde{M}_* \end{bmatrix}$ is right invertible in RH_∞ such that $\tilde{M}_*^{-1} \begin{bmatrix} \tilde{G}_1 & | & G_2 \end{bmatrix}$ is a coprime factorization. Moreover, by Property 5.4, one has

$$
\begin{aligned}
\mathrm{LFT_l}(P, K) &= \mathrm{CSD_l}(P_{*1}, \mathrm{CSD_r}(P_{*2}, K)) \\
&= \mathrm{CSD_l}\left(\tilde{M}_* P_{*1}, \mathrm{CSD_r}\left(\tilde{M}_* P_{*2}, K\right)\right) \\
&= \mathrm{CSD_l}\left(\tilde{G}_1, \mathrm{CSD_r}(G_2, K)\right).
\end{aligned} \tag{6.77}
$$

6.3 Normalized Coprime Factorization

As mentioned in the previous section, the coprime factorization of a given plant is not unique. There are a number of advantages in representing a possibly unstable transfer function in terms of two stable factors in this manner and will be exploited more in later chapters. There is however a special coprime factorization which is the *normalized coprime factorization* [4], and the normalized left and right coprime factors $\tilde{N}$, $\tilde{M}$, N, and M of a transfer function $P \in RH_\infty$ are unique [6]. Its properties are useful in the H_∞ loop-shaping control.

A normalized right coprime factorization is defined as that the right coprime factors satisfy

$$N^{\sim}(s)N(s) + M^{\sim}(s)M(s) = I, \tag{6.78}$$

i.e., $\begin{bmatrix} M(s) \\ N(s) \end{bmatrix}$ is inner. Similarly, its left counterpart is defined as

$$\tilde{N}(s)\tilde{N}^{\sim}(s) + \tilde{M}(s)\tilde{M}^{\sim}(s) = I, \tag{6.79}$$

i.e., $\begin{bmatrix} \tilde{M}(s) & \tilde{N}(s) \end{bmatrix}$ is co-inner.

Example 6.1 As shown in Fig. 6.9, given a controlled plant $G(s) \overset{s}{=} \left[\begin{array}{c|c} A & B \\ \hline C & 0 \end{array}\right]$, determine a normalized right coprime factorization of $G(s) = N(s)M^{-1}(s)$ in the state-space form.

Taking the LFT form as shown in Fig. 6.10, one has an augmented plant (i.e., the transfer function from r' to u, y)

$$P(s) = \begin{bmatrix} M(s) \\ N(s) \end{bmatrix} \overset{s}{=} \left[\begin{array}{c|c} A+BF & BW \\ \hline F & W \\ C & 0 \end{array}\right] \Rightarrow P^{\sim}(s) \overset{s}{=} \left[\begin{array}{c|cc} -(A+BF)^T & F^T & C^T \\ \hline -W^T B^T & W^T & 0 \end{array}\right]. \tag{6.80}$$

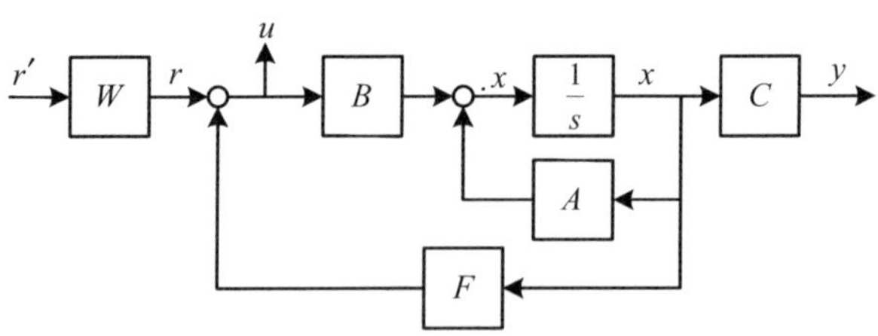

Fig. 6.9 State-space model of a control plant

Fig. 6.10 LFT formulation

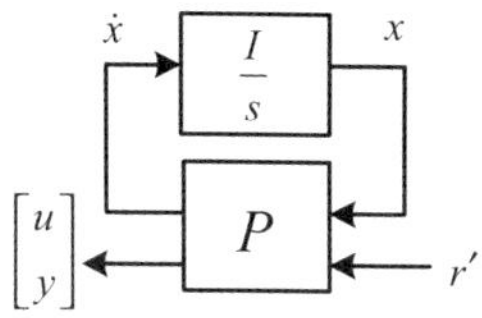

The state-feedback gain F stabilizes the controlled plant such that $A+BF$ is Hurwitz. By definition, the normalized right coprime factorization requires the augmented plant to be inner. Hence, it yields

$$P^{\sim}(s)P(s)=\begin{bmatrix} M^{\sim}(s) & N^{\sim}(s)\end{bmatrix}\begin{bmatrix} M(s) \\ N(s)\end{bmatrix}\stackrel{s}{=}\left[\begin{array}{cc|c} -(A+BF)^T & C^TC+F^TF & F^T \\ 0 & A+BF & BW \\ \hline -W^TB^T & F & W^TW \end{array}\right]. \tag{6.81}$$

By the similarity transformation $T=\begin{bmatrix} I & X \\ 0 & I\end{bmatrix}$, it yields

$$\begin{aligned} &\left[\begin{array}{cc|c} -(A+BF)^{\mathrm{T}} & C^{\mathrm{T}}C+F^{\mathrm{T}}F & F^{\mathrm{T}} \\ 0 & A+BF & BW \\ \hline -W^{\mathrm{T}}B^{\mathrm{T}} & F & W^{\mathrm{T}}W \end{array}\right] \quad (X \text{ row operation}, \ -X \text{ column operation}) \\ &\stackrel{s}{=}\left[\begin{array}{cc|c} -(A+BF)^{\mathrm{T}} & C^{\mathrm{T}}C+F^{\mathrm{T}}F+(A+BF)X+(A+BF)^{\mathrm{T}}X & F^{\mathrm{T}}+BWX \\ 0 & A+BF & BW \\ \hline -W^{\mathrm{T}}B^{\mathrm{T}} & F+W^{\mathrm{T}}B^{\mathrm{T}}X & W^{\mathrm{T}}W \end{array}\right] \end{aligned} \tag{6.82}$$

Therefore,

$$\begin{cases} W=I \\ C^{\mathrm{T}}C+F^{\mathrm{T}}F+(A+BF)X+(A+BF)^{\mathrm{T}}X=0 \\ F=-B^{\mathrm{T}}X. \end{cases} \tag{6.83}$$

Or equivalently, the state-feedback gain of F can be determined by the solution of the following equation:

$$A^{\mathrm{T}}X+XA-XBB^{\mathrm{T}}X+C^{\mathrm{T}}C=0. \tag{6.84}$$

In Example 6.1, one can find that the normalized right coprime factorization requires a state-feedback gain F such that $A+BF$ is Hurwitz and $\begin{bmatrix} M(s) \\ N(s)\end{bmatrix}$ is inner.

The algebraic Riccati equation of (6.84) needs to be solved to determine the solution of F. In the next chapter, properties of algebraic Riccati equations, the solutions, and applications will be further discussed. The state-space properties of normalized coprime factorization will then be introduced therein.

Exercises

1. Compute a coprime factorization of the following system with an inner denominator $T(s) \overset{s}{=} \left[\begin{array}{cc|c} 1 & 0 & 1 \\ 1 & 2 & 0 \\ \hline 0 & 1 & 0 \end{array}\right]$.

2. Let $G(s) = \frac{(s-1)(s+2)}{(s+3)(s-4)}$. Find a stable coprime factorization $G(s) = N(s)M^{-1}(s)$ and $X(s), Y(s) \in RH_\infty$ such that $X(s)N(s) + Y(s)M(s) = 1$.

References

1. Bézout É (1764) Cours de mathématiques: à l'usage des Gardes du Pavillon et de la Marine. avec un traité de navigation, Paris
2. Chen G, de Figueiredo RJP (1990) Construction of the left coprime fractional representation for a class of nonlinear control systems. Syst Control Lett 14:353–361
3. Maciejowski JM (1989) Multivariable feedback design. Addison-Wesley, Berkshire
4. McFarlane D, Glover K (1990) Robust controller design using normalized coprime factor plant descriptions. Springer, London
5. Tsai MC, Tsai CS (1993) A chain scattering matrix description approach to H_∞ control. IEEE Trans Autom Control 38:1416–1421
6. Vidyasagar M (1985) Control systems synthesis: a factorization approach. MIT Press, Cambridge, MA
7. Zhou K, Doyle JC, Glover K (1996) Robust and optimal control. Prentice Hall, Upper Saddle River
8. Zhou K, Doyle JC (1998) Essentials of robust control. Prentice Hall, Upper Saddle River

Chapter 7
Algebraic Riccati Equations and Spectral Factorizations

In the last chapter, it was discussed that the algebraic Riccati equation (ARE) need be solved in order to obtain the state-space solutions of the normalized coprime factorizations. In Chap. 2, the Lyapunov equation was employed to determine the controllability and observability gramians of a system. Both the algebraic Riccati and Lyapunov equations play prominent roles in the synthesis of robust and optimal control as well as in the stability analysis of control systems. In fact, the Lyapunov equation is a special case of the ARE. The ARE indeed has wide applications in control system analysis and synthesis. For example, the state-space formulation for particular coprime factorizations with a J-lossless (or dual J-lossless) *numerator* requires solving an ARE; in turn, the J-lossless and dual J-lossless systems are essential in the synthesis of robust controllers using the CSD approach. In this chapter, the ARE will be formally introduced. Solution procedures to AREs and their various properties will be discussed. Towards the end of this chapter, the coprime factorization approach to solve several spectral factorization problems is to be considered.

7.1 Algebraic Riccati Equations

The algebraic Riccati equation is useful for solving control synthesis problems such as the H_2/H_∞ (sub)optimal control problems [6, 10]. Let A, R, and Q be $n \times n$ real matrices with R and Q being symmetric. The following matrix equation is called an algebraic Riccati equation (ARE):

$$A^{\mathrm{T}}X + XA + XRX + Q = 0. \tag{7.1}$$

The matrix H defined in (7.2) is called a Hamiltonian matrix,

$$H := \begin{bmatrix} A & R \\ -Q & -A^{\mathrm{T}} \end{bmatrix} \in R^{2n\times 2n}. \tag{7.2}$$

M.-C. Tsai and D.-W. Gu, *Robust and Optimal Control: A Two-port Framework Approach*, Advances in Industrial Control, DOI 10.1007/978-1-4471-6257-5_7,
© Springer-Verlag London 2014

A Hamiltonian matrix H in this context is a $2n \times 2n$ real matrix and satisfies

$$J^{-1}HJ = -H^{\mathrm{T}}, \tag{7.3}$$

where $J = \begin{bmatrix} 0 & -I_n \\ I_n & 0 \end{bmatrix}$. Note that

$$J^2 = -I_{2n} \quad \text{and} \quad J^{\mathrm{T}} = -J = J^{-1}. \tag{7.4}$$

It is evident that a solution X to the ARE in (7.1) can be determined by the corresponding Hamiltonian matrix H in (7.2). In the context of control system analysis and synthesis, ARE solutions are further required to make $A + RX$ Hurwitz. Define an operator $\mathrm{Ric} : \mathbb{R}^{2n \times 2n} \to \mathbb{R}^{n \times n}$. Ric maps a Hamiltonian matrix H to such an ARE solution X, $X = \mathrm{Ric}(H)$. The domain of Ric is symbolized as dom(Ric). It can be directly verified that if λ is an eigenvalue of H, then $-\lambda$ is also an eigenvalue of H [8]. Also, the set of eigenvalues (the spectrum) of H is exactly the union of the eigenvalues of $A + RX$ and those of $-(A + RX)$. Hence, a necessary condition for $H \in \mathrm{dom(Ric)}$ is that H does not have any eigenvalues on the imaginary axis.

In summary, if $H \in \mathrm{dom(Ric)}$ and $X = \mathrm{Ric}(H)$, then

1. X is real symmetric.
2. X satisfies $XA + A^{\mathrm{T}}X + XRX + Q = 0$.
3. $A + RX$ is Hurwitz.
4. H does not have pure imaginary eigenvalues.

Given an ARE as (7.1), the Schur decomposition [7] offers a solution procedure to determine X. By the real Schur decomposition of H, there exists an orthonormal matrix $U_{2n \times 2n}$ (i.e., $U^{\mathrm{T}}U = I$), such that

$$H = UVU^{\mathrm{T}} = \begin{bmatrix} U_1 & U_2 \end{bmatrix} \begin{bmatrix} V_{11} & V_{12} \\ 0 & V_{22} \end{bmatrix} \begin{bmatrix} U_1 & U_2 \end{bmatrix}^{\mathrm{T}}, \tag{7.5}$$

where V_{11} contains all stable eigenvalues of H, and V_{22} contains all anti-stable eigenvalues. Multiplying U_1 from the right on both sides of (7.5) yields

$$HU_1 = \begin{bmatrix} U_1 & U_2 \end{bmatrix} \begin{bmatrix} V_{11} & V_{12} \\ 0 & V_{22} \end{bmatrix} \begin{bmatrix} I \\ 0 \end{bmatrix} = U_1 V_{11}. \tag{7.6}$$

Let

$$U_1 = \begin{bmatrix} X_1 \\ X_2 \end{bmatrix} = \begin{bmatrix} I \\ X_2 X_1^{-1} \end{bmatrix} X_1 = \begin{bmatrix} I \\ X \end{bmatrix} X_1, \tag{7.7}$$

where X_1 is assumed nonsingular and $X = X_2 X_1^{-1}$. Then, from (7.6) and (7.7),

$$HU_1 = \begin{bmatrix} A & R \\ -Q & -A^{\mathrm{T}} \end{bmatrix} \begin{bmatrix} I \\ X \end{bmatrix} X_1 = \begin{bmatrix} I \\ X \end{bmatrix} X_1 V_{11}, \tag{7.8}$$

and therefore,

$$\begin{bmatrix} A & R \\ -Q & -A^{\mathrm{T}} \end{bmatrix} \begin{bmatrix} I \\ X \end{bmatrix} = \begin{bmatrix} I \\ X \end{bmatrix} X_1 V_{11} X_1^{-1}. \tag{7.9}$$

This gives that, by multiplying $\begin{bmatrix} -X & I \end{bmatrix}$ from the left,

$$\begin{bmatrix} -X & I \end{bmatrix} \begin{bmatrix} A & R \\ -Q & -A^{\mathrm{T}} \end{bmatrix} \begin{bmatrix} I \\ X \end{bmatrix} = \begin{bmatrix} -X & I \end{bmatrix} \begin{bmatrix} I \\ X \end{bmatrix} X_1 V_{11} X_1^{-1} = 0. \tag{7.10}$$

The left-hand side of (7.10) is

$$\begin{bmatrix} -X & I \end{bmatrix} \begin{bmatrix} A + RX \\ -Q - A^{\mathrm{T}} X \end{bmatrix} = -\left(A^{\mathrm{T}} X + XA + XRX + Q\right), \tag{7.11}$$

which shows that such defined X solves the ARE (7.1). Furthermore, from (7.9),

$$\begin{bmatrix} A + RX \\ -Q - A^{\mathrm{T}} X \end{bmatrix} = \begin{bmatrix} X_1 V_{11} X_1^{-1} \\ X_2 V_{11} X_1^{-1} \end{bmatrix}. \tag{7.12}$$

Note that $A + RX = X_1 V_{11} X_1^{-1}$, which indicates that $A + RX$ shares the same eigenvalues of V_{11} and, hence, is Hurwitz. It can be proved that $X_1^T X_2$ is a Hermitian matrix [9, 11] (i.e., $M = M^{\mathrm{T}}$ in the real form or $M = M^*$ in the complex form) such that

$$X = X_2 X_1^{-1} = \left(X_1^{-1}\right)^{\mathrm{T}} X_1^{\mathrm{T}} X_2 X_1^{-1} \tag{7.13}$$

is Hermitian. By the following similarity transformation involving X, H is transformed into

$$\begin{bmatrix} I & 0 \\ -X & I \end{bmatrix} \begin{bmatrix} A & R \\ -Q & -A^{\mathrm{T}} \end{bmatrix} \begin{bmatrix} I & 0 \\ X & I \end{bmatrix} = \begin{bmatrix} A + RX & R \\ 0 & -(A + RX)^{\mathrm{T}} \end{bmatrix}. \tag{7.14}$$

This shows that these two matrices,

$$\begin{bmatrix} A & R \\ -Q & -A^{\mathrm{T}} \end{bmatrix} \quad \text{and} \quad \begin{bmatrix} A + RX & R \\ 0 & -(A + RX)^{\mathrm{T}} \end{bmatrix}, \tag{7.15}$$

have the same eigenvalues. Note that in most control synthesis problems, matrix $A + RX$ leads to the state matrix of the closed-loop system, which explains why $A + RX$ must be Hurwitz.

In general, if there exists a Hermitian matrix $\overline{X}$ and a square matrix W such that the following expression

$$\begin{bmatrix} A & R \\ -Q & -A^{\mathrm{T}} \end{bmatrix} \begin{bmatrix} I \\ \overline{X} \end{bmatrix} = \begin{bmatrix} I \\ \overline{X} \end{bmatrix} W \tag{7.16}$$

holds for a Hamiltonian matrix H as in (7.2) then $\overline{X} = \mathrm{Ric}(H)$ if the spectrum of W coincides with the set of stable eigenvalues of H.

In order to better understand the solution of an ARE, its scalar case (quadratic equation) is discussed in detail [5] in the following. Consider

$$rx^2 + 2ax + q = 0. \tag{7.17}$$

The corresponding Hamiltonian matrix is $H = \begin{bmatrix} a & r \\ -q & -a \end{bmatrix}$. Define the discriminant of (7.17) as

$$\Delta = 4\left(a^2 - qr\right). \tag{7.18}$$

If $\Delta > 0$, then Eq. (7.17) has two distinct real roots,

$$x_{1,2} = \frac{-a \pm \sqrt{a^2 - qr}}{r}, \tag{7.19}$$

which means that its graph, i.e., $y = rx^2 + 2ax + q$, will cross the x-axis twice. If $\Delta = 0$, (7.17) has two coincidental real roots,

$$x_1 = x_2 = -\frac{a}{r}, \tag{7.20}$$

and its graph is tangent to the x-axis at $x = -\frac{a}{r}$. If $\Delta < 0$, there is no intersection point on the x-axis, producing a graph which lies strictly either above or below the x-axis, i.e., no real roots. In addition, one can find that the determinant of the Hamiltonian matrix is related to the discriminant as

$$\Delta = 4\left(a^2 - qr\right) = -4\det(H). \tag{7.21}$$

Recall that if $r > 0$, the parabolic curve opens upward, which has the minimum value of $\frac{\det(H)}{r}$ at the vertex $\left(-\frac{a}{r}, \frac{\det(H)}{r}\right)$. The variable r controls the speed of increase/decrease of the quadratic function values from/to the vertex. A bigger positive r makes the function increase/decrease faster, and thus, the opening of the graph appears narrower. The variables a and r together indicate the x-coordinate of the vertex at which the axis of symmetry of the parabola lies. The variable $2a$ alone is the slope of the parabola as it crosses the y-axis. The variable q controls the "height"

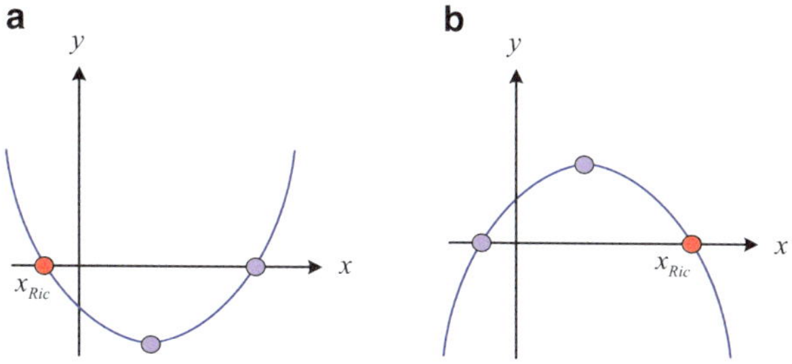

Fig. 7.1 Plot of quadratic curve and ARE solution (**a**) for $r > 0$ (**b**) for $r < 0$

of the parabola. More precisely, it is the point where the parabola crosses the y-axis. Similar discussions can be conducted for the case $r < 0$, while the function graph would be a straight line when $r = 0$.

To compare the quadratic equation with the Hamiltonian matrix, one can find from (7.18), for a solution x, that $\det(H) = -(a + rx)^2 \le 0$ in the scale case. If $\Delta > 0$, the eigenvalues of H are given by

$$\lambda = \pm (a + rx) = \pm\sqrt{-\det(H)}, \tag{7.22}$$

and the distance d between the two intersection points on the x-axis is equal to

$$d = \frac{2\sqrt{-\det(H)}}{r}. \tag{7.23}$$

When $\Delta > 0$, this indicates that H does not have any eigenvalues on the imaginary axis, i.e., $H \in \mathrm{dom(Ric)}$. Hence, there exists a required solution to the ARE. From (7.19), there are two solutions of the quadratic equation (7.16). For the purpose of a Hurwitz (stable) $A + RX$ (a negative number in the scalar case), the solution $x_2 = \frac{-a-\sqrt{a^2-qr}}{r}$ in (7.19) should be chosen, because it gives $a + rx = -\left(\sqrt{a^2 - qr}\right) < 0$. For the case $r > 0$, this required solution x_{Ric} is the negative (left) root in Fig. 7.1 and for $r < 0$, the positive (right) root. In fact, observation of Fig. 7.1 reveals that the ARE solution x_{Ric} is located at the branch of the parabolic curve which has a negative slope. This can be simply proven in the following. From (7.12), $a + rx = x_1 V_{11} x_1^{-1} = V_{11}$, for scalar x_1; therefore, $x = \frac{-a+V_{11}}{r}$. Now, the derivative of $y = rx^2 + 2ax + q$ is given by

$$y' = 2rx + 2a. \tag{7.24}$$

Then, the slope at the point $x = \frac{-a+V_{11}}{r}$ is $y' = 2V_{11}$, because the request $V_{11} < 0$, y' has to be negative. This concludes that the solution x_{Ric} to the ARE is always located at the branch of the parabolic curve which has a negative slope.

Although the above discussion is limited to the scalar case, there is some similarity in the matrix case. When the coefficient matrices R and Q are sign

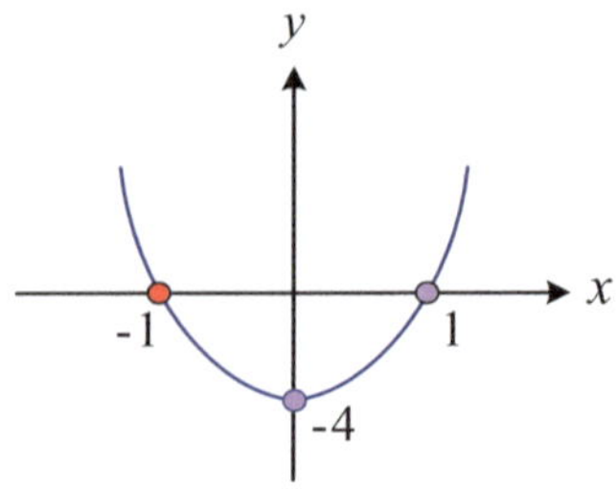

Fig. 7.2 Parabolic curve of $y=4x^2-4$

definite and both of the same sign, the required ARE solution X, if it exists, will be nonpositive definite, i.e., $X \leq 0$. On the other hand, if R and Q have opposite signs, the X is nonnegative definite, i.e., $X \geq 0$.

Example 7.1 Find the required solution of the quadratic equation (ARE) of $4x^2-4=0$.

It can be calculated that $\det(H)=-16<0$, where its corresponding Hamiltonian matrix is $H=\begin{bmatrix} a & r \\ -q & -a \end{bmatrix}=\begin{bmatrix} 0 & 4 \\ 4 & 0 \end{bmatrix}$. From the discussion above, one can conclude that $H \in \mathrm{dom(Ric)}$ with $\mathrm{eig}(H)=\pm 4$. Figure 7.2 shows its corresponding upward parabolic curve $y=4x^2-4$ $(r=4>0)$, and clearly, two solutions satisfying $4x^2-4=0$ can be found at $x=\pm 1$. Since at the point $x=-1$ has a negative slope as depicted in Fig.7.2, this means that $x=-1<0$ is the ARE solution. The following shows how to find the ARE solution step by step.

By the real Schur decomposition such as (7.5), one has

$$H=\begin{bmatrix} 0 & 4 \\ 4 & 0 \end{bmatrix}=\begin{bmatrix} -\frac{1}{\sqrt{2}} & \frac{1}{\sqrt{2}} \\ \frac{1}{\sqrt{2}} & \frac{1}{\sqrt{2}} \end{bmatrix}\begin{bmatrix} -4 & 0 \\ 0 & 4 \end{bmatrix}\begin{bmatrix} -\frac{1}{\sqrt{2}} & \frac{1}{\sqrt{2}} \\ \frac{1}{\sqrt{2}} & \frac{1}{\sqrt{2}} \end{bmatrix}^{\mathrm{T}}=\begin{bmatrix} U_1 & U_2 \end{bmatrix}\begin{bmatrix} V_{11} & V_{12} \\ 0 & V_{22} \end{bmatrix}\begin{bmatrix} U_1 & U_2 \end{bmatrix}^{\mathrm{T}},$$

where

$$U_1=\begin{bmatrix} x_1 \\ x_2 \end{bmatrix}=\begin{bmatrix} I \\ x_2 x_1^{-1} \end{bmatrix}x_1=\begin{bmatrix} I \\ x \end{bmatrix}x_1 \Rightarrow \begin{bmatrix} -\frac{1}{\sqrt{2}} \\ \frac{1}{\sqrt{2}} \end{bmatrix}=\begin{bmatrix} 1 \\ -1 \end{bmatrix}\frac{-1}{\sqrt{2}}.$$

Therefore, this obtains the solution $x=-1$. Furthermore, one gathers that

$$HU_1=\begin{bmatrix} A & R \\ -Q & -A^{\mathrm{T}} \end{bmatrix}\begin{bmatrix} I \\ x \end{bmatrix}x_1=\begin{bmatrix} I \\ x \end{bmatrix}x_1 V_{11} \Rightarrow \begin{bmatrix} 0 & 4 \\ 4 & 0 \end{bmatrix}\begin{bmatrix} -\frac{1}{\sqrt{2}} \\ \frac{1}{\sqrt{2}} \end{bmatrix}=\begin{bmatrix} 1 \\ -1 \end{bmatrix}\frac{-1}{\sqrt{2}}(-4),$$

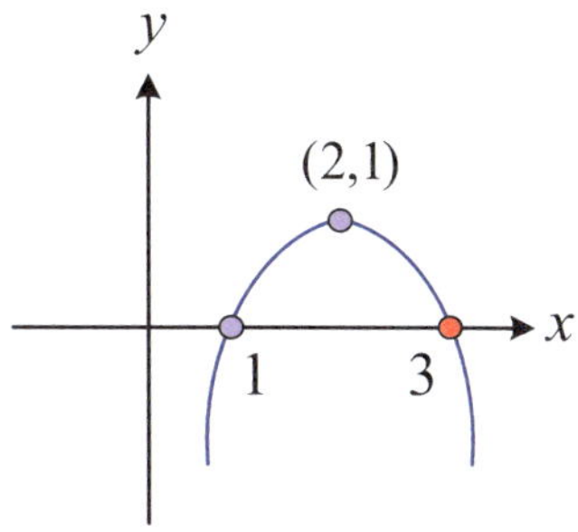

Fig. 7.3 Parabolic curve of $y = -x^2 + 4x - 3$

and then

$$\begin{bmatrix} 1 & 1 \end{bmatrix}\begin{bmatrix} -4 & 0 \\ 0 & 4 \end{bmatrix}\begin{bmatrix} 1 \\ -1 \end{bmatrix} = \begin{bmatrix} 1 & 1 \end{bmatrix}\begin{bmatrix} 1 \\ -1 \end{bmatrix}\left(-\frac{1}{\sqrt{2}}\right)(-4)\left(-\frac{1}{\sqrt{2}}\right)^{-1} = 0.$$

One concludes that

$$\begin{bmatrix} a + rx \\ -q - ax \end{bmatrix} = \begin{bmatrix} x_1 V_{11} x_1^{-1} \\ x_2 V_{11} x_1^{-1} \end{bmatrix} \Rightarrow \begin{bmatrix} -4 \\ 4 \end{bmatrix} = \begin{bmatrix} \left(-\frac{1}{\sqrt{2}}\right)(-4)\left(-\frac{1}{\sqrt{2}}\right)^{-1} \\ \frac{1}{\sqrt{2}}(-4)\left(-\frac{1}{\sqrt{2}}\right)^{-1} \end{bmatrix}.$$

This shows that $x = -1 < 0$ is the ARE solution, as depicted in Fig. 7.2, such that $V_{11} = -4$ and $a + rx = -1$. Note that $x = 1$ is a solution of $4x^2 - 4 = 0$, but it is not the ARE solution since $a + rx = 4$. ■

Example 7.2 Find the solution of the quadratic equation ARE of $-x^2 + 4x - 3 = 0$.

It can be calculated that $\det(H) = -1 < 0$ and $\mathrm{eig}(H) = \pm 1$, where the corresponding Hamiltonian matrix is $H = \begin{bmatrix} a & r \\ -q & -a \end{bmatrix} = \begin{bmatrix} 2 & -1 \\ 3 & -2 \end{bmatrix}$. One can deduce that $H \in \mathrm{dom(Ric)}$. Figure 7.3 shows its corresponding downward parabolic curve $y = -x^2 + 4x - 3$ $(r = -1 < 0)$, and evidently, two solutions satisfying $-x^2 + 4x - 3 = 0$ can be found at $x = 1$ and $x = 3$. Since the point $x = 3$ has a negative slope as depicted in Fig. 7.3, this means that $x = 3$ is the ARE solution. In fact, one can find the ARE solution step by step to obtain the ARE solution $x = 3$ such that $V_{11} = -1$ and $a + rx = -1$. Note that $x = 1 > 0$ is a solution of $-x^2 + 4x - 3 = 0$, but it is not the ARE solution since $a + rx = 2 + (-1) \times 1 = 1$.

The concept of the discriminant is useful in the scalar case to find the ARE solution of a quadratic equation. However, this property, unfortunately, cannot be directly extended to the general matrix ARE cases.

7.2 Similarity Transformation of Hamiltonian Matrices

Recall that the similarity transformation of a square matrix M is $M_T = TMT^{-1}$, where T is any nonsingular matrix. The similarity transformation preserves the eigenvalues of the matrix, i.e., {eigenvalues of M} = {eigenvalues of M_T}, the two matrices sharing the same spectrum. It would be interesting and useful to see if there is any invariance property of a Hamiltonian matrix under similarity transformations. Let a Hamiltonian matrix $H := \begin{bmatrix} A & R \\ -Q & -A^{\mathrm{T}} \end{bmatrix}$ and $X = \mathrm{Ric}(H)$. Given a nonsingular $2n \times 2n$ matrix T, would the matrix after the similarity transformation of T, $H_T = THT^{-1}$, still be Hamiltonian? If so, what kind of relations exist between the solutions $\mathrm{Ric}(H)$ and $\mathrm{Ric}(H_T)$, in terms of T? These questions will be answered in the following discussions.

By its definition, a Hamiltonian matrix H satisfies $J^{-1}HJ = -H^{\mathrm{T}}$. Hence, a sufficient and necessary condition for H to be Hamiltonian is

$$HJ + JH^{\mathrm{T}} = 0. \tag{7.25}$$

Let $T = \begin{bmatrix} T_{11} & T_{12} \\ T_{21} & T_{22} \end{bmatrix}$ be the $2n \times 2n$ nonsingular matrix, and its inverse

$$T^{-1} = \begin{bmatrix} \tilde{T}_{11} & \tilde{T}_{12} \\ \tilde{T}_{21} & \tilde{T}_{22} \end{bmatrix}. \tag{7.26}$$

For $H_T = THT^{-1}$ to be Hamiltonian, one has to show that $H_T J + JH_T^T = 0$. That is

$$THT^{-1}J + JT^{-\mathrm{T}}H^{\mathrm{T}}T^{\mathrm{T}} = 0$$

or

$$HT^{-1}JT^{-\mathrm{T}} + T^{-1}JT^{-\mathrm{T}}H^{\mathrm{T}} = 0. \tag{7.27}$$

One can easily see that H_T is Hamiltonian if and only if the matrix $HT^{-1}JT^{-\mathrm{T}}$ is skew symmetric. For the material presented in this book, however, just a sufficient condition is needed, which appears in the following lemma.

Lemma 7.1 *A Hamiltonian matrix H remains Hamiltonian under a similarity transformation, if the transformation matrix T satisfies the following conditions*:

(1) $\tilde{T}_{11}\tilde{T}_{12}^{\mathrm{T}}$ *and* $\tilde{T}_{22}\tilde{T}_{21}^{\mathrm{T}}$ *are both symmetric, where* $\tilde{T}_{ij}$ $(i, j = 1, 2)$ *are defined in* (7.26).
(2) $\tilde{T}_{22}\tilde{T}_{11}^{\mathrm{T}} - \tilde{T}_{21}\tilde{T}_{12}^{\mathrm{T}} = \alpha I$, *where* α *is a scalar constant.*

Proof When $T^{-1}JT^{-\mathrm{T}} = \alpha J$, (7.27) holds because H is Hamiltonian. Consequently, H_T is Hamiltonian from (7.27). Conditions (1) and (2) can be directly derived by block matrix manipulations in $T^{-1}JT^{-\mathrm{T}} = \alpha J$. ■

With Lemma 7.1, the following four cases are to be discussed, where H is assumed in dom(Ric) and $X = \mathrm{Ric}(H)$.

Case (I) $T = \begin{bmatrix} I & 0 \\ L & I \end{bmatrix}$ and $L = L^{\mathrm{T}}$.

In this case, $T^{-1} = \begin{bmatrix} I & 0 \\ -L & I \end{bmatrix}$. The two conditions in Lemma 7.1 hold by direct verification. Hence, $H_T = THT^{-1}$ is a Hamiltonian matrix, and $H_T T = TH$. Let $X = \mathrm{Ric}(H)$. By (7.14),

$$\begin{aligned} H_T T \begin{bmatrix} I & 0 \\ X & I \end{bmatrix} &= TH \begin{bmatrix} I & 0 \\ X & I \end{bmatrix} \\ &= T \begin{bmatrix} I & 0 \\ X & I \end{bmatrix} \begin{bmatrix} I & 0 \\ -X & I \end{bmatrix} H \begin{bmatrix} I & 0 \\ X & I \end{bmatrix} \\ &= T \begin{bmatrix} I & 0 \\ X & I \end{bmatrix} \begin{bmatrix} A+RX & R \\ 0 & -(A+RX)^{\mathrm{T}} \end{bmatrix}. \end{aligned}$$

The left-hand side is

$$H_T T \begin{bmatrix} I & 0 \\ X & I \end{bmatrix} = H_T \begin{bmatrix} I & 0 \\ L & I \end{bmatrix} \begin{bmatrix} I & 0 \\ X & I \end{bmatrix} = H_T \begin{bmatrix} I & 0 \\ L+X & I \end{bmatrix},$$

and the right-hand side is

$$\begin{aligned} T \begin{bmatrix} I & 0 \\ X & I \end{bmatrix} \begin{bmatrix} A+RX & R \\ 0 & -(A+RX)^{\mathrm{T}} \end{bmatrix} &= \begin{bmatrix} I & 0 \\ L+X & I \end{bmatrix} \begin{bmatrix} A+RX & R \\ 0 & -(A+RX)^{\mathrm{T}} \end{bmatrix} \\ &= \begin{bmatrix} A+RX & R \\ (L+X)(A+RX) & (L+X)R-(A+RX)^{\mathrm{T}} \end{bmatrix}. \end{aligned}$$

Quoting the first block column yields

$$H_T \begin{bmatrix} I \\ L+X \end{bmatrix} = \begin{bmatrix} A+RX \\ (L+X)(A+RX) \end{bmatrix} = \begin{bmatrix} I \\ L+X \end{bmatrix} (A+RX). \tag{7.28}$$

From (7.16), $\mathrm{Ric}(H_T) = L + X$.

Case (II) $T = \begin{bmatrix} I & 0 \\ 0 & \gamma^{-1} I \end{bmatrix}$.

In this case, $T^{-1} = \begin{bmatrix} I & 0 \\ 0 & \gamma I \end{bmatrix}$. Again, the conditions in Lemma 7.1 hold, and H_T is Hamiltonian. Similarly,

$$H_T T \begin{bmatrix} I & 0 \\ X & I \end{bmatrix} = T \begin{bmatrix} I & 0 \\ X & I \end{bmatrix} \begin{bmatrix} A + RX & R \\ 0 & -(A + RX)^{\mathrm{T}} \end{bmatrix},$$

$$H_T \begin{bmatrix} I & 0 \\ \gamma^{-1} X & \gamma^{-1} I \end{bmatrix} = \begin{bmatrix} I & 0 \\ \gamma^{-1} X & \gamma^{-1} I \end{bmatrix} \begin{bmatrix} A + RX & R \\ 0 & -(A + RX)^{\mathrm{T}} \end{bmatrix},$$

i.e.,

$$H_T \begin{bmatrix} I \\ \gamma^{-1} X \end{bmatrix} = \begin{bmatrix} I & 0 \\ \gamma^{-1} X & \gamma^{-1} I \end{bmatrix} \begin{bmatrix} A + RX \\ 0 \end{bmatrix} = \begin{bmatrix} I \\ \gamma^{-1} X \end{bmatrix} (A + RX).$$

Hence, $\mathrm{Ric}(H_T) = \gamma^{-1} X$.

Case (III) $T = \begin{bmatrix} U^{\mathrm{T}} & 0 \\ 0 & U^{\mathrm{T}} \end{bmatrix}$ and $U^{\mathrm{T}} = U^{-1}$, i.e., U is orthonormal.

Here, $T^{-1} = \begin{bmatrix} U & 0 \\ 0 & U \end{bmatrix}$. Straightforward manipulations show that the conditions in Lemma 7.1 are satisfactory, and thus, $H_T = THT^{-1}$ is a Hamiltonian matrix. One obtains accordingly that

$$H_T \begin{bmatrix} U^{\mathrm{T}} \\ U^{\mathrm{T}} X \end{bmatrix} = \begin{bmatrix} U^{\mathrm{T}} \\ U^{\mathrm{T}} X \end{bmatrix} (A + RX). \tag{7.29}$$

By further multiplying U from the right on both sides, it yields

$$H_T \begin{bmatrix} I \\ U^{\mathrm{T}} X U \end{bmatrix} = \begin{bmatrix} I \\ U^{\mathrm{T}} X U \end{bmatrix} U^{-1} (A + RX) U. \tag{7.30}$$

This concludes that $\mathrm{Ric}(H_T) = U^{\mathrm{T}} X U$ for $T = \begin{bmatrix} U^{\mathrm{T}} & 0 \\ 0 & U^{\mathrm{T}} \end{bmatrix}$.

Case (IV) $T = \begin{bmatrix} I & \tilde{L} \\ 0 & I \end{bmatrix}$ and $\tilde{L} = \tilde{L}^{\mathrm{T}}$.

It can be verified that $H_T = THT^{-1}$ is a Hamiltonian matrix, by using Lemma 7.1. One can obtain that

$$H_T \begin{bmatrix} I \\ X\left(I + \tilde{L}X\right)^{-1} \end{bmatrix} = \begin{bmatrix} I \\ X\left(I + \tilde{L}X\right)^{-1} \end{bmatrix} \left(I + \tilde{L}X\right)(A + RX)\left(I + \tilde{L}X\right)^{-1}. \tag{7.31}$$

Table 7.1 Four cases of similarity transformations

	$H_T = THT^{-1}$	Ric (H_T)
Case (I)	$T = \begin{bmatrix} I & 0 \\ L & I \end{bmatrix}$ and $L = L^{\mathrm{T}}$	$L + X$
Case (II)	$T = \begin{bmatrix} I & 0 \\ 0 & r^{-1}I \end{bmatrix}$	$r^{-1}X$
Case (III)	$T = \begin{bmatrix} U^{-1} & 0 \\ 0 & U^{\mathrm{T}} \end{bmatrix}$ and $U^{\mathrm{T}} = U^{-1}$	$U^{\mathrm{T}}XU$
Case (IV)	$T = \begin{bmatrix} I & \tilde{L} \\ 0 & I \end{bmatrix}$ and $\tilde{L} = \tilde{L}^{\mathrm{T}}$	$X(I + \tilde{L}X)^{-1}$

This concludes that $\mathrm{Ric}\,(H_T) = \mathrm{Ric}\,(THT^{-1}) = X(I + \tilde{L}X)^{-1}$. Table 7.1 summarizes these four cases.

It should be noted that in the above discussion, one neither requires the sign definitive nor nonsingular qualities of the ARE solution X for the sake of generality. In some applications, an ARE of the following form is considered:

$$A^{\mathrm{T}}X + XA - XRX + Q = 0, \tag{7.32}$$

where assumptions of $R \geq 0$ and $Q \geq 0$ are made. The solution $X = \mathrm{Ric}(H_X)$ is required to be nonnegative definite, and $A - RX$ should be made Hurwitz, where

$$H_X = \begin{bmatrix} A & -R \\ -Q & -A^{\mathrm{T}} \end{bmatrix}. \tag{7.33}$$

Such an ARE has a dual form equation:

$$AY + YA^{\mathrm{T}} - YQY + R = 0, \tag{7.34}$$

where $Y = \mathrm{Ric}(H_Y) \geq 0$ and

$$H_Y = \begin{bmatrix} A^{\mathrm{T}} & -Q \\ -R & -A \end{bmatrix}. \tag{7.35}$$

It is clear that $H_Y = H_X^T$. From

$$\begin{bmatrix} I & 0 \\ -X & I \end{bmatrix} H_X \begin{bmatrix} I & 0 \\ X & I \end{bmatrix} = \begin{bmatrix} A - RX & -R \\ 0 & -(A - RX)^{\mathrm{T}} \end{bmatrix}$$

$$\Rightarrow H_X = \begin{bmatrix} I & 0 \\ X & I \end{bmatrix} \begin{bmatrix} A - RX & -R \\ 0 & -(A - RX)^{\mathrm{T}} \end{bmatrix} \begin{bmatrix} I & 0 \\ -X & I \end{bmatrix}$$

and

$$\begin{bmatrix} I & 0 \\ -Y & I \end{bmatrix} H_Y \begin{bmatrix} I & 0 \\ Y & I \end{bmatrix} = \begin{bmatrix} (A-YQ)^{\mathrm{T}} & -Q \\ 0 & -(A-YQ) \end{bmatrix}$$

$$\Rightarrow H_Y = \begin{bmatrix} I & 0 \\ Y & I \end{bmatrix} \begin{bmatrix} (A-YQ)^{\mathrm{T}} & -Q \\ 0 & -(A-YQ) \end{bmatrix} \begin{bmatrix} I & 0 \\ -Y & I \end{bmatrix},$$

one has, from $H_X = H_Y^T$,

$$\begin{bmatrix} I & 0 \\ X & I \end{bmatrix} \begin{bmatrix} A-RX & -R \\ 0 & -(A-RX)^{\mathrm{T}} \end{bmatrix} \begin{bmatrix} I & 0 \\ -X & I \end{bmatrix}$$
$$= \begin{bmatrix} I & -Y \\ 0 & I \end{bmatrix} \begin{bmatrix} A-YQ & 0 \\ -Q & -(A-YQ)^{\mathrm{T}} \end{bmatrix} \begin{bmatrix} I & Y \\ 0 & I \end{bmatrix}.$$

Hence,

$$\begin{bmatrix} A-YQ & 0 \\ -Q & -(A-YQ)^{\mathrm{T}} \end{bmatrix} \begin{bmatrix} I & Y \\ 0 & I \end{bmatrix} \begin{bmatrix} I & 0 \\ X & I \end{bmatrix}$$
$$= \begin{bmatrix} I & Y \\ 0 & I \end{bmatrix} \begin{bmatrix} I & 0 \\ X & I \end{bmatrix} \begin{bmatrix} A-RX & -R \\ 0 & -(A-RX)^{\mathrm{T}} \end{bmatrix}.$$

That is,

$$\begin{bmatrix} A-YQ & 0 \\ -Q & -(A-YQ)^{\mathrm{T}} \end{bmatrix} \begin{bmatrix} I+YX & Y \\ X & I \end{bmatrix} = \begin{bmatrix} I+YX & Y \\ X & I \end{bmatrix} \begin{bmatrix} A-RX & -R \\ 0 & -(A-RX)^{\mathrm{T}} \end{bmatrix}.$$

Considering the first block column of the product matrices on both sides yields

$$\begin{bmatrix} A-YQ & 0 \\ -Q & -(A-YQ)^{\mathrm{T}} \end{bmatrix} \begin{bmatrix} I+YX \\ X \end{bmatrix} = \begin{bmatrix} I+YX & Y \\ X & I \end{bmatrix} \begin{bmatrix} A-RX \\ 0 \end{bmatrix}.$$

Comparing the first row on either side leads to

$$(A-YQ)(I+YX) = (I+YX)(A-RX).$$

Therefore,

$$(A-YQ) = (I+YX)(A-RX)(I+YX)^{-1}. \tag{7.36}$$

Equation (7.36) shows the similarity transformation relationship between $A-RX$ and $A-YQ$, which is useful for the solutions to H_∞ control.

7.3 Lyapunov Equation

Assuming that A is Hurwitz and P and Q are symmetric,

$$PA + A^{\mathrm{T}} P = -Q \tag{7.37}$$

is called a Lyapunov equation. When A is Hurwitz and Q is semi-positive, i.e., $Q = Q^{\mathrm{T}} \geq 0$, one can have that the solution P is positive definite, $P > 0$. On the other hand, if $Q = Q^{\mathrm{T}} \geq 0$ and $P > 0$, then matrix A is Hurwitz. These two results will be shown in a lemma below. In theory, a solution to the Lyapunov equation (7.37) can be found as

$$P = \int_0^{\infty} \mathrm{e}^{A^{\mathrm{T}} t} Q \mathrm{e}^{At} \mathrm{d}t. \tag{7.38}$$

It can be shown that such a P indeed solves (7.37) by

$$\begin{aligned} PA + A^{\mathrm{T}} P &= \int_0^{\infty} \mathrm{e}^{A^{\mathrm{T}} t} Q \mathrm{e}^{At} A \mathrm{d}t + \int_0^{\infty} A^{\mathrm{T}} \mathrm{e}^{A^{\mathrm{T}} t} Q \mathrm{e}^{At} \mathrm{d}t \\ &= \int_0^{\infty} \frac{\mathrm{d}}{\mathrm{d}t} \left(\mathrm{e}^{A^{\mathrm{T}} t} Q \mathrm{e}^{At} \right) \mathrm{d}t \\ &= \left[\mathrm{e}^{A^{\mathrm{T}} t} Q \mathrm{e}^{At} \right] |_0^{\infty} \\ &= -Q. \end{aligned} \tag{7.39}$$

The last equation comes from $\mathrm{e}^{At} \to 0$, as $t \to \infty$, due to A being Hurwitz. Note that a Lyapunov equation is a special case of the ARE (i.e., $R = 0$), $A^{\mathrm{T}} X + XA + Q = 0$, and

$$X = \mathrm{Ric} \begin{bmatrix} A & 0 \\ -Q & -A^{\mathrm{T}} \end{bmatrix}.$$

The observability and controllability gramians were previously mentioned in Chap. 2. In this section, they are defined in detail as follows. Given a control system, let the observability gramian of (C,A) be defined by

$$P_{\mathrm{o}} := \int_0^{\infty} \mathrm{e}^{A^{\mathrm{T}} t} C^{\mathrm{T}} C \mathrm{e}^{At} \mathrm{d}t, \tag{7.40}$$

which can be shown similarly as in the deduction of (7.39) to satisfy

$$A^{\mathrm{T}} P_{\mathrm{o}} + P_{\mathrm{o}} A = -C^{\mathrm{T}} C. \tag{7.41}$$

Observability gramian P_o determines the total energy in the system output, which is driven by a given initial state in the case of identical zero input.

In control system analysis and synthesis, the constant term on the right-hand side of a Lyapunov equation is usually not negative definite; rather it is nonpositive. For such cases, one has the following result.

Lemma 7.2 *Let (C, A) be observable. Then* $A^T P_o + P_o A = -C^T C$ *has a positive definite solution* P_o *if and only if A is Hurwitz.*

Proof (Sufficiency) Construct matrix P_o as in (7.40). If A is Hurwitz, it can be shown as in the deduction of (7.39) that such a P_o is indeed a solution to the Lyapunov equation. P_o is obviously nonnegative. Suppose that P_o is rank deficient. Let $\mathbb{N}$ be the null space of P_o and N_p be its matrix representation, i.e., $N_p = [\zeta_1, \ldots, \zeta_l]$, $1 \le l < n$, where n is the order of the system (dimension of A and thus of P_o), and $P_o N_p = O_{n*l}$. Multiplying ${N_p}^T$ and N_p from the left and right, respectively, on both sides of (7.41) concludes that $CN_p = O$. Then, multiplying N_p from the right on (7.41) leads to $P_o A N_p = O$. Hence, AN_p falls into $\mathbb{N}$, and there exists matrix L of $l \times l$ dimension such that

$$AN_p = N_p L. \tag{7.42}$$

From $CN_p = O$ and $CA^j N_p = CN_p L^j = O$, $j = 1, \ldots, n-1$, one concludes that $\begin{bmatrix} C \\ CA \\ \vdots \\ CA^{n-1} \end{bmatrix}$ is not of full rank, and therefore, (C,A) is not completely observable which contradicts the assumption. Hence, P_o is of full rank, i.e., $P_o > 0$.

(Necessity) If A is not Hurwitz, let $\lambda = \alpha + j\beta$ be an unstable eigenvalue of A and the corresponding eigenvector be $\xi(\neq 0)$, i.e., $\alpha \ge 0$ and $A\xi = \lambda\xi$. Multiplying ξ^* and ξ from the left and right, respectively, on both sides of (7.41), yields

$$\begin{aligned} \xi^* A^T P_0 \xi + \xi^* P_0 A \xi &= -\xi^* C^T C \xi \\ \bar{\lambda}\xi^* P_0 \xi + \lambda \xi^* P_0 \xi &= -\xi^* C^T C \xi \\ \left(\bar{\lambda} + \lambda\right) \xi^* P_0 \xi &= -(C\xi)^* (C\xi) \\ 2\alpha \xi^* P_0 \xi &= -(C\xi)^* (C\xi). \end{aligned} \tag{7.43}$$

For a positive definite P_o, the left-hand side of (7.43) is nonnegative while its right-hand side is nonpositive. Hence, $C\xi = 0$. Considering that ξ is also such that $A\xi = \lambda\xi$, this contradicts the assumption of (C,A) being completely observable. Hence, A is Hurwitz. ■

This property can be easily seen in the scalar case when (7.41) becomes a linear equation $2ap = -c^2$. Then, one can obtain $p = \frac{-c^2}{2a}$. It shows that $p > 0$ (i.e., a positive definite solution) if and only if $a < 0$ (i.e., Hurwitz).

Similarly, define controllability gramian as

$$P_c := \int_0^\infty e^{At} BB^T e^{A^T t} dt, \tag{7.44}$$

which satisfies

$$AP_c + P_c A^T = -BB^T. \tag{7.45}$$

In a physical engineering system with an "impulse" input and zero initial states, controllability gramian P_c determines the total energy in the states generated. A result dual to Lemma 7.2 is also available under the condition of complete controllability of (A,B).

7.4 State-Space Formulae for Spectral Factorizations Using Coprime Factorization Approach

In this section, three cases of spectral factorizations in the state-space form are to be introduced. They are obtained via a unified procedure, i.e., by employing coprime factorizations. State-space formulae are derived to find certain coprime factors for each spectral factorization. The factorization procedure is characterized by using the so-called *weighted all-pass* function, which is a generalization of all-pass functions introduced earlier in Chap. 6 and will be formally defined next.

Definition 7.1 Let $\Sigma = \Sigma^T$ and $\breve{\Sigma} = \breve{\Sigma}^T$ be constant matrices with compatible dimensions. Then, $P(s) \in RL_\infty$ satisfying

$$P^\sim(s)\Sigma P(s) = \breve{\Sigma} \tag{7.46}$$

is defined as *weighted all-pass*. Dually, the *weighted co-all-pass* is defined as

$$P(s)\Sigma P^\sim(s) = \breve{\Sigma}. \tag{7.47}$$

The following lemma gives conditions for a transfer function matrix $P(s)$ to be weighted all-pass.

Lemma 7.3 *Let* $P(s) \stackrel{s}{=} \left[\begin{array}{c|c} A & B \\ \hline C & D \end{array}\right] \in RL_\infty$. *Given* $\Sigma = \Sigma^T$ *and* $\breve{\Sigma} = \breve{\Sigma}^T$, *if*

$$D^{\mathrm{T}}\Sigma D = \breve{\Sigma}, \tag{7.48}$$

and there exists a matrix $X = X^{\mathrm{T}} \geq 0$ *such that*

$$XB + C^{\mathrm{T}}\Sigma D = 0, \tag{7.49}$$

$$A^{\mathrm{T}}X + XA + C^{\mathrm{T}}\Sigma C = 0, \tag{7.50}$$

then $P^{\sim}(s)\Sigma P(s) = \breve{\Sigma}$. *Dually, if*

$$D\Sigma D^{\mathrm{T}} = \breve{\Sigma}, \tag{7.51}$$

and there exists a matrix $Y = Y^{\mathrm{T}} \geq 0$ *such that*

$$YC^{\mathrm{T}} + B\Sigma D^{\mathrm{T}} = 0, \tag{7.52}$$

$$AY + YA^{\mathrm{T}} + B\Sigma B^{\mathrm{T}} = 0, \tag{7.53}$$

then $P(s)\Sigma P^{\sim}(s) = \breve{\Sigma}$.

Proof The weighted all-pass (and co-all-pass) proof follows the proof procedure of a correspondent result on *standard* inner (and co-inner) functions in [1, 3] and, therefore, is omitted here. ■

Note that for the case $P(s) \in RH_{\infty}^{n\times k}$, $\Sigma = I_n$, and $\breve{\Sigma} = I_k$, the weighted all-pass (or weighted co-all-pass) system $P(s)$ becomes an inner (or co-inner) function. Additionally, for the case that $P(s) \in RH_{\infty}^{(n_1+n_2)\times(k_1+k_2)}$, $\Sigma = \begin{bmatrix} I_{n_1} & 0 \\ 0 & -I_{n_2} \end{bmatrix}$, and $\breve{\Sigma} = \begin{bmatrix} I_{k_1} & 0 \\ 0 & -I_{k_2} \end{bmatrix}$, the weighted all-pass $P(s)$ becomes a J-lossless function. The properties of lossless two-port networks in the viewpoint of power wave propagation were discussed in Chap. 3. Recall that in Chap. 5, it demonstrated that properties of J-lossless and dual J-lossless both play an important role in CSD control systems.

For certain engineering systems, spectral factorization is a useful tool. Spectral factorization separates the causal and minimum-phase component from the rest, and this in turn reveals energy transformation involved with this system. In some sense, a spectral factor shows the *magnitude* of the system. Next, the formal definition of *standard* spectral factorization will be given first, and subsequently, several *spectral factorizations* that are often used in control system synthesis and analysis will be introduced. State-space formulae of these spectral factorizations will be described;

they are all obtained via weighted all-pass functions, which are constructed in a unified framework of coprime factorizations.

Definition 7.2 [2] Consider a square matrix $\Omega(s)$ having the properties

$$\Omega \in RL_{\infty}, \quad \Omega^{-1} \in RL_{\infty}, \quad \Omega^{\sim} = \Omega, \quad \text{and} \quad \Omega(\infty) > 0. \tag{7.54}$$

Then,

$$\Omega = \Phi^{\sim}\Phi \tag{7.55}$$

is called a spectral factorization of Ω, where Φ is a spectral factor and

$$\Phi \in GH_{\infty}, \tag{7.56}$$

i.e., both Φ and its inverse are stable.

Note that such a matrix $\Omega(s)$ has poles and zeros in symmetry about the imaginary axis and such a spectral factor $\Phi(s)$ is also called *outer* (stable with minimum phase).

Let $P(s) \stackrel{s}{=} \left[\begin{array}{c|c} A & B \\ \hline C & D \end{array}\right]$ and $P(s) = N(s)M^{-1}(s) = \tilde{M}^{-1}(s)\tilde{N}(s)$ be the right and left coprime factorizations, respectively. By Lemma 6.1, state-space formulae of the coprime factors are given by, for purposely chosen F and H,

$$\begin{bmatrix} M(s) \\ N(s) \end{bmatrix} \stackrel{s}{=} \left[\begin{array}{c|c} A+BF & BW \\ \hline F & W \\ C+DF & DW \end{array}\right] \in RH_{\infty}, \tag{7.57}$$

$$\begin{bmatrix} \tilde{N}(s) & \tilde{M}(s) \end{bmatrix} \stackrel{s}{=} \left[\begin{array}{c|cc} A+HC & B+HD & H \\ \hline \tilde{W}C & \tilde{W}D & \tilde{W} \end{array}\right] \in RH_{\infty}. \tag{7.58}$$

Since matrices F, W, H, and $\tilde{W}$ are *free* parameters, one can choose them to suit various requirements on the coprime factors in addition to stability. In the following, one shows how to find these matrices for the required spectral factorizations, where $P(s) \stackrel{s}{=} \left[\begin{array}{c|c} A & B \\ \hline C & D \end{array}\right] \in RH_{\infty}$ is assumed. Note that for a general $P(s) \in RL_{\infty}$, an extra coprime factorization can be applied first in order to obtain spectral factorizations.

7.4.1 Spectral Factorization Case I

The first type of spectral factorization is to find an outer matrix $\Phi(s)$ such that

$$R + P^{\sim}(s)QP(s) = \Phi^{\sim}(s)\Phi(s), \tag{7.59}$$

where $Q = Q^{\mathrm{T}}$ and $R = R^{\mathrm{T}}$ are given, constant matrices. Let $P(s) = N(s)M^{-1}(s)$ be a right coprime factorization. The left-hand side of (7.59) can be rewritten as

$$\begin{aligned} R + P^{\sim}(s)QP(s) &= R + \left(M^{-1}(s)\right)^{\sim} N^{\sim}(s)QN(s)M^{-1}(s) \\ &= \left(M^{-1}(s)\right)^{\sim} \left(M^{\sim}(s)RM(s) + N^{\sim}(s)QN(s)\right) M^{-1}(s) \\ &= \left(M^{-1}(s)\right)^{\sim} \begin{bmatrix} M(s) \\ N(s) \end{bmatrix}^{\sim} \begin{bmatrix} R & 0 \\ 0 & Q \end{bmatrix} \begin{bmatrix} M(s) \\ N(s) \end{bmatrix} M^{-1}(s). \end{aligned}$$

By defining $\Sigma = \begin{bmatrix} R & 0 \\ 0 & Q \end{bmatrix}$, $\breve{\Sigma} = I$, one may see that if the coprime factorization $P = NM^{-1}$ is such that $P_s = \begin{bmatrix} M \\ N \end{bmatrix}$ is a weighted all-pass function with M being outer, i.e., the following equation holds,

$$\begin{bmatrix} M(s) \\ N(s) \end{bmatrix}^{\sim} \begin{bmatrix} R & 0 \\ 0 & Q \end{bmatrix} \begin{bmatrix} M(s) \\ N(s) \end{bmatrix} = I, \tag{7.60}$$

then $M^{-1}(s)$ would be the required spectral factor $\Phi(s)$ provided that $M(s)$ is also outer. Next, it shows how to choose F and W in the coprime factorization of $P(s)$ to make $\begin{bmatrix} M(s) \\ N(s) \end{bmatrix}$ weighted all-pass, i.e., to satisfy (7.60) and to ensure $M(s)$ outer.

Herein, substituting the state-space realization of (7.57) and $\Sigma = \begin{bmatrix} R & 0 \\ 0 & Q \end{bmatrix}$, $\breve{\Sigma} = I$ into (7.48), (7.49) and (7.50) will obtain that

$$W^{\mathrm{T}}\left(R + D^{\mathrm{T}}QD\right)W = I, \tag{7.61}$$

$$XB + C^{\mathrm{T}}QD + F^{\mathrm{T}}\left(R + D^{\mathrm{T}}QD\right) = 0, \tag{7.62}$$

$$(A + BF)^{\mathrm{T}}X + X(A + BF) + F^{\mathrm{T}}RF + (C + DF)^{\mathrm{T}}Q(C + DF) = 0. \tag{7.63}$$

If one has $R + P^{\sim}(s)QP(s) > 0$, $\forall\, s = j\omega$, then $R_x = (R + D^{\mathrm{T}}QD) > 0$. Hence, define

$$W = R_x^{-1/2}, \tag{7.64}$$

$$F = -R_x^{-1}\left(B^{\mathrm{T}}X + D^{\mathrm{T}}QC\right), \tag{7.65}$$

where X is the solution to the following ARE,

$$\begin{aligned}&\left(A - BR_x^{-1}D^{\mathrm{T}}QC\right)^{\mathrm{T}}X + X\left(A - BR_x^{-1}D^{\mathrm{T}}QC\right) - XBR_x^{-1}B^{\mathrm{T}}X\\&\quad + C^{\mathrm{T}}\left(Q - QDR_x^{-1}D^{\mathrm{T}}Q\right)C = 0.\end{aligned} \tag{7.66}$$

That is,

$$X = \mathrm{Ric}\begin{bmatrix} A - BR_x^{-1}D^{\mathrm{T}}QC & -BR_x^{-1}B^{\mathrm{T}} \\ -C^{\mathrm{T}}\left(Q - QDR_x^{-1}D^{\mathrm{T}}Q\right)C & -\left(A - BR_x^{-1}D^{\mathrm{T}}QC\right)^{\mathrm{T}} \end{bmatrix} \geq 0. \tag{7.67}$$

Note that X is such that $A + BF$ is Hurwitz. By (7.57), the "denominator" is given by

$$M(s) \stackrel{s}{=} \left[\begin{array}{c|c} A + BF & BW \\ \hline F & W \end{array}\right] = \left[\begin{array}{c|c} A - BR_x^{-1/2}(B^T X + D^T QC) & BR_x^{1/2} \\ \hline -R_x^{-1/2}(B^T X + D^T QC) & R_x^{1/2} \end{array}\right]. \tag{7.68}$$

Note that $M^{-1}(s) \stackrel{s}{=} \left[\begin{array}{c|c} A & B \\ \hline -W^{-1}F & W^{-1} \end{array}\right] \in RH_\infty$. Hence, $\Phi(s) = M(s) \in GH_\infty$ is required.

Dually, the spectral factorizations of Case (I) are to find an outer matrix $\Phi(s)$ such that

$$R + P(s)QP^{\sim}(s) = \Phi(s)\Phi^{\sim}(s), \tag{7.69}$$

where $Q = Q^{\mathrm{T}}$ and $R = R^{\mathrm{T}}$. Similarly, a dual procedure can be followed starting from $P(s) = \tilde{M}^{-1}(s)\tilde{N}(s)$, $\Sigma = \begin{bmatrix} R & 0 \\ 0 & Q \end{bmatrix}$, and $\breve{\Sigma} = I$. Define $R_y = (R + DQD^{\mathrm{T}}) > 0$, and

$$\tilde{W} = R_y^{-1/2}, \tag{7.70}$$

$$H = -\left(YC^{\mathrm{T}} + BQD^{\mathrm{T}}\right)R_y^{-1}, \tag{7.71}$$

where Y solves

$$\begin{aligned}&Y\left(A - BD^{\mathrm{T}}R_y^{-1}QC\right)^{\mathrm{T}} + \left(A - BD^{\mathrm{T}}R_y^{-1}QC\right)Y\\&\quad - YC^{\mathrm{T}}\left(Q - QDR_y^{-1}D^{\mathrm{T}}Q\right)CY + BR_y^{-1}B^{\mathrm{T}} = 0.\end{aligned} \tag{7.72}$$

That is,

$$Y = \mathrm{Ric}\begin{bmatrix} A - BD^{\mathrm{T}}R_y{}^{-1}QC & -C^{\mathrm{T}}\left(Q - QDR_y{}^{-1}D^{\mathrm{T}}Q\right)C \\ -BR_y{}^{-1}B^{\mathrm{T}} & -\left(A - BD^{\mathrm{T}}R_y{}^{-1}QC\right)^{\mathrm{T}} \end{bmatrix} \geq 0. \quad (7.73)$$

Note that Y is chosen such that $A + HC$ is Hurwitz. By (7.58), $\tilde{M}(s)$ is given by

$$\tilde{M}(s) \stackrel{s}{=} \left[\begin{array}{c|c} A + HC & -(YC^{T} + BQD)R_y^{-1/2} \\ \hline R_y^{1/2}C & R_y^{1/2} \end{array}\right]. \quad (7.74)$$

Note that $\tilde{M}^{-1}(s) \stackrel{s}{=} \left[\begin{array}{c|c} A & H\tilde{W}^{-1} \\ \hline -C & \tilde{W}^{-1} \end{array}\right] \in RH_{\infty}$ and $\Phi(s) = \tilde{M}^{-1}(s) \in GH_{\infty}$ are required.

The above is summarized in the lemma below.

Lemma 7.4 *Let* $Q = Q^{\mathrm{T}}$ and $R = R^{\mathrm{T}}$. *There exists a right coprime factorization of* $P(s) = N(s)M^{-1}(s)$ *given by* $\begin{bmatrix} N(s) \\ M(s) \end{bmatrix} \stackrel{s}{=} \left[\begin{array}{c|c} A + BF & BW \\ \hline C + DF & DW \\ F & W \end{array}\right]$, *where*

$$W = R_x^{-1/2}, \quad (7.75)$$

$$R_x = \left(R + D^{\mathrm{T}}QD\right), \quad (7.76)$$

$$F = -R_x{}^{-1}\left(B^{\mathrm{T}}X + D^{\mathrm{T}}QC\right), \quad (7.77)$$

$$X = \mathrm{Ric}\begin{bmatrix} A - BR_x{}^{-1}D^{\mathrm{T}}QC & -BR_x{}^{-1}B^{\mathrm{T}} \\ -C^{\mathrm{T}}\left(Q - QDR_x{}^{-1}D^{\mathrm{T}}Q\right)C & -\left(A - BR_x{}^{-1}D^{\mathrm{T}}QC\right)^{\mathrm{T}} \end{bmatrix}. \quad (7.78)$$

Additionally, an outer function satisfying that $R + P^{\sim}(s)QP(s) = \Phi^{\sim}(s)\Phi(s)$ *is given by*

$$\Phi(s) = M^{-1}(s) \stackrel{s}{=} \left[\begin{array}{c|c} A & B \\ \hline R_x^{-1/2}(B^{T}X + D^{T}QC) & R_x^{1/2} \end{array}\right] \in GH_{\infty}. \quad (7.79)$$

Moreover, there exists a left coprime factorization of $P(s) = \tilde{M}^{-1}(s)\tilde{N}(s)$ *given by* $\begin{bmatrix} \tilde{N}(s) & \tilde{M}(s) \end{bmatrix} \stackrel{s}{=} \left[\begin{array}{c|cc} A + HC & B + HD & H \\ \hline \tilde{W}C & \tilde{W}D & \tilde{W} \end{array}\right]$, *where*

$$\tilde{W} = R_y^{-1/2}, \tag{7.80}$$

$$R_y = \left(R + DQD^{\mathrm{T}}\right), \tag{7.81}$$

$$H = -\left(YC^{\mathrm{T}} + BQD^{\mathrm{T}}\right) R_y{}^{-1}, \tag{7.82}$$

$$Y = \mathrm{Ric}\begin{bmatrix} A - BD^{\mathrm{T}}R_y{}^{-1}QC & -C^{\mathrm{T}}\left(Q - QDR_y{}^{-1}D^{\mathrm{T}}Q\right)C \\ -BR_y{}^{-1}B^{\mathrm{T}} & -\left(A - BD^{\mathrm{T}}R_y{}^{-1}QC\right)^{\mathrm{T}} \end{bmatrix}. \tag{7.83}$$

Furthermore, an outer function such that $R + P(s)QP^{\sim}(s) = \Phi(s)\Phi^{\sim}(s)$ *is given by*

$$\Phi(s) = \tilde{M}^{-1}(s) \stackrel{s}{=} \left[\begin{array}{c|c} A & (YC^T + BQD)R_y^{-1/2} \\ \hline C & R_y^{1/2} \end{array}\right]. \tag{7.84}$$

■

In Case (I), different R and Q correspond to different applications. Four applications of Case (I) are noted in the following.

7.4.1.1 Normalized Coprime Factorization

Let $P(s) = N(s)M^{-1}(s)\ \left(= \tilde{M}^{-1}(s)\tilde{N}(s)\right)$ be a right (left) coprime factorization. Here it should be noted that $P(s) \stackrel{s}{=} \left[\begin{array}{c|c} A & B \\ \hline C & D \end{array}\right]$ is general, not assumed to be stable. One needs specific coprime factors which satisfy

$$M^{\sim}(s)M(s) + N^{\sim}(s)N(s) = I \tag{7.85}$$

or

$$\tilde{M}(s)\tilde{M}^{\sim}(s) + \tilde{N}(s)\tilde{N}^{\sim}(s) = I. \tag{7.86}$$

Such factorizations are called normalized coprime factorizations. This is equivalent to Case (I) by setting $R = I$ and $Q = I$, i.e., $\Sigma = \begin{bmatrix} I & 0 \\ 0 & I \end{bmatrix}$ and $\breve{\Sigma} = I$. By Lemma 7.4, then there exists a normalized right coprime factorization of $P(s) = N(s)M^{-1}(s)$ given by $\begin{bmatrix} N(s) \\ M(s) \end{bmatrix} \stackrel{s}{=} \left[\begin{array}{c|c} A + BF & BW \\ \hline C + DF & DW \\ F & W \end{array}\right]$, where

$$F = -R_x{}^{-1}\left(B^{\mathrm{T}}X + D^{\mathrm{T}}C\right), \tag{7.87}$$

$$R_x = I + D^{\mathrm{T}}D, \tag{7.88}$$

$$W^{\mathrm{T}}\left(I + D^{\mathrm{T}}D\right)\, W = I, \tag{7.89}$$

and $X \geq 0$ is the solution of the following ARE:

$$\begin{aligned}&\left(A - BR_x^{-1}D^{\mathrm{T}}C\right)^{\mathrm{T}}X + X\left(A - BR_x^{-1}D^{\mathrm{T}}C\right) - XBR_x^{-1}B^{\mathrm{T}}X\\&\quad + C^{\mathrm{T}}\left(I - DR_x^{-1}D^{\mathrm{T}}\right)C = 0.\end{aligned} \tag{7.90}$$

In this case, the inverse of $M(s)$ is not necessarily stable for an unstable $P(s)$, which is not required in this application.

In addition, a normalized left coprime factorization $P(s) = \tilde{M}^{-1}(s)\tilde{N}(s)$ can be found by $\begin{bmatrix}\tilde{N}(s) & \tilde{M}(s)\end{bmatrix} \overset{s}{=} \left[\begin{array}{c|cc} A+HC & B+HD & H \\ \hline \tilde{W}C & \tilde{W}D & \tilde{W}\end{array}\right]$, where

$$H = -\left(YC^{\mathrm{T}} + BD^{\mathrm{T}}\right)R_y{}^{-1}, \tag{7.91}$$

$$R_y = I + DD^{\mathrm{T}}, \tag{7.92}$$

$$\tilde{W}^{\mathrm{T}}\left(I + DD^{\mathrm{T}}\right)\, \tilde{W} = I, \tag{7.93}$$

and $Y \geq 0$ is the solution of the following ARE:

$$\begin{aligned}&Y\left(A - BD^{\mathrm{T}}R_y^{-1}C\right)^{\mathrm{T}} + \left(A - BD^{\mathrm{T}}R_y^{-1}C\right)Y\\&\quad - YC^{\mathrm{T}}\left(I - DR_y^{-1}D^{\mathrm{T}}\right)CY + BR_y^{-1}B^{\mathrm{T}} = 0.\end{aligned} \tag{7.94}$$

For the case $D = 0$, the AREs of (7.90) will be

$$A^{\mathrm{T}}X + XA - XBB^{\mathrm{T}}X + C^{\mathrm{T}}C = 0. \tag{7.95}$$

Dually, the ARE of (7.94) will be

$$YA^{\mathrm{T}} + AY - YC^{\mathrm{T}}CY + BB^{\mathrm{T}} = 0. \tag{7.96}$$

Example 7.3 Consider $P(s) = \frac{s-1}{s+3}$, using the spectral factorizations to find the normalized right coprime factors of $G(s)$.

Applying the method discussed in Case (I) to solve it, one has

$$P(s)=\frac{s-1}{s+3}\overset{s}{=}\left[\begin{array}{c|c} A & B \\ \hline C & D \end{array}\right]=\left[\begin{array}{c|c} -3 & 1 \\ \hline -4 & 1 \end{array}\right] \text{ with } R=1 \text{ and } Q=1.$$

Hence,

$$R_x = \left(R + D^{\mathrm{T}}QD\right) = 2, \quad \text{and} \quad W = R_x^{-\frac{1}{2}} = \frac{1}{\sqrt{2}}.$$

From

$$\left(A - BR_x^{-1}D^{\mathrm{T}}QC\right)^{\mathrm{T}}X + X\left(A - BR_x^{-1}D^{\mathrm{T}}QC\right) - XBR_x^{-1}B^{\mathrm{T}}X$$
$$+ C^{\mathrm{T}}\left(Q - QDR_x^{-1}D^{\mathrm{T}}Q\right)C = 0,$$

one has $X = 2.4721$ and $F = -R_x^{-1}(B^{\mathrm{T}}X + D^{\mathrm{T}}QC) = 0.7639$. Then,

$$A + BF = -2.2361 < 0$$

is Hurwitz. Consequently, the coprime factors of (7.57) yield

$$\begin{bmatrix} M(s) \\ N(s) \end{bmatrix} \overset{s}{=} \left[\begin{array}{c|c} -2.236 & 1/\sqrt{2} \\ \hline 0.7639 & 1/\sqrt{2} \\ -3.2361 & 1/\sqrt{2} \end{array}\right].$$

It can be seen that $M^{\sim}(s)M(s) + N^{\sim}(s)N(s) = 1$. ■

Example 7.4 Using MATLAB, determine the state-space realization of the normalized coprime factorizations of a given plant $G(s)\overset{s}{=}\left[\begin{array}{c|c} A & B \\ \hline C & D \end{array}\right]$.

The following MATLAB code is an example for readers' reference.

```
clear all;clc;
disp('Normalized Coprime Factorization')
syms A B C D;
A=input('A:');
B=input('B:');
C=input('C:');
D=input('D:');
```

```
R=1;
Q=1;
sys=ss(A,B,C,D);
%Right coprime
Rx=(R+D'*Q*D);
W=Rx^(-1/2);
Hx_A = (A-B*Rx^(-1)*D'*Q*C);
Hx_B = B';
Hx_C = (C'*(Q-Q*D*Rx^(-1)*D'*Q)*C);
[x,l,g]=care(Hx_A,Hx_B,Hx_C,Rx);
F=-Rx^(-1)*(B'*x+D'*Q*C);
disp('eigenvalues of A+B*F')
eig(A+B*F)
disp('Value of X')
x
M_inv=ss(A,B,Rx^(-1/2)*(B'*x+D'*Q*C),Rx^(1/2));
disp('State-space of M')
M=inv(M_inv)
disp('State-space of N')
N=ss(A+B*F,B*W,C+D*F,D*W)
%Left coprime
Ry=(R+D*Q*D');
W_w=Ry^(-1/2);
Hy_A = (A-B*D'*Ry^(-1)*Q*C);
Hy_B = (C'*(Q-Q*D*Ry^(-1)*D'*Q)*C);
Hy_C = B*inv(Ry)*B';
[y,ll,gg]=care(Hy_A,Hy_B,Hy_C);
H=-(y*C'+B*Q*D')*inv(Ry);
M_w_inv=ss(A,(y*C'+B*Q*D)*Ry^(-1/2),C,Ry^(1/2));
disp('State-space of Mw')
M=inv(M_inv)
disp('State-space of Nw')
N=ss(A+H*C,B+H*D,W_w*C,W_w*D)
```
■

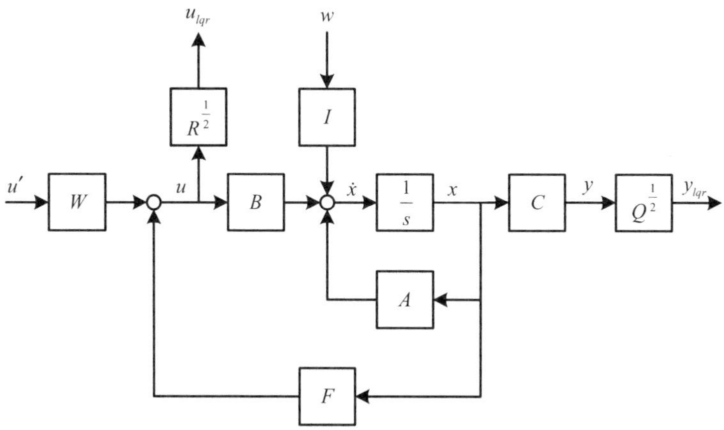

Fig. 7.4 LQR problem with D=0

7.4.1.2 Optimal Linear Quadratic Regulation

As shown in Fig. 7.4, the linear quadratic regulation (LQR) is to find a stabilizing state feedback gain F to minimize the deterministic cost function

$$J_{\text{lqr}} = \int_0^{\infty} \left(y^{\text{T}}(t)Qy(t) + u^{\text{T}}(t)Ru(t)\right) \text{d}t, \tag{7.97}$$

where $R > 0$ and $Q \geq 0$ are the weights. From Fig. 7.4, the right coprime factorization of $\begin{bmatrix} u \\ y \end{bmatrix} = \begin{bmatrix} M \\ N \end{bmatrix} u'$ yields $\begin{bmatrix} M(s) \\ N(s) \end{bmatrix} \overset{s}{=} \left[\begin{array}{c|c} A+BF & BW \\ \hline F & W \\ C & 0 \end{array}\right]$. Let

$$\begin{bmatrix} u_{\text{lqr}} \\ y_{\text{lqr}} \end{bmatrix} = \begin{bmatrix} R^{1/2}M \\ Q^{1/2}N \end{bmatrix} u'. \tag{7.98}$$

It can be seen in Chap. 8 that the optimal LQR control problem is equivalent to solving for particular coprime factors such that

$$M^{\sim}(s)RM(s) + N^{\sim}(s)QN(s) = I. \tag{7.99}$$

Thus, the optimal state feedback gain F and feedforward gain W, which minimize the cost function, can be obtained such that $\begin{bmatrix} R^{1/2}M \\ Q^{1/2}N \end{bmatrix}$ is inner. Actually, (7.99) is equivalent to forming the weighted all-pass function (7.46) by setting

$\Sigma = \begin{bmatrix} R & 0 \\ 0 & Q \end{bmatrix}$ and $\breve{\Sigma} = I$. This optimal LQR problem for any initial state is thus, in fact, an application of Lemma 7.4. Again, $P(s) \stackrel{s}{=} \left[\begin{array}{c|c} A & B \\ \hline C & D \end{array}\right]$ is general and not assumed to be stable.

7.4.1.3 Inner-Outer Factorization

For $P(s) \in RH_\infty$, the inner-outer factorization $P(s) = N(s)\Phi(s)$ is to find $\Phi(s) \in GH_\infty$ such that $P^\sim(s)P(s) = \Phi^\sim(s)\Phi(s)$. Let $P(s) = N(s)M^{-1}(s)$ be a coprime factorization of $P(s)$. Solving the inner-outer factorization problem is equivalent to setting $R = 0$ and $Q = I$ in (7.59), i.e., $\Sigma = \begin{bmatrix} 0 & 0 \\ 0 & I \end{bmatrix}$ and $\breve{\Sigma} = I$. By Lemma 7.4 with $R = 0$ and $Q = I$, one can have the state-space solution of the coprime factorization $P(s) = N(s)M^{-1}(s)$ such that

$$N^\sim(s)N(s) = I. \tag{7.100}$$

So, $N(s)$ is the inner part and $M^{-1}(s) = \Phi(s) \in GH_\infty$, where $M(s)$ is the outer part of $P(s) \in RH_\infty$. That is, $M(s)$ is an outer function.

Example 7.5 Given $P(s) = \frac{(s-2)}{(s+5)}$, compute the inner-outer factorization such that $P^\sim(s)P(s) = \Phi^\sim(s)\Phi(s)$, using spectral factorizations. Implement the solution to construct the right coprime factors of $G(s)$.

(1) Compute $\Phi(s) \in GH_\infty$ directly. It is evident that

$$\begin{aligned} P^\sim(s)P(s) &= \left(\frac{-s-2}{-s+5}\right)\left(\frac{s-2}{s+5}\right) = \left(\frac{s-2}{s-5}\right)\left(\frac{s+2}{s+5}\right) \\ &= \left(\frac{-s+2}{-s+5}\right)\left(\frac{s+2}{s+5}\right) = \Phi^\sim(s)\Phi(s), \end{aligned}$$

where $\Phi(s) = \frac{s+2}{s+5}$.

(2) Apply the method noted in Case (I) to solve it. It yields

$$P(s) = \frac{s-2}{s+5} \stackrel{s}{=} \left[\begin{array}{c|c} A & B \\ \hline C & D \end{array}\right] = \left[\begin{array}{c|c} -5 & 1 \\ \hline -7 & 1 \end{array}\right], \quad R = 0, \text{ and } Q = 1.$$

Hence,

$$R_x = \left(R + D^{\mathrm{T}}QD\right) = 1, \text{ and } W = R_x^{-\frac{1}{2}} = 1.$$

From

$$\left(A - BR_x^{-1}D^{\mathrm{T}}QC\right)^{\mathrm{T}}X + X\left(A - BR_x^{-1}D^{\mathrm{T}}QC\right) - XBR_x^{-1}B^{\mathrm{T}}X$$
$$- C^{\mathrm{T}}\left(Q - QDR_x^{-1}D^{\mathrm{T}}Q\right)C = 0,$$

one has $X = 4$ and

$$F = -R_x^{-1}\left(B^{\mathrm{T}}X + D^{\mathrm{T}}QC\right) = -1\left(1 \times (4) + 1 \times 1 \times (-7)\right) = 3.$$

Then,

$$A + BF = -5 + 1 \times (3) = -2 < 0$$

is Hurwitz. Then, by (7.57), an outer function such that $(M^{-1}(s))^{\sim}M^{-1}(s) = \Phi^{\sim}(s)\Phi(s)$ is given by

$$\Phi(s) = M^{-1}(s) \stackrel{s}{=} \left[\begin{array}{c|c} A & B \\ \hline R_x^{-1/2}(B^T X + D^T QC) & R_x^{1/2} \end{array}\right] = \left[\begin{array}{c|c} A & B \\ \hline -W^{-1}F & W^{-1} \end{array}\right] = \left[\begin{array}{c|c} -5 & 1 \\ \hline -3 & 1 \end{array}\right] \stackrel{s}{=} \frac{s+2}{s+5}.$$

It is apparent that results (1) and (2) are the same. ∎

7.4.1.4 Bounded Real Lemma

Suppose that $P(s) \stackrel{s}{=} \left[\begin{array}{c|c} A & B \\ \hline C & D \end{array}\right] \in RH_\infty$ and $\|P(s)\|_\infty < \gamma$. Consider the spectral factorization

$$\gamma^2 I - P^{\sim}(s)P(s) = \Phi^{\sim}(s)\Phi(s). \tag{7.101}$$

This problem is equivalent to setting $R = \gamma^2 I$ and $Q = -I$ in (7.59), i.e., $\Sigma = \begin{bmatrix} \gamma^2 I & 0 \\ 0 & -I \end{bmatrix}$ and $\breve{\Sigma} = I$. By Lemma 7.4 with $R = \gamma^2 I$ and $Q = -I$, one can obtain the state-space solution of the coprime factorization $P(s) = N(s)M^{-1}(s)$ such that

$$\gamma^2 M^{\sim}(s)M(s) - N^{\sim}(s)N(s) = I. \tag{7.102}$$

Then, $\Phi(s)$ is given by (7.68),

$$\Phi(s) = M^{-1}(s) \stackrel{s}{=} \left[\begin{array}{c|c} A & B \\ \hline R_x^{-\frac{1}{2}}(B^T X + D^T QC) & R_x^{\frac{1}{2}} \end{array}\right] \in GH_\infty, \tag{7.103}$$

where

$$R_x = \left(\gamma^2 I - D^{\mathrm{T}} D\right), \text{ and} \tag{7.104}$$

$$X = \mathrm{Ric}\begin{bmatrix} A + BR_x{}^{-1}D^{\mathrm{T}}C & -BR_x{}^{-1}B^{\mathrm{T}} \\ C^{\mathrm{T}}\left(I + DR_x{}^{-1}D^{\mathrm{T}}\right)C & -\left(A + BR_x{}^{-1}D^{\mathrm{T}}C\right)^{\mathrm{T}} \end{bmatrix} \geq 0. \tag{7.105}$$

The spectral factorization of (7.101) is actually the well-known and widely applied *bounded real lemma* (BRL) [4]. The above deduction shows that the BRL can be proved via a spectral factorization, which is solved by a unified approach using the weighed all-pass concept and coprime factorization.

7.4.2 Spectral Factorization Case II

The second case of spectral factorization problems is to find an outer matrix $\Phi(s)$ such that, for $R = R^{\mathrm{T}} > 0$,

$$R + P^{\sim}(s) + P(s) = \Phi^{\sim}(s)\Phi(s). \tag{7.106}$$

For $P(s) = N(s)M^{-1}(s)$, $\Sigma = \begin{bmatrix} R & I \\ I & 0 \end{bmatrix}$, and $\breve{\Sigma} = I$, one finds that if the coprime factorization is such that $P_s(s) = \begin{bmatrix} M \\ N \end{bmatrix}$ is weighted all-pass, with M being outer, with regard to such defined Σ and $\breve{\Sigma}$, i.e., the following equation holds,

$$\begin{bmatrix} M \\ N \end{bmatrix}^{\sim} \begin{bmatrix} R & I \\ I & 0 \end{bmatrix} \begin{bmatrix} M \\ N \end{bmatrix} = I. \tag{7.107}$$

Then, because

$$\begin{aligned} R + P^{\sim}(s) + P(s) &= R + \left(M^{-1}(s)\right)^{\sim} N^{\sim}(s) + N(s)M^{-1}(s) \\ &= \left(M^{\sim}(s)\right)^{-1}\left(M^{\sim}(s)RM(s) + N^{\sim}(s)M(s) + M^{\sim}(s)N(s)\right)M^{-1}(s) \\ &= \left(M^{-1}(s)\right)^{\sim} M^{-1}(s), \end{aligned}$$

$\Phi(s) = M^{-1}(s)$ is a required spectral factor provided that $M(s)$ is outer.

Hence, similar to Case (I), one can define,

$$R_x = \left(R + D^{\mathrm{T}} + D\right) > 0, \tag{7.108}$$

and

$$W = R_x^{-1/2}, \tag{7.109}$$

$$F = -R_x^{-1}\left(B^{\mathrm{T}}X + C\right), \tag{7.110}$$

where

$$X = \mathrm{Ric}\begin{bmatrix} A - BR_x^{-1}C & -BR_x^{-1}B^{\mathrm{T}} \\ C^{\mathrm{T}}R_x^{-1}C & -\left(A - BR_x^{-1}C\right)^{\mathrm{T}} \end{bmatrix} \geq 0. \tag{7.111}$$

Then, the "denominator" which makes $P_s(s)$ weighted all-pass is

$$M(s) \stackrel{s}{=} \left[\begin{array}{c|c} A - BR_x^{-1}(B^T X + C) & BR_x^{1/2} \\ \hline -R_x^{-1/2}(B^T X + C) & R_x^{1/2} \end{array}\right] \tag{7.112}$$

and

$$M^{-1}(s) \stackrel{s}{=} \left[\begin{array}{c|c} A & B \\ \hline R_x^{-1/2}(B^T X + C) & R_x^{1/2} \end{array}\right]. \tag{7.113}$$

Because both $M(s)$ and $M^{-1}(s)$ are stable, $\Phi(s) = M^{-1}(s)$ is a solution to (7.106). Note that for the case $R = 0$, $P^{\sim}(s) + P(s) = \Phi^{\sim}(s)\Phi(s)$ is the spectral factorization of a strictly positive real matrix. Readers can refer to the definitions of positive real functions in Chap. 3.

Also, from $P(s) = \tilde{M}^{-1}(s)\tilde{N}(s)$, $\Sigma = \begin{bmatrix} R & I \\ I & 0 \end{bmatrix}$, and $\breve{\Sigma} = I$, one finds

$$\begin{bmatrix} \tilde{M} & \tilde{N} \end{bmatrix}\begin{bmatrix} R & I \\ I & 0 \end{bmatrix}\begin{bmatrix} \tilde{M} & \tilde{N} \end{bmatrix}^{\sim} = I. \tag{7.114}$$

H and $\tilde{W}$ can be found to make the above viable, i.e.,

$$\tilde{M}(s)R\tilde{M}^{\sim}(s) + \tilde{N}(s)\tilde{M}^{\sim}(s) + \tilde{M}(s)\tilde{N}^{\sim}(s) = I. \tag{7.115}$$

Hence,

$$\tilde{M}^{-1}(s)\left(\tilde{M}(s)R\tilde{M}^{\sim}(s) + \tilde{N}(s)\tilde{M}^{\sim}(s) + \tilde{M}(s)\tilde{N}^{\sim}(s)\right)\left(\tilde{M}^{-1}(s)\right)^{\sim} = \tilde{M}^{-1}(s)\left(\tilde{M}^{-1}(s)\right)^{\sim}$$

$$\Rightarrow R + G(s) + G^{\sim}(s) = \tilde{M}^{-1}(s)\left(\tilde{M}^{-1}(s)\right)^{\sim} = \Phi(s)\Phi^{\sim}(s). \tag{7.116}$$

Suppose that $R + P(s) + P^{\sim}(s) > 0$ (positive real), $\forall\, s = j\omega$. With $\Sigma = \begin{bmatrix} R & I \\ I & 0 \end{bmatrix}$ and the same manipulation, one gathers

$$\tilde{W} = R_y^{-1/2}, \tag{7.117}$$

$$H = -\left(YC^{\mathrm{T}} + B\right) R_y{}^{-1}, \tag{7.118}$$

$$R_y = \left(R + D + D^{\mathrm{T}}\right) > 0, \tag{7.119}$$

and

$$Y = \mathrm{Ric}\begin{bmatrix} A - BR_y{}^{-1}C & -C^{\mathrm{T}}R_y{}^{-1}C \\ -BR_y{}^{-1}B^{\mathrm{T}} & -\left(A - BR_y{}^{-1}C\right)^{\mathrm{T}} \end{bmatrix} \geq 0. \tag{7.120}$$

Then, an outer function such that

$$R + P^{\sim}(s) + P(s) = \Phi^{\sim}(s)\Phi(s) \tag{7.121}$$

is given by

$$\Phi(s) = \tilde{M}^{-1}(s) \stackrel{s}{=} \left[\begin{array}{c|c} A & (YC^T + B)R_y^{-1/2} \\ \hline C & R_y^{1/2} \end{array}\right]. \tag{7.122}$$

The above is summarized in the next lemma for the convenience of future reference.

Lemma 7.5 *For* $R > 0$, *there exists a right coprime factorization of* $P(s) = N(s)M^{-1}(s)$ *given by* $\begin{bmatrix} N(s) \\ M(s) \end{bmatrix} \stackrel{s}{=} \left[\begin{array}{c|c} A + BF & BW \\ \hline C + DF & DW \\ F & W \end{array}\right]$, *where*

$$W = R_x{}^{-1/2}, \tag{7.123}$$

$$F = -R_x{}^{-1}\left(B^{\mathrm{T}}X + C\right), \tag{7.124}$$

$$R_x = \left(R + D^{\mathrm{T}} + D\right) > 0, \tag{7.125}$$

and

$$X = \mathrm{Ric}\begin{bmatrix} A - BR_x{}^{-1}C & -BR_x{}^{-1}B^{\mathrm{T}} \\ C^{\mathrm{T}}R_x{}^{-1}C & -\left(A - BR_x{}^{-1}C\right)^{\mathrm{T}} \end{bmatrix} \geq 0. \tag{7.126}$$

Additionally, an outer function such that $R + G^{\sim}(s) + G(s) = \Phi^{\sim}(s)\Phi(s)$ *is given by*

$$\Phi(s) = M^{-1}(s) \stackrel{s}{=} \left[\begin{array}{c|c} A & B \\ \hline R_x^{-1/2}(B^T X + C) & R_x^{1/2} \end{array}\right]. \tag{7.127}$$

Furthermore, there exists a left coprime factorization of $P(s) = \tilde{M}^{-1}(s)\tilde{N}(s)$ *given by* $\left[\tilde{N}(s) \quad \tilde{M}(s)\right] \stackrel{s}{=} \left[\begin{array}{c|cc} A + HC & B + HD & H \\ \hline \tilde{W}C & \tilde{W}D & \tilde{W} \end{array}\right]$*, where*

$$\tilde{W} = R_y^{-1/2}, \tag{7.128}$$

$$H = -\left(YC^{\mathrm{T}} + B\right)R_y{}^{-1}, \tag{7.129}$$

$$R_y = \left(R + D + D^{\mathrm{T}}\right) > 0, \tag{7.130}$$

and

$$Y = \mathrm{Ric}\begin{bmatrix} A - BR_y{}^{-1}C & -C^{\mathrm{T}}R_y{}^{-1}C \\ -BR_y{}^{-1}B^{\mathrm{T}} & -\left(A - BR_y{}^{-1}C\right)^{\mathrm{T}} \end{bmatrix} \geq 0. \tag{7.131}$$

Moreover, an outer function $\Phi(s)$ *such that* $R + G^{\sim}(s) + G(s) = \Phi^{\sim}(s)\Phi(s)$ *is given by*

$$\Phi(s) = \tilde{M}^{-1}(s) \stackrel{s}{=} \left[\begin{array}{c|c} A & (YC^T + B)R_y^{-1/2} \\ \hline C & R_y^{1/2} \end{array}\right]. \tag{7.132}$$

■

Example 7.6 Given $P(s) = \frac{s-4}{s+8}$, $R=2$, use spectral factorization to find an outer matrix $\Phi(s)$ such that $R + P^{\sim}(s) + P(s) = \Phi^{\sim}(s)\Phi(s)$.

(1) Compute $\Phi(s) \in GH_\infty$ directly. It is evident that

$$\begin{aligned} R + P^{\sim}(s) + P(s) &= 2 + \frac{-s-4}{-s+8} + \frac{s-4}{s+8} = \frac{4s^2-64}{s^2-64} \\ &= 2\frac{(s-4)}{(s-8)}2\frac{(s+4)}{(s+8)} = 2\frac{(-s+4)}{(-s+8)}2\frac{(s+4)}{(s+8)} \\ &= \Phi^{\sim}(s)\Phi(s), \end{aligned}$$

where $\Phi(s) = \frac{2(s+4)}{s+8}$.

(2) Apply the method noted in Case II to solve it. It yields

$$P(s) = \frac{s-4}{s+8} \overset{s}{=} \left[\begin{array}{c|c} A & B \\ \hline C & D \end{array}\right] = \left[\begin{array}{c|c} -8 & 1 \\ \hline -12 & 1 \end{array}\right].$$

Let $R = 2 = R^{\mathrm{T}} \geq 0$, then

$$R_x = \left(R + D^{\mathrm{T}} + D\right) = (2+1+1) = 4 > 0,$$

$$W = R_x^{-1/2} = 4^{-1/2} = \frac{1}{2}.$$

From

$$\left(A - BR_x^{-1}C\right)^{\mathrm{T}} X + X\left(A - BR_x^{-1}C\right) - XBR_x^{-1}B^{\mathrm{T}}X - C^{\mathrm{T}}R_x^{-1}C = 0,$$

one finds $X = (-36, -4)$. When $X = -4$,

$$F = -R_x^{-1}\left(B^{\mathrm{T}}X + C\right) = -\frac{1}{4}\left(1 \times (-4) + (-12)\right) = 4.$$

Then,

$$A + BF = -8 + 1 \times 4 = -4 < 0$$

is Hurwitz. An outer function such that $2 + \frac{-s-4}{-s+8} + \frac{s-4}{s+8} = \Phi^{\sim}(s)\Phi(s)$ is given by

$$\Phi(s) = M^{-1}(s) \overset{s}{=} \left[\begin{array}{c|c} A & B \\ \hline R_x^{-1/2}(B^T X + C) & R_x^{1/2} \end{array}\right] = \left[\begin{array}{c|c} A & B \\ \hline -W^{-1}F & W^{-1} \end{array}\right] = \left[\begin{array}{c|c} -8 & 1 \\ \hline -8 & 2 \end{array}\right] \overset{s}{=} \frac{2(s+4)}{s+8}.$$

Consequently, results (1) and (2) are the same. ■

7.4.3 Spectral Factorization Case III

Let $P(s) \stackrel{s}{=} \left[\begin{array}{c|c} A & B \\ \hline C & D \end{array}\right]$ be partitioned as $\begin{bmatrix} P_{11}(s) & P_{12}(s) \\ P_{21}(s) & P_{22}(s) \end{bmatrix} \in RH_\infty^{(n_1+n_2)\times(k_1+k_2)}$ with $n_1 \geq k_1$ and $n_2 = k_2$. The final case, which is so-called *J-spectral factorization*, is to find a matrix function $\Phi(s) \in GH_\infty^{(k_1+k_2)\times(k_1+k_2)}$ such that

$$P^{\sim}(s) J_1 P(s) = \Phi^{\sim}(s) J_2 \Phi(s), \tag{7.133}$$

where $J_1 = \begin{bmatrix} I_{n_1} & 0 \\ 0 & -I_{n_2} \end{bmatrix}$ and $J_2 = \begin{bmatrix} I_{k_1} & 0 \\ 0 & -I_{k_2} \end{bmatrix}$.

By defining $\Sigma = \begin{bmatrix} 0 & 0 \\ 0 & J_1 \end{bmatrix}$, $\breve{\Sigma} = J_2$, one may see that the factorization can be solved if the coprime factorization $P = NM^{-1}$ is such that $\begin{bmatrix} M \\ N \end{bmatrix}$ is a weighted all-pass function and M is outer, i.e., the following equation maintains

$$\begin{bmatrix} M(s) \\ N(s) \end{bmatrix}^{\sim} \begin{bmatrix} 0 & 0 \\ 0 & J_1 \end{bmatrix} \begin{bmatrix} M(s) \\ N(s) \end{bmatrix} = J_2. \tag{7.134}$$

From (7.134), one has, for $P = NM^{-1}$,

$$\left(M^{-1}\right)^{\sim} N^{\sim} J_1 N M^{-1} = \left(M^{-1}\right)^{\sim} J_2 M^{-1}. \tag{7.135}$$

It is clear that if the coprime factorization of $P(s)$ is found such that $M(s)$ is outer, then $\Phi = M^{-1}$ would be a solution to this spectral factorization problem. Using the state-space formula in (7.57), $\begin{bmatrix} M(s) \\ N(s) \end{bmatrix}$ is weighted all-pass with a specified Σ and $\breve{\Sigma}$ if and only if there exists a nonsingular matrix W satisfying $W^{\mathrm{T}} R_x W = J_2$, where $R_x = D^{\mathrm{T}} J_1 D$, and the corresponding ARE is

$$\begin{aligned} &\left(A - BR_x^{-1} D^{\mathrm{T}} J_1 C\right)^{\mathrm{T}} X + X\left(A - BR_x^{-1} D^{\mathrm{T}} J_1 C\right) - XBR_x^{-1} B^{\mathrm{T}} X \\ &\quad + C^{\mathrm{T}}\left(J_1 - J_1 D R_x^{-1} D^{\mathrm{T}} J_1\right) C = 0. \end{aligned} \tag{7.136}$$

Hence, with the stabilizing solution

$$X = \mathrm{Ric}\begin{bmatrix} A - BR_x^{-1} D^{\mathrm{T}} J_1 C & -BR_x^{-1} B^{\mathrm{T}} \\ -C^{\mathrm{T}}\left(J_1 - J_1 D R_x^{-1} D^{\mathrm{T}} J_1\right) C & -\left(A - BR_x^{-1} D^{\mathrm{T}} J_1 C\right)^{\mathrm{T}} \end{bmatrix}, \tag{7.137}$$

an outer function satisfying that

$$P^{\sim}(s)J_1P(s) = \Phi^{\sim}(s)J_2\Phi(s) \tag{7.138}$$

is given by

$$\Phi(s) = M^{-1}(s) \overset{s}{=} \left[\begin{array}{c|c} A & B \\ \hline -W^{-1}F & W^{-1} \end{array}\right], \tag{7.139}$$

where

$$F = -R_x^{-1}\left(B^{\mathrm{T}}X + D^{\mathrm{T}}J_1C\right). \tag{7.140}$$

Note that a solution W of $W^{\mathrm{T}}R_xW = J_2$ is given by $W = \begin{bmatrix} W_{11} & 0 \\ W_{21} & W_{22} \end{bmatrix}$, where

$$W_{22} = \left(D_{22}^{\mathrm{T}}D_{22} - D_{12}^{\mathrm{T}}D_{12}\right)^{-1/2}, \tag{7.141}$$

$$W_{11} = \left[D_{11}^{\mathrm{T}}D_{11} - D_{21}^{\mathrm{T}}D_{21} + \left(D_{12}^{\mathrm{T}}D_{11} - D_{22}^{\mathrm{T}}D_{21}\right)^{\mathrm{T}}W_{22}^2\left(D_{12}^{\mathrm{T}}D_{11} - D_{22}^{\mathrm{T}}D_{21}\right)\right]^{-1/2}, \text{ and} \tag{7.142}$$

$$W_{21} = W_{22}^2\left(D_{12}^{\mathrm{T}}D_{11} - D_{22}^{\mathrm{T}}D_{21}\right)W_{11}. \tag{7.143}$$

Clearly, if $D_{21} = 0$ and $D_{22}^TD_{22} > D_{12}^TD_{12}$, then W is nonsingular.

Dually, let $P(s) \overset{s}{=} \left[\begin{array}{c|c} A & B \\ \hline C & D \end{array}\right]$ be partitioned as $\begin{bmatrix} P_{11}(s) & P_{12}(s) \\ P_{21}(s) & P_{22}(s) \end{bmatrix} \in RH_\infty^{(n_1+n_2)\times(k_1+k_2)}$ with $n_1 = k_1$ and $n_2 \le k_2$. One needs to find a matrix function $\Phi(s) \in GH_\infty^{(n_1+n_2)\times(n_1+n_2)}$ satisfying

$$P(s)J_2P^{\sim}(s) = \Phi(s)J_1\Phi^{\sim}(s). \tag{7.144}$$

Also, by defining $\Sigma = \begin{bmatrix} 0 & 0 \\ 0 & J_2 \end{bmatrix}$, $\breve{\Sigma} = J_1$, one may see that if the left coprime factorization $P(s) = \tilde{M}^{-1}(s)\tilde{N}(s)$ is found such that $\begin{bmatrix} \tilde{M} & \tilde{N} \end{bmatrix}$ is a weighted all-pass function with M being outer, i.e., the following equation keeps

$$\begin{bmatrix} \tilde{M} & \tilde{N} \end{bmatrix}\begin{bmatrix} 0 & 0 \\ 0 & J_2 \end{bmatrix}\begin{bmatrix} \tilde{M} & \tilde{N} \end{bmatrix}^{\sim} = J_1. \tag{7.145}$$

From (7.144), one obtains

$$\tilde{M}^{-1}\tilde{N}J_2\tilde{N}^{\sim}\left(\tilde{M}^{-1}\right)^{\sim} = \Phi J_1\Phi^{\sim}. \tag{7.146}$$

$P = \tilde{M}^{-1}\tilde{N}$ can be found such that $\tilde{N}(s)J_2\tilde{N}^{\sim}(s) = J_1$ (dual J-lossless) if and only if there exists a nonsingular matrix $\tilde{W}$ satisfying $\tilde{W}R_y\tilde{W}^{\mathrm{T}} = J_1$, where $R_y = DJ_2D^{\mathrm{T}}$, and the ARE is

$$\begin{aligned}&\left(A - BJ_2D^{\mathrm{T}}R_y{}^{-1}C\right)Y + Y\left(A - BJ_2D^{\mathrm{T}}R_y{}^{-1}C\right)^{\mathrm{T}} - YC^{\mathrm{T}}R_y{}^{-1}CY\\&\quad - B\left(J_2 - J_2D^{\mathrm{T}}R_y{}^{-1}DJ_2\right)B^{\mathrm{T}} = 0.\end{aligned} \tag{7.147}$$

Hence,

$$Y = \mathrm{Ric}\begin{bmatrix}\left(A - BJ_2D^{\mathrm{T}}R_y{}^{-1}C\right)^{\mathrm{T}} & -C^{\mathrm{T}}R_y{}^{-1}C\\ B\left(J_2 - J_2D^{\mathrm{T}}R_y{}^{-1}DJ_2\right)B^{\mathrm{T}} & -\left(A - BJ_2D^{\mathrm{T}}R_y{}^{-1}C\right)\end{bmatrix} \tag{7.148}$$

has a stabilizing solution. Then, an outer function satisfying

$$P(s)J_2P^{\sim}(s) = \Phi(s)J_1\Phi^{\sim}(s) \tag{7.149}$$

is given by

$$\Phi(s) = \tilde{M}^{-1}(s) \stackrel{s}{=} \left[\begin{array}{c|c} A & H\tilde{W}^{-1}\\ \hline -C & \tilde{W}^{-1}\end{array}\right], \tag{7.150}$$

where

$$H = \left(YC^{\mathrm{T}} - BJ_2D^{\mathrm{T}}\right)R_y^{-1}. \tag{7.151}$$

Note that a solution $\tilde{W}R_y\tilde{W}^{\mathrm{T}} = J_2$ is given by $\tilde{W} = \begin{bmatrix}\tilde{W}_{11} & 0\\ \tilde{W}_{21} & \tilde{W}_{22}\end{bmatrix}$, where

$$\tilde{W}_{22} = \left[D_{21}\left(I - D_{11}^{\mathrm{T}}D_{11}\right)^{-1}D_{21}^{\mathrm{T}}\right]^{-1/2}, \tag{7.152}$$

$$\tilde{W}_{11} = \left[D_{12}^{\mathrm{T}}\left(I - D_{11}D_{11}^{\mathrm{T}}\right)^{-1}D_{12}\right]^{1/2}, \quad \text{and} \tag{7.153}$$

$$\tilde{W}_{21} = -\left[D_{21}\left(I - D_{11}^{\mathrm{T}}D_{11}\right)^{-1}D_{21}^{\mathrm{T}}\right]^{-1/2}D_{21}\left(I - D_{11}^{\mathrm{T}}D_{11}\right)^{-1}D_{11}^{\mathrm{T}}D_{12}. \tag{7.154}$$

Furthermore, the proposed methods can be directly applied to the discrete-time spectral factorization. Equations (7.134) and (7.145) show that $N^{\sim}J_1N = J_2$ ($\tilde{N}J_2\tilde{N}^{\sim} = J_1$). Hence, the coprime factor N $(\tilde{N})$ is J-lossless (dual J-lossless). The factorization of $P(s)$ into a J-lossless and outer parts has important applications. Therefore, the results are summarized in the following theorem.

Theorem 7.1 *Let* $P(s) \overset{s}{=} \left[\begin{array}{c|c} A & B \\ \hline C & D \end{array}\right] \in RH_{\infty}^{(n_1+n_2)\times(k_1+k_2)}$.

1. *There exists an rcf* $P(s) = \Theta(s)\Pi^{-1}(s)$ *such that* $\Theta(s)$ *is J-lossless, and* $\Pi(s)$ *is outer if there exists a nonsingular matrix* $W = \begin{bmatrix} W_{11} & 0 \\ W_{21} & W_{22} \end{bmatrix} \in R^{(k_1+k_2)\times(k_1+k_2)}$ *satisfying* $W^{\mathrm{T}}D^{\mathrm{T}}J_1DW = J_2$, *and* $A_{H_x} \in \mathrm{dom}\,(\mathrm{Ric})$, $F = -\left(D^{\mathrm{T}}J_1D\right)^{-1}\left(B^{\mathrm{T}}X + D^{\mathrm{T}}J_1C\right)$ *where* $X = \mathrm{Ric}\,(A_{H_x}) \geq 0$,

$$A_{H_x} = \begin{bmatrix} A - B\left(D^{\mathrm{T}}J_1D\right)^{-1}D^{\mathrm{T}}J_1C & -B\left(D^{\mathrm{T}}J_1D\right)^{-1}B^{\mathrm{T}} \\ -C^{\mathrm{T}}\left(J_1 - J_1D\left(D^{\mathrm{T}}J_1D\right)^{-1}D^{\mathrm{T}}J_1\right)C & -\left(A - B\left(D^{\mathrm{T}}J_1D\right)^{-1}D^{\mathrm{T}}J_1C\right)^{\mathrm{T}} \end{bmatrix}. \tag{7.155}$$

2. *There exists an lcf* $P(s) \overset{s}{=} \left[\begin{array}{c|c} A & B \\ \hline C & D \end{array}\right] \overset{s}{=} \tilde{\Pi}(s)^{-1}\tilde{\Theta}(s)$ *such that* $\tilde{\Theta}(s)$ *is dual J-lossless if there exists a nonsingular matrix* $\tilde{W} = \begin{bmatrix} \tilde{W}_{11} & 0 \\ \tilde{W}_{21} & \tilde{W}_{22} \end{bmatrix} \in R^{(n_1+n_2)\times(n_1+n_2)}$ *satisfying* $\tilde{W}DJ_2D^{\mathrm{T}}\tilde{W}^{\mathrm{T}} = J_1$, *and* $A_{H_y} \in \mathrm{dom}\,(\mathrm{Ric})$, $H = \left(YC^{\mathrm{T}} - BJ_2D^{\mathrm{T}}\right)\left(DJ_2D^{\mathrm{T}}\right)^{-1}$, *where* $Y = \mathrm{Ric}\left(A_{H_y}\right) \geq 0$,

$$A_{H_y} = \begin{bmatrix} \left(A - BJ_2D^{\mathrm{T}}(DJ_2D^{\mathrm{T}})^{-1}C\right)^{\mathrm{T}} & -C^{\mathrm{T}}(DJ_2D^{\mathrm{T}})^{-1}C \\ B\left(J_2 - J_2D^{\mathrm{T}}(DJ_2D^{\mathrm{T}})^{-1}DJ_2\right)B^{\mathrm{T}} & -\left(A - BJ_2D^{\mathrm{T}}(DJ_2D^{\mathrm{T}})^{-1}C\right) \end{bmatrix}. \tag{7.156}$$

■

Example 7.7 Let $P(s) = \begin{bmatrix} \frac{s-1}{s+2} & 0 \\ 0 & \frac{s+3}{s+4} \end{bmatrix} \in RH_{\infty}^{(1+1)\times(1+1)}$. Use the J-spectral factorization, which is defined earlier in this section, to find an outer matrix function $\Phi(s) \in GH_{\infty}^{(1+1)\times(1+1)}$ such that $P^{\sim}(s)J_1P(s) = \Phi^{\sim}(s)J_2\Phi(s)$.

Here, one can determine the state-space realization of $P(s)$ as

$$P(s) = \begin{bmatrix} \dfrac{s-1}{s+2} & 0 \\ 0 & \dfrac{s+3}{s+4} \end{bmatrix} \overset{s}{=} \left[\begin{array}{cc|cc} -2 & 0 & 1 & 0 \\ 0 & -4 & 0 & 1 \\ \hline -3 & 0 & 1 & 0 \\ 0 & -1 & 0 & 1 \end{array}\right].$$

By Case (III), $R_x = D^{\mathrm{T}} J_1 D = \begin{bmatrix} 1 & 0 \\ 0 & -1 \end{bmatrix}$ leads to

$$\begin{aligned}&\left(A - BR_x{}^{-1}D^{\mathrm{T}}J_1C\right)^{\mathrm{T}} X + X\left(A - BR_x{}^{-1}D^{\mathrm{T}}J_1C\right) - XBR_x{}^{-1}B^{\mathrm{T}}X \\ &\quad + C^{\mathrm{T}}\left(J_1 - J_1DR_x{}^{-1}D^{\mathrm{T}}J_1\right)C = 0\end{aligned}$$

and

$$X = \begin{bmatrix} 2 & 0 \\ 0 & 0 \end{bmatrix},$$

$$F = -R_x^{-1}\left(B^{\mathrm{T}}X + D^{\mathrm{T}}J_1C\right) = \begin{bmatrix} 1 & 0 \\ 0 & 1 \end{bmatrix}.$$

Thus,

$$A + BF = \begin{bmatrix} -2 & 0 \\ 0 & -4 \end{bmatrix} + \begin{bmatrix} 1 & 0 \\ 0 & 1 \end{bmatrix}\begin{bmatrix} 1 & 0 \\ 0 & 1 \end{bmatrix} = \begin{bmatrix} -1 & 0 \\ 0 & -3 \end{bmatrix} < 0$$

is Hurwitz. From

$$W_{22} = \left(D_{22}^{\mathrm{T}}D_{22} - D_{12}^{\mathrm{T}}D_{12}\right)^{-1/2},$$

$$W_{11} = \left[D_{11}^{\mathrm{T}}D_{11} - D_{21}^{\mathrm{T}}D_{21} + \left(D_{12}^{\mathrm{T}}D_{11} - D_{22}^{\mathrm{T}}D_{21}\right)^{\mathrm{T}} W_{22}^2\left(D_{12}^{\mathrm{T}}D_{11} - D_{22}^{\mathrm{T}}D_{21}\right)\right]^{-1/2},$$

$$W_{21} = W_{22}^2\left(D_{12}^{\mathrm{T}}D_{11} - D_{22}^{\mathrm{T}}D_{21}\right)W_{11},$$

it yields

$$W = \begin{bmatrix} W_{11} & 0 \\ W_{21} & W_{22} \end{bmatrix} = \begin{bmatrix} 1 & 0 \\ 0 & 1 \end{bmatrix}.$$

Then,

$$\Phi(s) = M^{-1}(s) \overset{s}{=} \left[\begin{array}{c|c} A & B \\ \hline -W^{-1}F & W^{-1} \end{array}\right] = \left[\begin{array}{cc|cc} -2 & 0 & 1 & 0 \\ 0 & -4 & 0 & 1 \\ \hline -1 & 0 & 1 & 0 \\ 0 & -1 & 0 & 1 \end{array}\right] \overset{s}{=} \begin{bmatrix} \dfrac{s+1}{s+2} & 0 \\ 0 & \dfrac{s+3}{s+4} \end{bmatrix}.$$

■

In this chapter, the matrix equation of ARE was introduced. In determination of stable coprime factors of a given system, solutions to AREs are required. The AREs are also employed in the formulation of a CSD for determining the conditions of J-lossless and dual J-lossless. This is the reason why one introduces the ARE in a seemingly "*independent*" chapter, before introducing controller synthesis. As mentioned in Chap. 1, this book aims to propose a unified controller synthesis approach based on the CSDs. In the next chapters, the issues on control systems synthesis will be discussed.

Exercises

1. Find a J-lossless spectral coprime factorization of $G(s) = \begin{bmatrix} \frac{s-1}{s+4} & 0 \\ 0 & \frac{s-1}{s+2} \end{bmatrix}$.
2. Given $G(s) = \frac{s-1}{s-4}$, compute spectral factor $\Phi(s)$ using spectral factorizations Case (I) and $R = 0, Q = 1$.
3. Given $G(s) = \begin{bmatrix} \frac{3(s-3)(s-4)}{(s+1)(s+2)} \\ \frac{4(s-3)(s-4)}{(s+1)(s+2)} \end{bmatrix}$, compute spectral factor $\Phi(s)$ using the spectral factorization Case (I) and $R = 0, Q = 1$.
4. Given $G(s) = \frac{s-3}{s+4} = N(s)M^{-1}(s)$, compute the normalized coprime factors $N(s)$ and $M(s)$ using spectral factorizations.
5. Given $G(s) = \frac{2(s-1)(s-3)}{(s+1)(s+2)}$, compute inner-outer factorization using coprime factorizations.
6. Compute the J-lossless factorization of the following systems:

(a)

$$G_1(s) = \begin{bmatrix} \dfrac{s}{s+2} & \dfrac{-2}{s+1} \\ \dfrac{-6}{s+2} & \dfrac{s-2}{s+2} \end{bmatrix} \stackrel{s}{=} \left[\begin{array}{ccc|cc} -2 & 0 & 0 & 2 & 0 \\ 0 & -1 & 0 & 0 & 1 \\ 0 & 0 & -2 & 0 & 2 \\ \hline -1 & -2 & 0 & 1 & 0 \\ -3 & 0 & -2 & 0 & 1 \end{array}\right];$$

(b)

$$G_2(s) = \begin{bmatrix} \dfrac{s}{s+2} & \dfrac{s-1}{s+1} \\ \dfrac{s-4}{s+2} & \dfrac{s-2}{s+2} \end{bmatrix} \stackrel{s}{=} \left[\begin{array}{ccc|cc} -2 & 0 & 0 & 2 & 0 \\ 0 & -1 & 0 & 0 & 1 \\ 0 & 0 & -2 & 0 & 2 \\ \hline -1 & -2 & 0 & 1 & 1 \\ -3 & 0 & -2 & 1 & 1 \end{array}\right].$$

7. Given $G(s) = \frac{2}{s-3}$, use coprime factorization to find an outer matrix $\Phi(s)$ such that $R + G^{\sim}(s)QG(s) = \Phi^{\sim}(s)\Phi(s)$, where $R = 1/4$ and $Q = 10$.
8. Given $G(s) = \frac{(s-1)}{(s+2)(s+3)}$, use spectral factorization to find an outer matrix $\Phi(s)$ such that $1 - G^{\sim}(s)G(s) = \Phi^{\sim}(s)\Phi(s)$.
9. Show that system an LFT SCC P matrix is unitary if $G^{\sim}JG = J$, where $\mathrm{LFT_l}(P,\Phi) = \mathrm{CSD_r}(\Theta,\Phi)$.
10. Provided that $\Phi \in BH_\infty$ and Θ is J-lossless, prove that $\mathrm{CSD_r}(\Theta,\Phi) \in BH_\infty$.

References

1. Doyle JC (1984) Lecture notes in advanced multivariable control. Lecture notes at ONR/Honeywell workshop, Minneapolis
2. Francis BA (1987) A course in H_∞ control theory. Springer, Berlin
3. Glover K (1984) All optimal Hankel-norm approximations of linear multivariable systems and their L_∞-error bounds. Int J Control 39:1115–1193
4. Green M, Limebeer DJN (1995) Linear robust control. Prentice Hall, Englewood Cliffs
5. Jan CW (1974) On the existence of a non-positive solution to the Riccati equation. IEEE Trans Autom Control 19(5):592–593
6. John D, Keith G, Promad K, Bruce F (1989) State-space solutions to standard H_2 and H_∞ control problems. IEEE Trans Autom Control 34(8):831–847
7. Kailath T (1980) Linear systems. Prentice-Hall, Englewood Cliffs
8. Maciejowski JM (1989) Multivariable feedback design. Addison-Wesley, Berkshire
9. Rajendra B (1997) Matrix analysis. Springer, New York
10. Sergio B, Alan JL, Jan CW (1991) The Riccati equation. Springer, Berlin
11. Zhou K, Doyle JC, Glover K (1996) Robust and optimal control. Prentice Hall, Upper Saddle River

Chapter 8
CSD Approach to Stabilization Control and H_2 Optimal Control

For a control system under investigation, an important design objective is to find a controller such that the closed-loop system satisfies a set of desired specifications. Furthermore, it is desirable to characterize all such controllers which achieve the required specifications. To accomplish this task, the required specifications can be first formulated into an adequate performance index, and then one can minimize the index to obtain sought controllers. This chapter presents a unified approach for solving a fairly general class of control synthesis problems by employing chain scattering-matrix descriptions and coprime factorizations. In particular, robust stabilization problems and the optimal H_2 control will be discussed in this chapter, respectively. The H_∞ (sub)optimal control problem can also be solved along the same approach which will be presented in the next chapter.

It is shown that general control synthesis problems, including those for stabilizing controllers or further for H_2/H_∞ controllers, can be first represented in terms of two associated chain scattering-matrix descriptions and then be solved by means of two coprime factorizations. The proposed method is virtually different from conventional approaches [8, 9, 12] in that it follows a straightforward development without introducing any pseudo signals for augmentation. Another major benefit of the proposed method is in its unified framework, which directly characterizes all stabilizing controllers and then leads to the optimal H_2 solution by finding a pair of special coprime factorizations. The corresponding formulae in the form of state-space realizations are included to show how the solutions are constructed with numerical advantages. Related graphic network representations are also presented to help explaining the whole concept which one believes will benefit engineering readers.

M.-C. Tsai and D.-W. Gu, *Robust and Optimal Control: A Two-port Framework Approach*, Advances in Industrial Control, DOI 10.1007/978-1-4471-6257-5_8,
© Springer-Verlag London 2014

8.1 Introduction

The issue of controller design has always played a major role in dealing with dynamic systems. Abundant techniques have been developed based on numerous mathematical tools and various requirements. For the first controller, one can refer to the governor for steam engines investigated by J. Watt in 1787, which significantly initiated the studies of control systems. So far the development of corresponding research could roughly be traced back into four periods. In 1932, Nyquist [14] used the theory of complex variables to build the stability criterion for single-input and single-output systems based on the frequency response. By late 1930s, Bode and Nichols progressed the analysis of frequency domain and led to great innovations followed by the root locus developed by Evans in 1948 [4], in which the importance of pole/zero of a system was characterized. All the above mentioned methods mainly compose the kernel of classical control. During the second period, Kalman and others probed into the optimal control theory by the state-space description in 1960s. With the assumption that the plant model is accurately representative, the so-called linear quadratic regulator (LQR) controller was derived, in which an infinity gain margin and at least 60° phase margin were congenitally obtained, proved by Safonov and Athans [16].

The third stage, led by Rosenbrock and MacFarlane in 1970s, has turned back to frequency analysis, due to its good robustness property, and much focus on the multidimensional systems. Even Rosenbrock [15] and MacFarlane [13] tried to promote an integration design for overcoming the stability problem caused by the altering of loop gain, there was not a complete framework with respect to minimize the influence of uncertainties. As a summary, model uncertainty and disturbance were not considered formally in controller design within these three periods.

Since the end of 1980s, robust control theory has become more and more well known because the issue of model uncertainty increases its recognition over the years, which significantly turned the research into the fourth period. In robust control theory, H_2 and H_∞ controls provide powerful tools to deal with model uncertainty, which also have been tremendously developed for linear time-invariant systems since the work of Zames [5, 19, 20]. In 1992, Smith and Doyle [17] optimized a perturbed coprime factor plant and connected the robust control issues to coprime factorizations. The contribution is significant in control synthesis and has received considerable attention in later developments [7]. In Smith and Doyle's method, noise is utilized as an exogenous input of a generalized plant, in which the uncertainty block is connected as surrounding feedback. Furthermore, the general model consistency problem was solved in the frequency domain for uncertain discrete-time plants represented by linear fractional transformations (LFTs) in [2]. Chen showed that these general consistency problems can be transformed into a set of convex optimization problems, which can be efficiently evaluated by mathematical techniques. Moreover, Benoit and Bruce [1] simplified the consistency examination, which made this method more easily applicable.

In 1995, Kimura used a single chain scattering description (CSD) to carry out the general H_∞ synthesis in which the so-called four-block problem can be solved by augmenting with some fictitious signals [9, 10]. The method proposed by Kimura greatly simplifies the sophisticated mathematics encountered in the H_∞ optimization synthesis. However, signal augmentation and additional calculations are still required for constructing a controller generator for the general H_∞ control synthesis problem. For coping with this shortcoming, an alternative CSD approach which significantly differs from Kimura's method was proposed in that the framework involves constructing two coupled (right and left) CSD matrices and just coprime factorizations [18]. The two coupled CSD approach does not require any fictitious signals for the augmentation so that it is fairly straightforward. This chapter proposes a comprehensive treatment of control synthesis via the latter CSD framework which is based on two coupled coprime factorizations of the SCC plant P to construct a general Bezout identity and then to characterize all the stabilization solutions. Based on these results, the H_2 optimal solution is readily obtained. In addition to the complete state-space formulae given, an outstanding feature of this approach is to use intensively graphic representations and manipulations which will hopefully help readers understand and grasp the synthesis procedures without requiring too much mathematics.

8.2 Characterization of All Stabilizing Controllers

Recall the general feedback control framework discussed in Chap. 4, as shown in Fig. 8.1 where the two-port transfer function matrix $P = \begin{bmatrix} P_{11} & P_{12} \\ P_{21} & P_{22} \end{bmatrix}$ denotes the standard control configuration (SCC) and K is the stabilizing controller to be designed. The closed-loop transfer function from w to z is given as

$$\mathrm{LFT_l}\left(\begin{bmatrix} P_{11} & P_{12} \\ P_{21} & P_{22} \end{bmatrix}, K\right) = P_{11} + P_{12}K(I - P_{22}K)^{-1}P_{21} \tag{8.1}$$

In contrast to the linear fractional transformation (LFT), the CSD developed in Chap. 5, originally for use of the network circuits, provides a straightforward interconnection in a cascaded manner.

It was shown in Chap. 5 that if P_{21} (or P_{12}) in Fig. 8.1 is square, then there exists $\mathrm{CSD_r}(G,K) = \mathrm{LFT_l}(P,K)$ (or $\mathrm{CSD_l}\left(\tilde{G}, K\right) = \mathrm{LFT_l}(P, K)$). It can be found

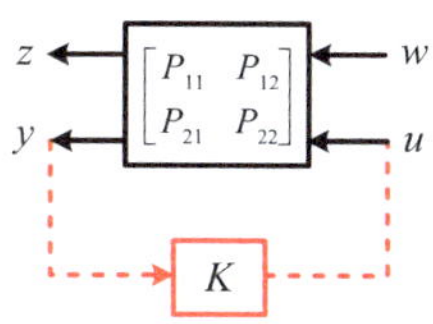

Fig. 8.1 Linear fractional transformation

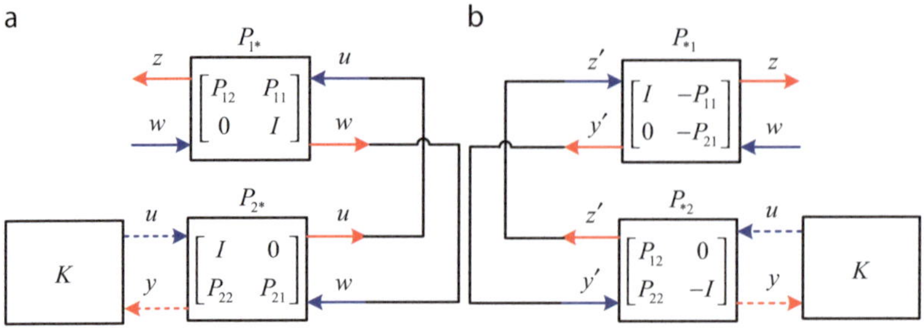

Fig. 8.2 Two coupled CSDs for an LFT matrix. (**a**) $\mathrm{CSD_r} \rightleftarrows \mathrm{CSD_l}$, (**b**) $\mathrm{CSD_l} \rightleftarrows \mathrm{CSD_r}$

that if both P_{21} and P_{12} are not square (i.e., the most general four-block problem), then a single equivalent CSD does not exist for the purpose. Therefore, utilizing pseudo signals is a possible approach for satisfying the square augmentation as in [9, 11]. However, as mentioned early in Chap. 2, the P matrix in SCC can always be represented by two coupled CSDs, i.e., right + left or left + right, as follows:

$$\mathrm{LFT}\left(\begin{bmatrix} P_{11} & P_{12} \\ P_{21} & P_{22} \end{bmatrix}, K\right) = \mathrm{CSD_r}\left(\begin{bmatrix} P_{12} & P_{11} \\ 0 & I \end{bmatrix}, \mathrm{CSD_l}\left(\begin{bmatrix} I & 0 \\ P_{22} & P_{21} \end{bmatrix}, K\right)\right) \quad (8.2)$$

or

$$\mathrm{LFT}\left(\begin{bmatrix} P_{11} & P_{12} \\ P_{21} & P_{22} \end{bmatrix}, K\right) = \mathrm{CSD_l}\left(\begin{bmatrix} I & -P_{11} \\ 0 & -P_{21} \end{bmatrix}, \mathrm{CSD_r}\left(\begin{bmatrix} P_{12} & 0 \\ P_{22} & -I \end{bmatrix}, K\right)\right). \quad (8.3)$$

This shows the equivalence of LFT and a pair of CSD representations. In other words, the P matrix in Fig. 8.1 can always be represented either by a right CSD matrix associated with a left CSD matrix or by a left CSD matrix associated with a right CSD matrix, as illustrated in Fig. 8.2. Note that $\mathrm{CSD_r} \rightleftarrows \mathrm{CSD_l}$ will be used in the following to denote the framework of a right CSD associated with a left one and $\mathrm{CSD_l} \rightleftarrows \mathrm{CSD_r}$ refers to its dual part. In the following, a new approach to directly characterize stabilization solutions for a given SCC plant is presented. The approach follows the framework of constructing two coupled CSDs and solving two coprime factorizations. Two methods are considered, corresponding to the two CSD representations in Fig. 8.2, respectively.

8.2.1 Method I: $\mathrm{CSD}_r \rightleftarrows \mathrm{CSD}_l$ Using a Right CSD Coupled with a Left CSD

To solve the stabilizing problem of Fig. 8.2a, let

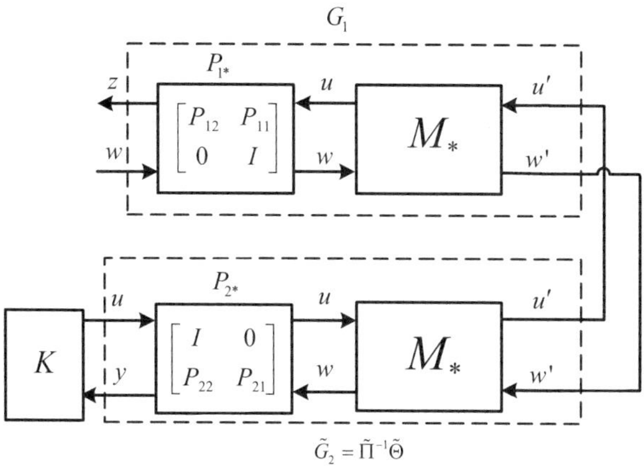

Fig. 8.3 Right coprime factorization over $\text{CSD}_r \rightleftarrows \text{CSD}_l$

$$\begin{bmatrix} P_{1*} \\ P_{2*} \end{bmatrix} = \begin{bmatrix} P_{12} & P_{11} \\ 0 & I \\ \hline I & 0 \\ P_{22} & P_{21} \end{bmatrix} = \begin{bmatrix} G_1 \\ \tilde{G}_2 \end{bmatrix} M_*^{-1} \tag{8.4}$$

be a right coprime factorization (*rcf*) over RH_∞. Consequently, as illustrated in Fig. 8.3, the stabilization problem of Fig. 8.1 could be realized by multiplying the denominator M_* on the right side of the CSD representations, as $\begin{bmatrix} P_{1*} \\ P_{2*} \end{bmatrix} M_*$. The resultant is the $\begin{bmatrix} G_1 \\ \tilde{G}_2 \end{bmatrix}$ from (8.4). Given an invertible $M_* \in RH_\infty$, by Property 5.3, one has

$$\begin{aligned} \text{LFT}_l(P, K) &= \text{CSD}_r(P_{1*}, \text{CSD}_l(P_{2*}, K)) \\ &= \text{CSD}_r(P_{1*}M_*, \text{CSD}_l(P_{2*}M_*, K)) \\ &= \text{CSD}_r\left(G_1, \text{CSD}_l\left(\tilde{G}_2, K\right)\right). \end{aligned} \tag{8.5}$$

Also let $\tilde{G}_2 = \tilde{\Pi}^{-1}\tilde{\Theta}$ be a left coprime factorization (*lcf*) over RH_∞, where $\tilde{\Pi}$ is a unit in RH_∞. By letting $K = \text{CSD}_l\left(\tilde{\Pi}, \Phi\right)$, $\forall \Phi \in RH_\infty$, the stabilization problem of Fig. 8.1 is represented in Fig. 8.4 where the closed-loop transfer function of $w \mapsto z$ is given as

$$z = \text{CSD}_r\left(\begin{bmatrix} G_{11} & G_{12} \\ G_{21} & G_{22} \end{bmatrix}, \text{CSD}_l\left(\begin{bmatrix} \tilde{\Theta}_{11} & \tilde{\Theta}_{12} \\ \tilde{\Theta}_{21} & \tilde{\Theta}_{22} \end{bmatrix}, \Phi\right)\right) w. \tag{8.6}$$

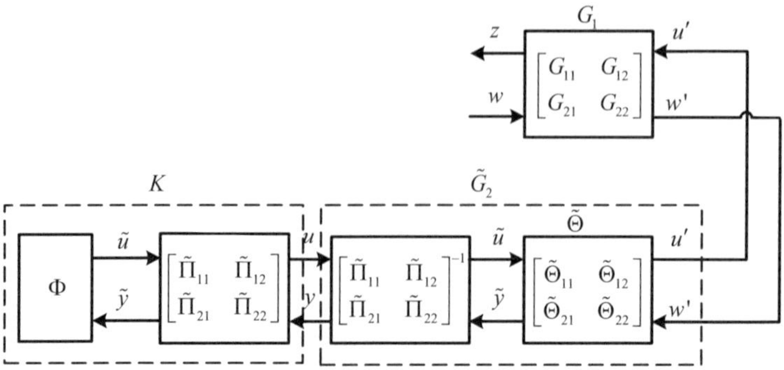

Fig. 8.4 $\tilde{G}_2 = \tilde{\Pi}^{-1}\tilde{\Theta}$ and $K = \mathrm{CSD}_{\mathrm{l}}\left(\tilde{\Pi}, \Phi\right)$

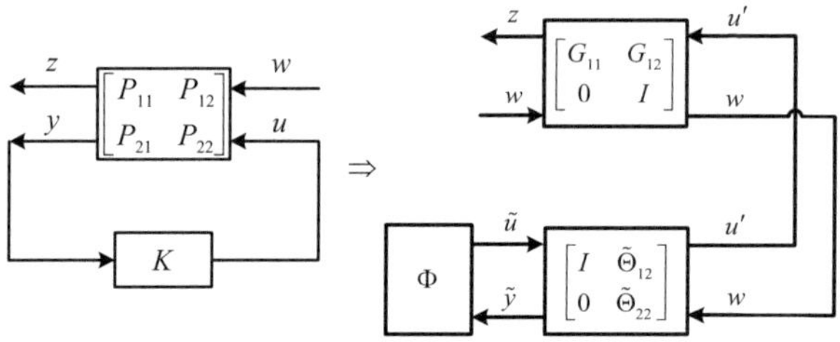

Fig. 8.5 Equivalent framework

As can be seen, the overall stability of the feedback system in Fig. 8.4 is not obvious, and it would be difficult to verify the stability with a general form of G_1 and $\tilde{\Theta}$. However, if G_1 in the *rcf* of (8.4) can be found to be in the triangular form of $G_1 = \begin{bmatrix} G_{11} & G_{12} \\ 0 & I \end{bmatrix}$, then one concludes $w' = w$ as shown in Fig. 8.5. Furthermore, in the *lcf* of $\tilde{G}_2 = \tilde{\Pi}^{-1}\tilde{\Theta}$, $\tilde{\Theta}$ can be found to be an upper triangular form of $\tilde{\Theta} = \begin{bmatrix} I & \tilde{\Theta}_{12} \\ 0 & \tilde{\Theta}_{22} \end{bmatrix}$. (Detailed formulae of such factorizations in the state-space form will be given in Sect. 8.3.) With such G_1 and $\tilde{\Theta}$, the input/output relation between w and z can be easily determined as, by direct manipulations,

$$z = \mathrm{CSD}_{\mathrm{r}}\left(\begin{bmatrix} G_{11} & G_{12} \\ 0 & I \end{bmatrix}, \mathrm{CSD}_{\mathrm{l}}\left(\begin{bmatrix} I & \tilde{\Theta}_{12} \\ 0 & \tilde{\Theta}_{22} \end{bmatrix}, \Phi\right)\right) = \left[G_{12} - G_{11}\left(\tilde{\Theta}_{12} - \Phi\tilde{\Theta}_{22}\right)\right] w. \tag{8.7}$$

Therefore, because the coprime factors $G_1, \tilde{\Theta}$ and the free parameter Φ are all in RH_∞, the closed-loop transfer function of $w \mapsto z$ is naturally stable.

The above procedure of finding a set of stabilizing controllers is characterized by solving two coprime factorization problems as summarized below:

Step 1: Find a right coprime factorization of $\begin{bmatrix} P_{1*} \\ P_{2*} \end{bmatrix} = \begin{bmatrix} G_1 \\ \tilde{G}_2 \end{bmatrix} M_*^{-1}$ over RH_∞ such that G_1 is the upper triangular form of $\begin{bmatrix} G_{11} & G_{12} \\ 0 & I \end{bmatrix}$.

Step 2: Find a left coprime factorization of $\tilde{G}_2 = \tilde{\Pi}^{-1}\tilde{\Theta}$ over RH_∞ such that $\tilde{\Pi} \in GH_\infty$ and $\tilde{\Theta}$ is of the upper triangular form of $\begin{bmatrix} I & \tilde{\Theta}_{12} \\ 0 & \tilde{\Theta}_{22} \end{bmatrix}$.

Step 3: All stabilizing controllers are determined by

$$K = \mathrm{CSD}_l\left(\tilde{\Pi}, \Phi\right) = -\left(\tilde{\Pi}_{11} - \Phi\tilde{\Pi}_{21}\right)^{-1}\left(\tilde{\Pi}_{12} - \Phi\tilde{\Pi}_{22}\right) \tag{8.8}$$

where $\Phi \in RH_\infty$ can also be called the Youla parameter. ■

Lemma 8.1 *The controller described in (8.8) stabilizes the transfer function from w to z for any Φ belonging to RH_∞.*

Proof The result can be proved directly from (8.7). ■

Furthermore, we may have (8.8) characterizing the set of *all* stabilizing controllers as described in the following theorem.

Theorem 8.1 *A controller K stabilizes the SCC in Fig. 8.1 if and only if it is given by (8.8) for any Φ in RH_∞.*

Proof The sufficiency is given by Lemma 8.1. The necessary part will be proved in Sect. 8.3.1, after the derivation of state-space formulae. ■

8.2.2 Method II: $CSD_l \rightleftarrows CSD_r$ Using a Left CSD Coupled with a Right CSD

Dually, the two-port SCC plant is described by a left CSD associated with a right one as shown in Fig. 8.2b. To solve this stabilization problem, let

$$\left[P_{*1} \mid P_{*2}\right] = \left[\begin{array}{cc|cc} I & -P_{11} & P_{12} & 0 \\ 0 & -P_{21} & P_{22} & -I \end{array}\right] = \tilde{M}_*^{-1}\left[\tilde{G}_1 \mid G_2\right] \tag{8.9}$$

be an *lcf* over RH_∞. Also let $G_2 = \Theta\Pi^{-1}$ be an *rcf* over RH_∞, where $\Pi \in GH_\infty$ is a unit in RH_∞. Dually, given an invertible $\tilde{M}_* \in RH_\infty$, by Property 5.4, one has

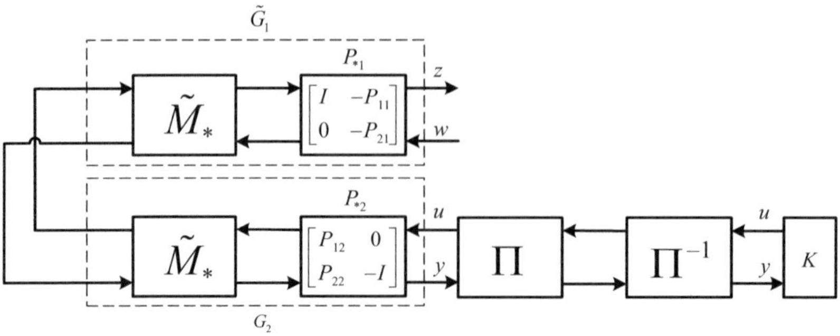

Fig. 8.6 Multiplying $\tilde{M}_*$ and inserting $\Pi^{-1}\Pi = I$ in Fig. 8.2b

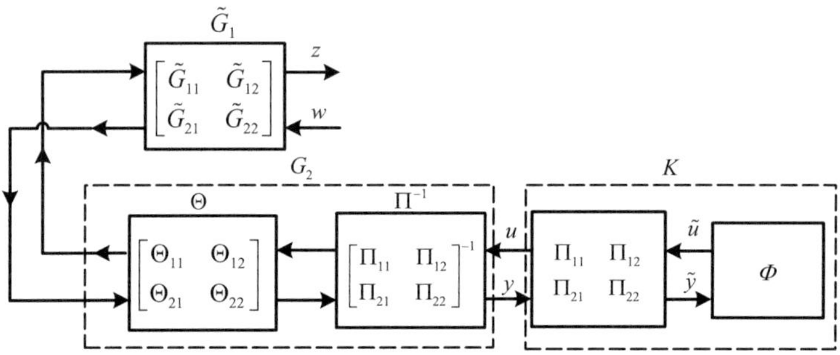

Fig. 8.7 $G_2 = \Theta\Pi^{-1}$ and $K = \mathrm{CSD_r}(\Pi,\Phi)$

$$\begin{aligned}\mathrm{LFT_l}(P, K) &= \mathrm{CSD_l}(P_{*1}, \mathrm{CSD_r}(P_{*2}, K)) \\ &= \mathrm{CSD_l}\left(\tilde{M}_* P_{*1}, \mathrm{CSD_r}\left(\tilde{M}_* P_{*2}, K\right)\right) \\ &= \mathrm{CSD_l}\left(\tilde{G}_1, \mathrm{CSD_r}(G_2, K)\right). \end{aligned} \tag{8.10}$$

As a dual form, this method can be realized by multiplying the denominator $\tilde{M}_*$ on the left side and by inserting the identity of $\Pi^{-1}\Pi = I$ between K and G_2 as illustrated in Fig. 8.6. By letting $K = \mathrm{CSD_r}(\Pi,\Phi)$ for $\Phi \in RH_\infty$, the stabilization problem of Fig. 8.1 will be depicted as in Fig. 8.7 where the closed-loop transfer function of $w \mapsto z$ is given by

$$z = \mathrm{CSD_l}\left(\begin{bmatrix} \tilde{G}_{11} & \tilde{G}_{12} \\ \tilde{G}_{21} & \tilde{G}_{22} \end{bmatrix}, \mathrm{CSD_l}\left(\begin{bmatrix} \Theta_{11} & \Theta_{12} \\ \Theta_{21} & \Theta_{22} \end{bmatrix}, \Phi\right)\right) w. \tag{8.11}$$

Apparently, if one can characterize the particular coprime factors $\tilde{G}_1 = \begin{bmatrix} I & \tilde{G}_{12} \\ 0 & \tilde{G}_{22} \end{bmatrix}$ and $\Theta = \begin{bmatrix} \Theta_{11} & \Theta_{12} \\ 0 & I \end{bmatrix}$, then the input/output relation between w and z, from the interconnection system in Fig. 8.8, can be easily determined as

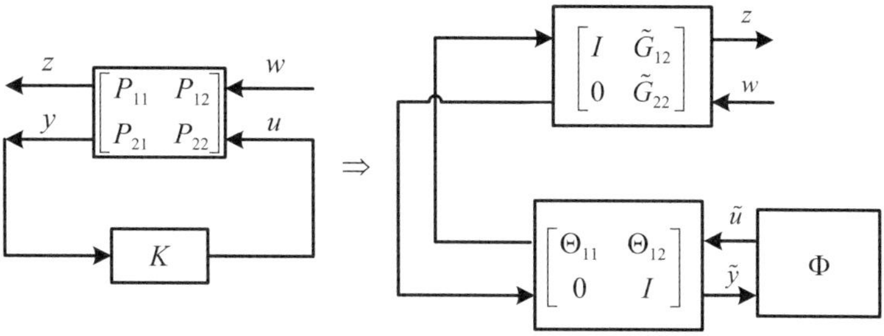

Fig. 8.8 Equivalent framework

$$z = \mathrm{CSD_l}\left(\begin{bmatrix} I & \tilde{G}_{12} \\ 0 & \tilde{G}_{22} \end{bmatrix}, \mathrm{CSD_r}\left(\begin{bmatrix} \Theta_{11} & \Theta_{12} \\ 0 & I \end{bmatrix}, \Phi\right)\right) = \left[(\Theta_{11}\Phi + \Theta_{12})\,\tilde{G}_{22} - \tilde{G}_{12}\right] w. \tag{8.12}$$

Dually, characterizing a set of stabilizing controllers is required for solving two related coprime factorization problems as summarized below:

Step 1: Find a left coprime factorization of $\begin{bmatrix} I & -P_{11} & P_{12} & 0 \\ 0 & -P_{21} & P_{22} & -I \end{bmatrix} = \tilde{M}_*^{-1}\begin{bmatrix} \tilde{G}_1 & G_2 \end{bmatrix}$ over RH_∞ such that $\tilde{G}_1$ is the upper triangular form of $\begin{bmatrix} I & \tilde{G}_{12} \\ 0 & \tilde{G}_{22} \end{bmatrix}$.

Step 2: Find a right coprime factorization of $G_2 = \Theta\Pi^{-1}$ over RH_∞ such that Θ is the upper triangular form of $\begin{bmatrix} \Theta_{11} & \Theta_{12} \\ 0 & I \end{bmatrix}$.

Step 3: All stabilizing controllers are determined by

$$K = \mathrm{CSD_r}(\Pi, \Phi) = (\Pi_{11}\Phi + \Pi_{12})(\Pi_{21}\Phi + \Pi_{22})^{-1} \tag{8.13}$$

where $\Phi \in RH_\infty$ can also be called the Youla parameter. ■

So far two alternative methods for finding all stabilizing controllers have been characterized. It should be emphasized that the approach of the right CSD associated with a left one is equivalent to the dual topology of the left CSD associated with a right one, in the sense that the stabilizing controller sets generated by (8.8) and (8.13) are the same. This result will be demonstrated at the end of Sect. 8.3.

In the next section, state-space formulae for the required coprime factorizations used in the above procedures will be derived and so those for the stabilizing controllers.

8.3 State-Space Formulae of Stabilizing Controllers

8.3.1 Method I: $CSD_r \rightleftarrows CSD_l$

In the following, the set of stabilizing controllers in the state-space form are to be constructed via three steps by a right CSD which is coupled with a left CSD formation. In the first step, the SCC plant is rearranged into a column stacked matrix form, which directly characterizes the transfer function in terms of a coupled CSD representation, and then a right coprime factorization is found of which the "numerator" matrix of the top half is upper triangular. In the second step, a left coprime factorization of the left CSD matrix (which is stable) is found such that the "denominator matrix" is outer and the "numerator" matrix is also upper triangular. Finally, all admissible controllers can be generated by the denominator which is a square bistable two-port matrix, i.e., in GH_∞. The details of the construction in the state-space form for each step are described below.

Consider the LFT interconnected system of Fig. 8.1, where $P(s)$ is a $(q_1+q_2)\times(m_1+m_2)$ proper, real-rational, transfer function matrix given by

$$P(s)=\begin{bmatrix} P_{11}(s) & P_{12}(s) \\ P_{21}(s) & P_{22}(s) \end{bmatrix} \overset{s}{=} \left[\begin{array}{c|cc} A & B_1 & B_2 \\ \hline C_1 & D_{11} & D_{12} \\ C_2 & D_{21} & 0 \end{array}\right] \tag{8.14}$$

with $(A,\, B_2)$ stabilizable and $(C_2,\, A)$ detectable. One can assume $D_{22}=0$ without loss of generality concerning the stabilization problem [6]. Recall that the solid lines in a matrix as in (8.14) show the matrix is a compact expression for a transfer function matrix, while dotted line in a matrix indicates usual matrix partition for the sake of clarity. In fact, (8.14) has the notation of state-space matrix form as

$$\begin{bmatrix} \dot{x} \\ z \\ y \end{bmatrix} = \begin{bmatrix} A & B_1 & B_2 \\ C_1 & D_{11} & D_{12} \\ C_2 & D_{21} & 0 \end{bmatrix} \begin{bmatrix} x \\ w \\ u \end{bmatrix} \tag{8.15}$$

with initial condition $x(0)=0$. As presented in Chap. 6, the following introduces the state-space procedures for the required coprime factorizations to parameterize a set of proper, real-rational controllers which stabilizes the SCC plant P.

An advantage of using the expression of a control system state-space model in (8.15) is that usual (constant) matrix manipulations can be applied when there are changes of system variables, input and output variables in particular. For example, for the state feedback control of $u=Fx+Wu'$, the state-space form of the closed-loop system can be readily obtained from (8.15), with $\begin{bmatrix} x \\ w \\ u \end{bmatrix} = \begin{bmatrix} I & 0 & 0 \\ 0 & I & 0 \\ F & 0 & W \end{bmatrix} \begin{bmatrix} x \\ w \\ u' \end{bmatrix}$, as

$$\begin{bmatrix} \dot{x} \\ z \\ y \end{bmatrix} = \begin{bmatrix} A & B_1 & B_2 \\ C_1 & D_{11} & D_{12} \\ C_2 & D_{21} & 0 \end{bmatrix} \begin{bmatrix} I & 0 & 0 \\ 0 & I & 0 \\ F & 0 & W \end{bmatrix} \begin{bmatrix} x \\ w \\ u' \end{bmatrix} = \begin{bmatrix} A+B_2F & B_1 & B_2W \\ C_1+D_{12}F & D_{11} & D_{12}W \\ C_2 & D_{21} & 0 \end{bmatrix} \begin{bmatrix} x \\ w \\ u' \end{bmatrix}. \tag{8.16}$$

This simple and intuitive approach is believed to be welcomed by engineers and will be extensively used in manipulations in the rest of this book.

Consider the coupled CSD representation depicted in Fig. 8.2a. Then (8.15) could be rewritten by exchanging the order of rows and columns to give the state-space realization of the coupled CSD representation as

$$\begin{bmatrix} \dot{x} \\ \hline z \\ y \\ \hline w \\ u \end{bmatrix} = \left[\begin{array}{c|cc} A & B_1 & B_2 \\ \hline C_1 & D_{11} & D_{12} \\ C_2 & D_{21} & 0 \\ \hline 0 & I & 0 \\ 0 & 0 & I \end{array}\right] \begin{bmatrix} x \\ w \\ u \end{bmatrix} \Rightarrow \begin{bmatrix} \dot{x} \\ \hline z \\ w \\ \hline u \\ y \end{bmatrix} = \left[\begin{array}{c|cc} A & B_2 & B_1 \\ \hline C_1 & D_{12} & D_{11} \\ 0 & 0 & I \\ \hline 0 & I & 0 \\ C_2 & 0 & D_{21} \end{array}\right] \begin{bmatrix} x \\ u \\ w \end{bmatrix}. \tag{8.17}$$

The following describes the state-space realization of each step in Method I.

Step 1: Find an *rcf* of a SCC plant P with an upper block triangular numerator G_1.

As stated in Chap. 6, the state-space form of the coprime factors $\begin{bmatrix} G_1 \\ \tilde{G}_2 \end{bmatrix}$ and M_* can be constructed by the state feedback of

$$\begin{bmatrix} x \\ u \\ w \end{bmatrix} = \begin{bmatrix} I & 0 & 0 \\ F_u & W_{uu} & 0 \\ F_w & W_{wu} & W_{ww} \end{bmatrix} \begin{bmatrix} x \\ u' \\ w' \end{bmatrix}. \tag{8.18}$$

Then it can be found from (8.17) that

$$\begin{bmatrix} \dot{x} \\ \hline z \\ w \\ \hline u \\ y \\ \hline u \\ w \end{bmatrix} = \left[\begin{array}{c|cc} A+B_1F_w+B_2F_u & B_1W_{wu}+B_2W_{uu} & B_1W_{ww} \\ \hline C_1+D_{11}F_w+D_{12}F_u & D_{11}W_{wu}+D_{12}W_{uu} & D_{11}W_{ww} \\ F_w & W_{wu} & W_{ww} \\ F_u & W_{uu} & 0 \\ C_2+D_{21}F_w & D_{21}W_{wu} & D_{21}W_{ww} \\ \hline F_u & W_{uu} & 0 \\ F_w & W_{wu} & W_{ww} \end{array}\right] \begin{bmatrix} x \\ u' \\ w' \end{bmatrix} \tag{8.19}$$

where $F = \begin{bmatrix} F_u \\ F_w \end{bmatrix}$ is chosen such that $A + B_1F_w + B_2F_u$ is Hurwitz and both W_{uu} and W_{ww} are nonsingular. The order of the vectors on the left and right sides of the first equation above is arranged according to $\begin{bmatrix} P_{1*} \\ P_{2*} \\ I \end{bmatrix}$ which is the transfer function matrix from the open-loop system input $\begin{bmatrix} u \\ w \end{bmatrix}$ to the output $\begin{bmatrix} z \\ w \\ u \\ y \end{bmatrix}$ stacking above the input itself. To find the right coprime factors with the required triangular form, let $F_w = 0$, $W_{wu} = 0$ and $W_{ww} = I$. Then the coprime factors are readily derived from the above equation as

$$\begin{bmatrix} G_1(s) & \tilde{G}_2(s) \mid M_*(s) \end{bmatrix}^T \overset{s}{=} \left[\begin{array}{c|cc} A+B_2F_u & B_2W_{uu} & B_1 \\ \hline C_1+D_{12}F_u & D_{12}W_{uu} & D_{11} \\ 0 & 0 & I \\ F_u & W_{uu} & 0 \\ C_2 & 0 & D_{21} \\ \hdashline F_u & W_{uu} & 0 \\ 0 & 0 & I \end{array}\right] \in RH_\infty \tag{8.20}$$

where

$$G_1(s) = \begin{bmatrix} G_{11}(s) & G_{12}(s) \\ 0 & I \end{bmatrix} \overset{s}{=} \left[\begin{array}{c|cc} A+B_2F_u & B_2W_{uu} & B_1 \\ \hline C_1+D_{12}F_u & D_{12}W_{uu} & D_{11} \\ 0 & 0 & I \end{array}\right] \in RH_\infty, \tag{8.21}$$

$$\tilde{G}_2(s) = \begin{bmatrix} \tilde{G}_{11}(s) & \tilde{G}_{12}(s) \\ \tilde{G}_{21}(s) & \tilde{G}_{22}(s) \end{bmatrix} \overset{s}{=} \left[\begin{array}{c|cc} A+B_2F_u & B_2W_{uu} & B_1 \\ \hline F_u & W_{uu} & 0 \\ C_2 & 0 & D_{21} \end{array}\right] \in RH_\infty, \tag{8.22}$$

$$M_*(s) = \begin{bmatrix} M_{11}(s) & M_{12}(s) \\ 0 & I \end{bmatrix} \overset{s}{=} \left[\begin{array}{c|cc} A+B_2F_u & B_2W_{uu} & B_1 \\ \hline F_u & W_{uu} & 0 \\ 0 & 0 & I \end{array}\right] \in RH_\infty. \tag{8.23}$$

Note that $A + B_2F_u$ is Hurwitz, with the selection request set earlier in the present choice of $F_w = 0$. As can be seen above, multiplying the denominator M_* of (8.23) on the right side of the CSD representations, as shown in Fig. 8.3, does not change the signal flow of the overall system but leads to a stable block triangular right CSD matrix G_1 and a stable left CSD matrix $\tilde{G}_2$.

Step 2: Find an *lcf* of $\tilde{G}_2$ with an upper block triangular numerator.

Moreover, for a stable $\tilde{G}_2$ of (8.22), a particular *lcf* of $\tilde{G}_2 = \tilde{\Pi}^{-1}\tilde{\Theta} \in RH_\infty$ is sought such that $\tilde{\Pi}$ is outer and $\tilde{\Theta}(s) = \begin{bmatrix} I & \tilde{\Theta}_{12}(s) \\ 0 & \tilde{\Theta}_{22}(s) \end{bmatrix} \in RH_\infty$. Since (C_2, A) is detectable, there exists an injection gain matrix H_y such that $A + H_yC_2$ is Hurwitz. Similar to Step 1, a general *lcf* of $\tilde{G}_2$ can be derived by the output injection of

$$\begin{bmatrix} \tilde{x} \\ \tilde{u} \\ \tilde{y} \end{bmatrix} = \begin{bmatrix} I & H_u & H_y \\ 0 & \tilde{W}_{uu} & \tilde{W}_{uy} \\ 0 & 0 & -\tilde{W}_{yy} \end{bmatrix} \begin{bmatrix} x \\ u \\ y \end{bmatrix}. \tag{8.24}$$

To find the required triangular form of the numerator, let $H_u = -B_2$, $\tilde{W}_{uy} = 0$ and $\tilde{W}_{uu} = W_{uu}^{-1}$. Note that the negative term of $-\tilde{W}_{yy}$ is defined such that Method I and Method II could be the dual cases, as derived in the end of Sect. 8.3 later. Then the coprime factors are readily derived from (8.22) as

$$\begin{aligned} &\left[\begin{array}{c|cc} I & -B_2 & H_y \\ \hline 0 & W_{uu}^{-1} & 0 \\ 0 & 0 & -\tilde{W}_{yy} \end{array}\right] \times \left[\begin{array}{c|cc|cc} A + B_2F_u & B_2W_{uu} & B_1 & 0 & 0 \\ \hline F_u & W_{uu} & 0 & I & 0 \\ C_2 & 0 & D_{21} & 0 & I \end{array}\right] \Rightarrow \\ &\left[\tilde{\Theta}(s) \mid \tilde{\Pi}(s)\right] \stackrel{s}{=} \left[\begin{array}{c|cc|cc} A + H_yC_2 & 0 & B_1 + H_yD_{21} & -B_2 & H_y \\ \hline W_{uu}^{-1}F_u & I & 0 & W_{uu}^{-1} & 0 \\ -\tilde{W}_{yy}C_2 & 0 & -\tilde{W}_{yy}D_{21} & 0 & -\tilde{W}_{yy} \end{array}\right] \end{aligned} \tag{8.25}$$

where $\tilde{\Theta}(s) \in RH_\infty$ is given by

$$\tilde{\Theta}(s) = \begin{bmatrix} I & \tilde{\Theta}_{12}(s) \\ 0 & \tilde{\Theta}_{22}(s) \end{bmatrix} \stackrel{s}{=} \left[\begin{array}{c|cc} A + H_yC_2 & 0 & B_1 + H_yD_{21} \\ \hline W_{uu}^{-1}F_u & I & 0 \\ -\tilde{W}_{yy}C_2 & 0 & -\tilde{W}_{yy}D_{21} \end{array}\right]. \tag{8.26}$$

Step 3: Find the stabilizing controllers.

Consequently, a set of all stabilizing controllers can be generated by $K = \mathrm{CSD}_l(\tilde{\Pi}, \Phi)$, $\forall \Phi \in RH_\infty$ where $\tilde{\Pi}(s)$, from Step 2 above, is given by

$$\tilde{\Pi}(s) = \begin{bmatrix} \tilde{\Pi}_{11}(s) & \tilde{\Pi}_{12}(s) \\ \tilde{\Pi}_{21}(s) & \tilde{\Pi}_{22}(s) \end{bmatrix} \stackrel{s}{=} \left[\begin{array}{c|cc} A + H_yC_2 & -B_2 & H_y \\ \hline W_{uu}^{-1}F_u & W_{uu}^{-1} & 0 \\ -\tilde{W}_{yy}C_2 & 0 & -\tilde{W}_{yy} \end{array}\right]. \tag{8.27}$$

■

It can be verified that the A-matrix of $\tilde{\Pi}^{-1}(s)$ is $A + B_2F_u$ and, therefore, the denominator of (8.27) belongs to GH_∞.

Fig. 8.9 Equivalent framework

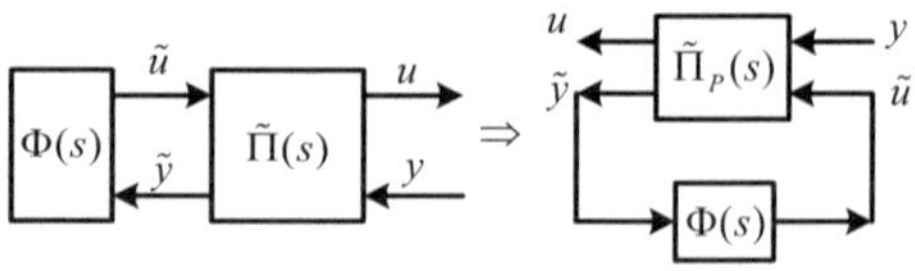

As illustrated in Fig. 8.9 which was presented in Chap. 5, the left CSD characterization of (8.27) can be equivalently transformed into its LFT form by (5.123), denoted by $\tilde{\Pi}_P$, such that $K = \mathrm{CSD_l}\left(\tilde{\Pi}, \Phi\right) = \mathrm{LFT_l}\left(\tilde{\Pi}_P, \Phi\right)$ where

$$\tilde{\Pi}_P(s) \overset{s}{=} \left[\begin{array}{c|cc} A + H_y C_2 + B_2 F_u & H_y & -B_2 W_{uu} \\ \hline -F_u & 0 & W_{uu} \\ -\tilde{W}_{yy} C_2 & -\tilde{W}_{yy} & 0 \end{array}\right]. \tag{8.28}$$

Thus, the central stabilizing controller is given by

$$K_0(s) = LFT_l(\tilde{\Pi}_P(s), 0) \overset{s}{=} \left[\begin{array}{c|c} A + H_y C_2 + B_2 F_u & H_y \\ \hline -F_u & 0 \end{array}\right]. \tag{8.29}$$

One may now complete the proof of Theorem 8.1.

Proof of Theorem 8.1 (continued) To prove the necessity part, one considers that a controller K stabilizes $P(s)$ in the SCC as in Fig. 8.1 if and only if K stabilizes P_{22}. As in the book [21], if K is a stabilizing controller, there exists a $Q(s) \in RH_\infty$ such that $K = \mathrm{LFT_l}(J,Q)$, where

$$J(s) \overset{s}{=} \left[\begin{array}{c|cc} A + B_2 F_u + H_y C_2 + H_y D_{22} F_u & H_y & -B_2 + H_y D_{22} \\ \hline -F_u & 0 & I \\ C_2 + D_{22} F_u & I & -D_{22} \end{array}\right]. \tag{8.30}$$

Without loss of generality, one may assume $D_{22} = 0$. In Step 3 above, it is obvious by letting $W_{uu} = I$, $\tilde{W}_{yy} = -I$ that $\tilde{\Pi}_P(s)$ of (8.28) is identical to $J(s)$, which shows that set of controllers generated by (8.8) (i.e., the state-space form of (8.27)) indeed includes all stabilizing controllers. ■

8.3.2 Method II: $CSD_l \rightleftarrows CSD_r$

A dual scheme of Method I can also be developed as well in which a left CSD is coupled with a right CSD. The procedure is similar to Method I. The details are summarized as follows.

Step 1: Find an *lcf of* a SCC plant P with an upper block triangular numerator $\tilde{G}_1$.

In the dual approach of stabilization, the state-space formulae can be derived as follows. A state-space realization of the row stacked matrix for (8.9) is given by

$$\tilde{P}_*(s) = \left[P_{*1}(s) \;\vdots\; P_{*2}(s) \right] \stackrel{s}{=} \left[\begin{array}{c|cc:cc} A & 0 & -B_1 & B_2 & 0 \\ \hline C_1 & I & -D_{11} & D_{12} & 0 \\ C_2 & 0 & -D_{21} & 0 & -I \end{array} \right]. \tag{8.31}$$

As presented in Chap. 6, by setting

$$\begin{bmatrix} \dot{\tilde{x}} \\ \tilde{z} \\ \tilde{y} \end{bmatrix} = \left[\begin{array}{c:cc} I & H_z & H_y \\ \hdashline 0 & \tilde{W}_{zz} & 0 \\ 0 & \tilde{W}_{yz} & \tilde{W}_{yy} \end{array} \right] \begin{bmatrix} \dot{x} \\ z_e \\ y_e \end{bmatrix}, \tag{8.32}$$

the *rcf* of $\left[P_{*1} \;\vdots\; P_{*2} \right] = \tilde{M}_*^{-1} \left[\tilde{G}_1 \;\vdots\; G_2 \right]$ can be constructed as

$$\left[\begin{array}{c:cc} I & H_z & H_y \\ \hdashline 0 & \tilde{W}_{zz} & 0 \\ 0 & \tilde{W}_{yz} & \tilde{W}_{yy} \end{array} \right] \times \left[\begin{array}{c:cccc:cc} A & 0 & -B_1 & B_2 & 0 & 0 & 0 \\ \hdashline C_1 & I & -D_{11} & D_{12} & 0 & I & 0 \\ C_2 & 0 & -D_{21} & 0 & -I & 0 & I \end{array} \right] \Rightarrow \left[\tilde{G}_1(s) \quad G_2(s) \;\vdots\; \tilde{M}_*(s) \right] \tag{8.33}$$

where $H = \left[H_z \; H_y \right]$ is chosen such that $A + H_z C_1 + H_y C_2$ is Hurwitz and $\tilde{W}_{zz}$ and $\tilde{W}_{yy}$ are nonsingular. Letting $H_z = 0$, $\tilde{W}_{yz} = 0$, and $\tilde{W}_{zz} = I$ in the above equation will yield the required coprime factors as

$$\tilde{G}_1(s) = \begin{bmatrix} I & \tilde{G}_{12}(s) \\ 0 & \tilde{G}_{22}(s) \end{bmatrix} \stackrel{s}{=} \left[\begin{array}{c|cc} A + H_y C_2 & 0 & -\left(B_1 + H_y D_{21} \right) \\ \hline C_1 & I & -D_{11} \\ \tilde{W}_{yy} C_2 & 0 & -\tilde{W}_{yy} D_{21} \end{array} \right] \in RH_\infty, \tag{8.34}$$

$$G_2(s) \stackrel{s}{=} \left[\begin{array}{c|cc} A + H_y C_2 & B_2 & -H_y \\ \hline C_1 & D_{12} & 0 \\ \tilde{W}_{yy} C_2 & 0 & -\tilde{W}_{yy} \end{array} \right] \in RH_\infty, \tag{8.35}$$

$$\tilde{M}_*(s) = \begin{bmatrix} I & \tilde{M}_{12}(s) \\ 0 & \tilde{M}_{22}(s) \end{bmatrix} \stackrel{s}{=} \left[\begin{array}{c|cc} A + H_y C_2 & 0 & H_y \\ \hline C_1 & I & 0 \\ \tilde{W}_{yy} C_2 & 0 & \tilde{W}_{yy} \end{array} \right] \in RH_\infty. \tag{8.36}$$

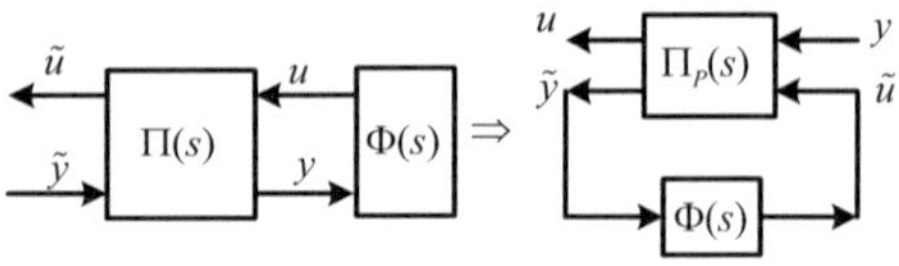

Fig. 8.10 $K = \mathrm{CSD_r}(\Pi,\Phi) = \mathrm{LFT_l}(\Pi_p,\Phi)$

Step 2: Find an *rcf* of G_2 with an upper block triangular numerator.

Moreover, given G_2 in (8.35), a state-space realization of $G_2 = \Theta\Pi^{-1}$ is given by

$$\left[\begin{array}{c|cc} A+H_yC_2 & B_2 & -H_y \\ \hline C_1 & D_{12} & 0 \\ \tilde{W}_{yy}C_2 & 0 & -\tilde{W}_{yy} \\ \hline 0 & I & 0 \\ 0 & 0 & I \end{array}\right] \times \left[\begin{array}{c|cc} I & 0 & 0 \\ \hline F_u & W_{uu} & 0 \\ C_2 & 0 & -\tilde{W}_{yy}^{-1} \end{array}\right] \Rightarrow \begin{bmatrix} \Theta(s) \\ \hline \Pi(s) \end{bmatrix} \stackrel{s}{=} \left[\begin{array}{c|cc} A+B_2F_u & B_2W_{uu} & H_y\tilde{W}_{yy}^{-1} \\ \hline C_1+D_{12}F_u & D_{12}W_{uu} & 0 \\ 0 & 0 & I \\ \hline F_u & W_{uu} & 0 \\ C_2 & 0 & -\tilde{W}_{yy}^{-1} \end{array}\right], \tag{8.37}$$

where

$$\begin{bmatrix} \Theta_{11}(s) & \Theta_{12}(s) \\ 0 & I \end{bmatrix} \stackrel{s}{=} \left[\begin{array}{c|cc} A+B_2F_u & B_2W_{uu} & H_y\tilde{W}_{yy}^{-1} \\ \hline C_1+D_{12}F_u & D_{12}W_{uu} & 0 \\ 0 & 0 & I \end{array}\right]. \tag{8.38}$$

Step 3: Find the stabilizing controller.

Then a set of all stabilizing controllers can be generated by $K = \mathrm{CSD_r}(\Pi,\Phi)$, $\forall\, \Phi \in RH_\infty$ where, from (8.37),

$$\Pi(s) = \begin{bmatrix} \Pi_{11}(s) & \Pi_{12}(s) \\ \Pi_{21}(s) & \Pi_{22}(s) \end{bmatrix} \stackrel{s}{=} \left[\begin{array}{c|cc} A+B_2F_u & B_2W_{uu} & H_y\tilde{W}_{yy}^{-1} \\ \hline F_u & W_{uu} & 0 \\ C_2 & 0 & -\tilde{W}_{yy}^{-1} \end{array}\right]. \tag{8.39}$$

As illustrated in Fig. 8.10, the right CSD characterization can also be transformed into its LFT form, denoted by Π_P, such that $K = \mathrm{CSD_r}(\Pi,\Phi) = \mathrm{LFT_l}(\Pi_P,\Phi)$ where

$$\Pi_P(s) \stackrel{s}{=} \left[\begin{array}{c|cc} A+B_2F_u+H_yC_2 & -H_y & B_2W_{uu} \\ \hline F_u & 0 & W_{uu} \\ \tilde{W}_{yy}C_2 & -\tilde{W}_{yy} & 0 \end{array}\right]. \tag{8.40}$$

Thus the central solution of the stabilizing controller is given by

$$K_0(s) = LFT_l(\tilde{\Pi}_P(s),0) \stackrel{s}{=} \left[\begin{array}{c|c} A+B_2F_u+H_yC_2 & -H_y \\ \hline F_u & 0 \end{array}\right]. \tag{8.41}$$

Note that the central solution is the same as that obtained by (8.29).

So far two alternative methods for finding all stabilizing controllers are characterized. It should be emphasized that the right CSD associated with a left one is indeed the dual topology of the left CSD associated with a right one. It can be verified from the inverse of (8.27) as

$$\tilde{\Pi}^{-1}(s) \overset{s}{=} \left[\begin{array}{c|cc} A+B_2F_u & B_2W_{uu} & H_y\tilde{W}_{yy}^{-1} \\ \hline F_u & W_{uu} & 0 \\ C_2 & 0 & -\tilde{W}_{yy}^{-1} \end{array}\right] \overset{s}{=} \Pi(s). \tag{8.42}$$

Then from (8.39), one concludes that $\tilde{\Pi}\Pi = I$, i.e., $\mathrm{CSD_r}(\Pi, \Phi) = \mathrm{CSD_l}(\tilde{\Pi}, \Phi)$. This implies that Method I and Method II are virtually the dual approach to each other for the general case. Note that the notation introduced in Method I for coprime factorizations has been employed purposely in Method II such that a general Bezout identity (i.e., $\tilde{\Pi}(s)\Pi(s) = I$) can be constructed from these two methods. The parameters of F_u, H_y, W_{uu}, and $\tilde{W}_{yy}$ in the controller generator formulae of Π (or $\tilde{\Pi}$) will be characterized by the H_2 control optimization later in Sect. 8.5.

8.4 Example of Finding Stabilizing Controllers

The following gives an example to show how to characterize all internally stabilizing solutions for a typical control problem using the proposed CSD approach. Consider the feedback stabilization problem in Fig. 8.11, where G is the controlled plant and K is the controller to be designed. Then the SCC plant of $\begin{bmatrix} w_1 \\ w_2 \end{bmatrix} \mapsto \begin{bmatrix} z_1 \\ z_2 \end{bmatrix}$ from Fig. 8.11 can be found as

$$P = \begin{bmatrix} P_{11} & P_{12} \\ P_{21} & P_{22} \end{bmatrix} = \begin{bmatrix} \begin{pmatrix} 0 & 0 \\ 0 & G \end{pmatrix} & \begin{pmatrix} I \\ G \end{pmatrix} \\ \begin{pmatrix} I & G \end{pmatrix} & G \end{bmatrix}. \tag{8.43}$$

The closed-loop transfer function of $w \mapsto z$ with $u = Ky$ is

$$\begin{bmatrix} z_1 \\ z_2 \end{bmatrix} = \mathrm{LFT_l}(P, K) \begin{bmatrix} w_1 \\ w_2 \end{bmatrix}, \tag{8.44}$$

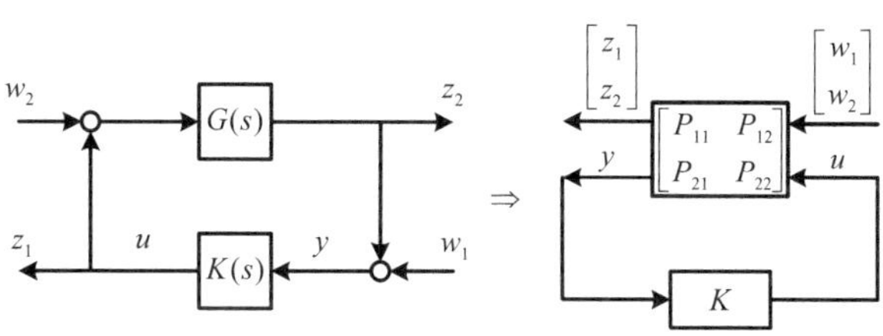

Fig. 8.11 Feedback control system

where

$$\mathrm{LFT}_l(P,K)=\begin{bmatrix} K(I-GK)^{-1} & K(I-GK)^{-1}G \\ GK(I-GK)^{-1} & G\left[I+K(I-GK)^{-1}G\right] \end{bmatrix}. \tag{8.45}$$

Assume that the state-space realization of $G(s)\overset{s}{=}\left[\begin{array}{c|c} A & B \\ \hline C & 0 \end{array}\right]$ is a minimal realization. Then the state-space realization of the augmented SCC plant (8.43) is given by

$$P(s)\overset{s}{=}\left[\begin{array}{c|cc} A & B_1 & B_2 \\ \hline C_1 & 0 & D_{12} \\ C_2 & D_{21} & 0 \end{array}\right]\Rightarrow\begin{bmatrix} \begin{pmatrix} 0 & 0 \\ 0 & G(s) \end{pmatrix} & \begin{pmatrix} I \\ G(s) \end{pmatrix} \\ \begin{pmatrix} I & G(s) \end{pmatrix} & G(s) \end{bmatrix}\overset{s}{=}\left[\begin{array}{c|cc} A & \begin{pmatrix} 0 & B \end{pmatrix} & B \\ \hline \begin{pmatrix} 0 \\ C \end{pmatrix} & \begin{pmatrix} 0 & 0 \\ 0 & 0 \end{pmatrix} & \begin{pmatrix} I \\ 0 \end{pmatrix} \\ C & \begin{pmatrix} I & 0 \end{pmatrix} & 0 \end{array}\right]. \tag{8.46}$$

8.4.1 Method I: $\mathrm{CSD}_r \rightleftarrows \mathrm{CSD}_l$ Using a Right CSD Associated with a Left CSD

Step 1: By (8.46), a state-space realization of the column stacked matrix $\begin{bmatrix} P_{1*} \\ P_{2*} \end{bmatrix}$ can be found from (8.17) with $D_{11}=0$ as

$$\left[\begin{array}{c} \dot{x} \\ \hline \begin{pmatrix} z_1 \\ z_2 \end{pmatrix} \\ \begin{pmatrix} w_1 \\ w_2 \end{pmatrix} \\ \hdashline u \\ y \end{array}\right]=\left[\begin{array}{c|cc} A & B & \begin{pmatrix} 0 & B \end{pmatrix} \\ \hline \begin{pmatrix} 0 \\ C \end{pmatrix} & \begin{pmatrix} I \\ 0 \end{pmatrix} & \begin{pmatrix} 0 & 0 \\ 0 & 0 \end{pmatrix} \\ \begin{pmatrix} 0 \\ 0 \end{pmatrix} & \begin{pmatrix} 0 \\ 0 \end{pmatrix} & \begin{pmatrix} I & 0 \\ 0 & I \end{pmatrix} \\ \hdashline 0 & I & \begin{pmatrix} 0 & 0 \end{pmatrix} \\ C & 0 & \begin{pmatrix} I & 0 \end{pmatrix} \end{array}\right]\begin{bmatrix} x \\ u \\ \begin{pmatrix} w_1 \\ w_2 \end{pmatrix} \end{bmatrix}. \tag{8.47}$$

The coupled CSD form of Fig. 8.1 becomes Fig. 8.12.

Step 2: Right coprime factors $\begin{bmatrix} G_1 \\ \tilde{G}_2 \end{bmatrix}$ and M_*, by (8.20) with $F_u=F$ and $W_{uu}=I$, can be constructed as

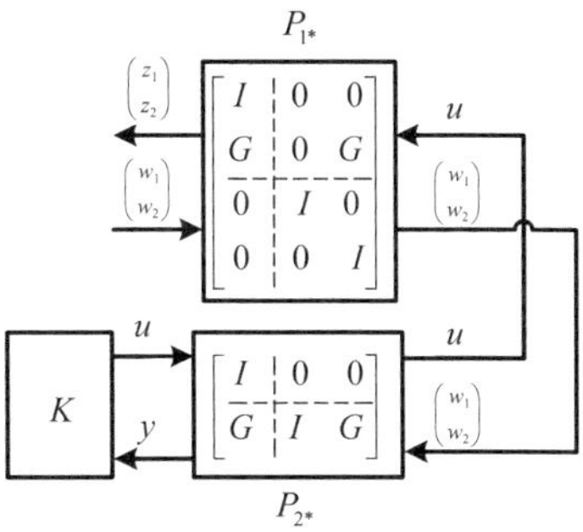

Fig. 8.12 Equivalent $CSD_r \rightleftarrows CSD_l$

$$G_1(s) \stackrel{s}{=} \left[\begin{array}{c|c|c} A+BF & B & \begin{pmatrix} 0 & B \end{pmatrix} \\ \hline \begin{pmatrix} F \\ C \end{pmatrix} & \begin{pmatrix} I \\ 0 \end{pmatrix} & \begin{pmatrix} 0 & 0 \\ 0 & 0 \end{pmatrix} \\ \begin{pmatrix} 0 \\ 0 \end{pmatrix} & \begin{pmatrix} 0 \\ 0 \end{pmatrix} & \begin{pmatrix} I & 0 \\ 0 & I \end{pmatrix} \end{array}\right], \tag{8.48}$$

$$\tilde{G}_2(s) \stackrel{s}{=} \left[\begin{array}{c|cc} A+BF & B & \begin{pmatrix} 0 & B \end{pmatrix} \\ \hline F & I & \begin{pmatrix} 0 & 0 \end{pmatrix} \\ C & 0 & \begin{pmatrix} I & 0 \end{pmatrix} \end{array}\right], \tag{8.49}$$

$$M_*(s) \stackrel{s}{=} \left[\begin{array}{c|c|c} A+BF & B & \begin{pmatrix} 0 & B \end{pmatrix} \\ \hline F & I & \begin{pmatrix} 0 & 0 \end{pmatrix} \\ \begin{pmatrix} 0 \\ 0 \end{pmatrix} & \begin{pmatrix} 0 \\ 0 \end{pmatrix} & \begin{pmatrix} I & 0 \\ 0 & I \end{pmatrix} \end{array}\right]. \tag{8.50}$$

To derive the CSD representations in terms of the transfer function level, let $G = NM^{-1} = \tilde{M}^{-1}\tilde{N}$ denote the doubly coprime factorizations of the controlled plant. By Lemma 6.2, there exists the Bezout identity of (6.40) depicted as

$$\begin{bmatrix} \tilde{X}(s) & \tilde{Y}(s) \\ \tilde{N}(s) & \tilde{M}(s) \end{bmatrix} \begin{bmatrix} M(s) & -Y(s) \\ -N(s) & X(s) \end{bmatrix} = \begin{bmatrix} I & 0 \\ 0 & I \end{bmatrix}. \tag{8.51}$$

From (6.22), (6.23), (6.31), and (6.35) with $W = \tilde{W} = I$, the state-space form of (8.51) can be represented by

$$\begin{bmatrix} \tilde{X}(s) & \tilde{Y}(s) \\ \tilde{N}(s) & \tilde{M}(s) \end{bmatrix} \stackrel{s}{=} \left[\begin{array}{c|cc} A+HC & -B & -H \\ \hline F & I & 0 \\ -C & 0 & I \end{array}\right] \text{ and } \begin{bmatrix} M(s) & -Y(s) \\ -N(s) & X(s) \end{bmatrix} \stackrel{s}{=} \left[\begin{array}{c|cc} A+BF & B & H \\ \hline F & I & 0 \\ -C & 0 & I \end{array}\right]. \tag{8.52}$$

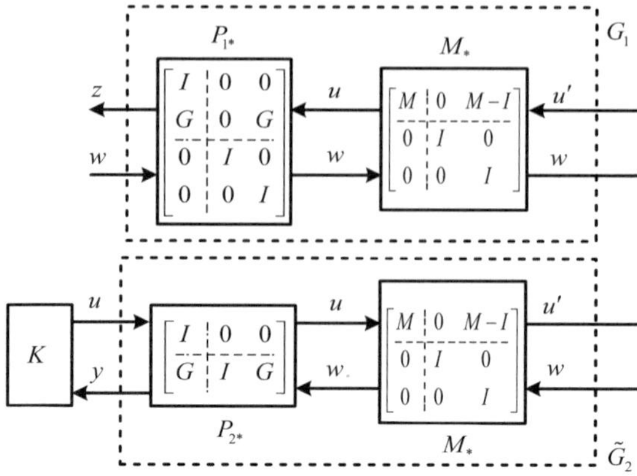

Fig. 8.13 Multiplying M_* at the right terminals

On the other hand, one can represent (8.48), (8.49) and (8.50) in the transfer function level as (Fig. 8.13)

$$M_* = \begin{bmatrix} M & 0 & M-I \\ 0 & I & 0 \\ 0 & 0 & I \end{bmatrix}, \tag{8.53}$$

$$G_1 = \begin{bmatrix} G_{11} & G_{12} \\ 0 & I \end{bmatrix} = P_{1*}M_* = \left[\begin{array}{c|cc} I & 0 & 0 \\ G & 0 & G \\ \hline 0 & I & 0 \\ 0 & 0 & I \end{array}\right]\left[\begin{array}{c|cc} M & 0 & M-I \\ \hline 0 & I & 0 \\ 0 & 0 & I \end{array}\right] = \left[\begin{array}{c|cc} M & 0 & M-I \\ N & 0 & N \\ \hline 0 & I & 0 \\ 0 & 0 & I \end{array}\right]_\infty, \tag{8.54}$$

$$\tilde{G}_2 = P_{2*}M_* = \left[\begin{array}{c|cc} I & 0 & 0 \\ \hline G & I & G \end{array}\right]\left[\begin{array}{c|cc} M & 0 & M-I \\ \hline 0 & I & 0 \\ 0 & 0 & I \end{array}\right] = \left[\begin{array}{c|cc} M & 0 & M-I \\ \hline N & I & N \end{array}\right]. \tag{8.55}$$

Step 3: Given (8.49), the required coprime factors $\tilde{\Theta}$ and $\tilde{\Pi}$ can be found from (8.25) with $H_y = H$, $\tilde{W}_{yy} = I$, and $W_{uu} = I$ as

$$\tilde{\Theta}(s) = \begin{bmatrix} I & \tilde{\Theta}_{12}(s) \\ 0 & \tilde{\Theta}_{22}(s) \end{bmatrix} \stackrel{s}{=} \left[\begin{array}{c|cc} A+HC & 0 & (H \quad B) \\ \hline F & I & (0 \quad 0) \\ -C & 0 & -(I \quad 0) \end{array}\right], \tag{8.56}$$

$$\tilde{\Pi}(s) \stackrel{s}{=} \left[\begin{array}{c|cc} A+HC & -B & H \\ \hline F & I & 0 \\ -C & 0 & -I \end{array}\right]. \tag{8.57}$$

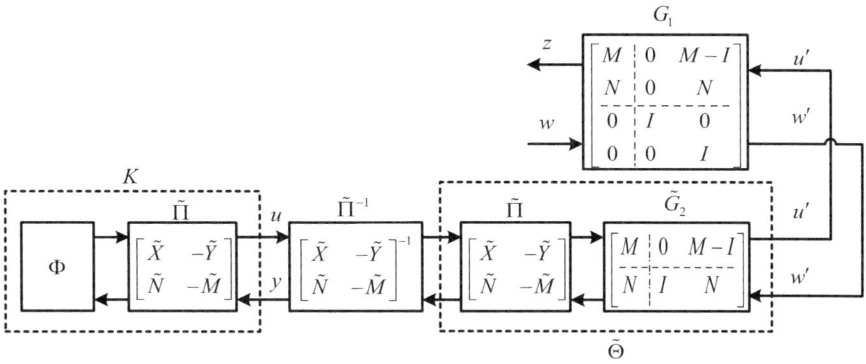

Fig. 8.14 Insertion of $\tilde{\Pi}^{-1}\tilde{\Pi}$

Therefore, $\tilde{\Pi}(s)$ above can be written in transfer function level such as

$$\tilde{\Pi}(s) = \begin{bmatrix} \tilde{X}(s) & -\tilde{Y}(s) \\ \tilde{N}(s) & -\tilde{M}(s) \end{bmatrix} \overset{s}{=} \left[\begin{array}{c|cc} A+HC & -B & H \\ \hline F & I & 0 \\ -C & 0 & -I \end{array}\right]. \tag{8.58}$$

It should be emphasized that the state-space form defined in (8.58) is identical to (8.52); therefore, one concludes that $\tilde{X}M - \tilde{Y}N = I$ from (8.51), and

$$\begin{bmatrix} I & \tilde{\Theta}_{12} \\ 0 & \tilde{\Theta}_{22} \end{bmatrix} = \tilde{\Pi}\tilde{G}_2 = \left[\begin{array}{c|c} \tilde{X} & -\tilde{Y} \\ \hline \tilde{N} & -\tilde{M} \end{array}\right]\left[\begin{array}{c|cc} M & 0 & M-I \\ \hline N & I & N \end{array}\right] = \left[\begin{array}{c|cc} I & -\tilde{Y} & I-\tilde{X} \\ \hline 0 & -\tilde{M} & -\tilde{N} \end{array}\right]. \tag{8.59}$$

Finally, let $K = \mathrm{CSD}_l\left(\tilde{\Pi}, \Phi\right)$, $\forall \Phi \in RH_\infty$ as shown in Fig. 8.14. Then the closed-loop transfer function of $w \mapsto z$ can be found from (8.54) and (8.59) as

$$\begin{aligned}
&\mathrm{LFT}_l\,(P, K) \\
&\quad = G_{12} - G_{11}\left(\tilde{\Theta}_{12} - \Phi\tilde{\Theta}_{22}\right) \\
&\quad = \begin{bmatrix} 0 & M-I \\ 0 & N \end{bmatrix} + \begin{bmatrix} M \\ N \end{bmatrix}\left[\tilde{Y} + \Phi\tilde{M} \;\; \Phi\tilde{N} + \left(\tilde{X} - I\right)\right] \\
&\quad = \begin{bmatrix} 0 & M-I \\ 0 & N \end{bmatrix} + \begin{bmatrix} M\left(\tilde{Y} + \Phi\tilde{M}\right) & M\left(\Phi\tilde{N} + \left(\tilde{X} - I\right)\right) \\ N\left(\tilde{Y} + \Phi\tilde{M}\right) & N\left(\Phi\tilde{N} + \left(\tilde{X} - I\right)\right) \end{bmatrix} \\
&\quad = \begin{bmatrix} M\left(\tilde{Y} + \Phi\tilde{M}\right) & M\Phi\tilde{N} + M\tilde{X} - I \\ N\tilde{Y} + N\Phi\tilde{M} & N\left(\Phi\tilde{N} + \tilde{X}\right) \end{bmatrix} \\
&\quad = \begin{bmatrix} (Y + M\Phi)\,\tilde{M} & (M\Phi + Y)\,\tilde{N} \\ N\left(\tilde{Y} + \Phi\tilde{M}\right) & N\left(\Phi\tilde{N} + \tilde{X}\right) \end{bmatrix}.
\end{aligned} \tag{8.60}$$

It can be concluded that the closed-loop transfer function of $w \mapsto z$ is stable for any $\Phi \in RH_\infty$ and the stabilizing controllers determined by (8.57) are then given by

$$K = \mathrm{CSD_l}\left(\tilde{\Pi}, \Phi\right) = \left(\tilde{X} - \Phi\tilde{N}\right)^{-1}\left(\tilde{Y} + \Phi\tilde{M}\right), \quad \forall\Phi \in RH_\infty. \tag{8.61}$$

8.4.2 Method II: $CSD_l \rightleftarrows CSD_r$ Using a Left CSD Associated with a Right CSD

Dually left coprime factors of $[P_{*1} \mid P_{*2}] = \tilde{M}_*^{-1}[\tilde{G}_1 \mid G_2]$ can be constructed by (8.33) with $H_y = H$ and $\tilde{W}_{yy} = I$. From (8.34), (8.35), and (8.36), one has, respectively,

$$\tilde{G}_1(s) \overset{s}{=} \left[\begin{array}{c|cc} A+HC & \begin{pmatrix} 0 & 0 \end{pmatrix} & \begin{pmatrix} -H & -B \end{pmatrix} \\ \hline \begin{pmatrix} 0 \\ C \end{pmatrix} & \begin{pmatrix} I & 0 \\ 0 & I \end{pmatrix} & \begin{pmatrix} 0 & 0 \\ 0 & 0 \end{pmatrix} \\ C & \begin{pmatrix} 0 & 0 \end{pmatrix} & -\begin{pmatrix} I & 0 \end{pmatrix} \end{array}\right], \tag{8.62}$$

$$G_2(s) \overset{s}{=} \left[\begin{array}{c|cc} A+HC & B & -H \\ \hline \begin{pmatrix} 0 \\ C \end{pmatrix} & \begin{pmatrix} I \\ 0 \end{pmatrix} & \begin{pmatrix} 0 \\ 0 \end{pmatrix} \\ C & 0 & -I \end{array}\right], \tag{8.63}$$

$$\tilde{M}_*(s) \overset{s}{=} \left[\begin{array}{c|cc} A+HC & \begin{pmatrix} 0 & 0 \end{pmatrix} & H \\ \hline \begin{pmatrix} 0 \\ C \end{pmatrix} & \begin{pmatrix} I & 0 \\ 0 & I \end{pmatrix} & \begin{pmatrix} 0 \\ 0 \end{pmatrix} \\ C & \begin{pmatrix} 0 & 0 \end{pmatrix} & I \end{array}\right]. \tag{8.64}$$

Dually, it can also be found in the transfer function level as (Fig. 8.15)

$$\tilde{M}_* = \begin{bmatrix} I & 0 & 0 \\ 0 & I & \tilde{M} - I \\ 0 & 0 & \tilde{M} \end{bmatrix}, \tag{8.65}$$

$$\tilde{G}_1 = \begin{bmatrix} I & \tilde{G}_{12} \\ 0 & \tilde{G}_{22} \end{bmatrix} = \left[\begin{array}{c|cc} I & 0 & 0 \\ \hline 0 & I & \tilde{M} - I \\ 0 & 0 & \tilde{M} \end{array}\right] \left[\begin{array}{cc|cc} I & 0 & 0 & 0 \\ 0 & I & 0 & -G \\ \hline 0 & 0 & -I & -G \end{array}\right] = \left[\begin{array}{cc|cc} I & 0 & 0 & 0 \\ 0 & I & I - \tilde{M} & -\tilde{N} \\ \hline 0 & 0 & -\tilde{M} & -\tilde{N} \end{array}\right], \tag{8.66}$$

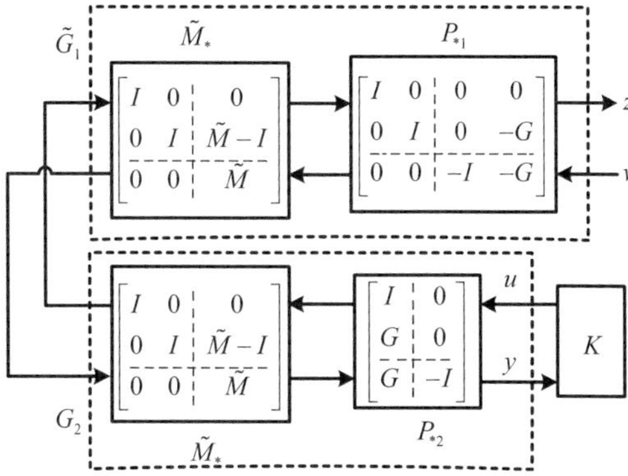

Fig. 8.15 Multiplying $\tilde{M}_*$ at the left terminal

$$G_2 = \left[\begin{array}{c|cc} I & 0 & 0 \\ \hline 0 & I & \tilde{M}-I \\ 0 & 0 & \tilde{M} \end{array}\right]\left[\begin{array}{c|c} I & 0 \\ G & 0 \\ \hline G & -I \end{array}\right] = \left[\begin{array}{c|c} I & 0 \\ \tilde{N} & I-\tilde{M} \\ \hline \tilde{N} & -\tilde{M} \end{array}\right]. \tag{8.67}$$

Furthermore, the right coprime factors Θ and Π, given by (8.38) and (8.39) with $F_u = F,\ W_{uu} = I,\ \tilde{W}_{yy} = I$, can be found as

$$\Pi(s) \stackrel{s}{=} \left[\begin{array}{c|cc} A+BF & B & H \\ \hline F & I & 0 \\ C & 0 & -I \end{array}\right], \tag{8.68}$$

$$\Theta(s) \stackrel{s}{=} \left[\begin{array}{c|cc} A+BF & B & H \\ \hline \begin{pmatrix} F \\ C \end{pmatrix} & \begin{pmatrix} I \\ 0 \end{pmatrix} & \begin{pmatrix} 0 \\ 0 \end{pmatrix} \\ 0 & 0 & I \end{array}\right]. \tag{8.69}$$

Similarly, $\Pi(s)$ above can be written in transfer function level as

$$\Pi(s) = \begin{bmatrix} M(s) & -Y(s) \\ N(s) & -X(s) \end{bmatrix} \stackrel{s}{=} \left[\begin{array}{c|cc} A+BF & B & H \\ \hline F & I & 0 \\ C & 0 & -I \end{array}\right], \tag{8.70}$$

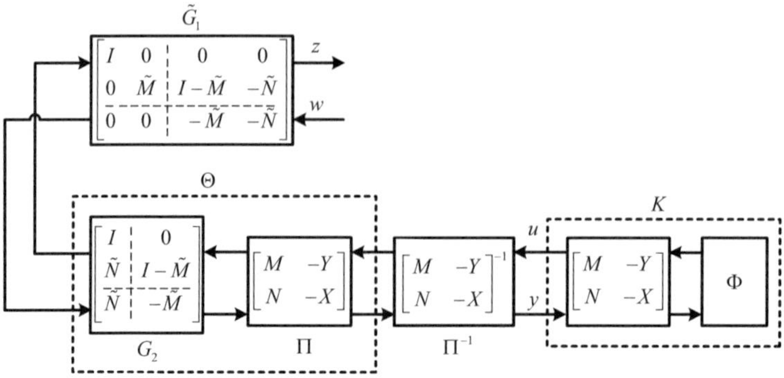

Fig. 8.16 Insertion of $\Pi\Pi^{-1}$

Since $\tilde{M}X - \tilde{N}Y = I$, one concludes that (Fig. 8.16)

$$\Theta = G_2\Pi = \begin{bmatrix} I & 0 \\ \tilde{N} & I-\tilde{M} \\ \hline \tilde{N} & -\tilde{M} \end{bmatrix}\begin{bmatrix} M & -Y \\ \hline N & -X \end{bmatrix} = \begin{bmatrix} M & Y \\ -N & I-X \\ \hline 0 & I \end{bmatrix} = \begin{bmatrix} \Theta_{11} & \Theta_{12} \\ 0 & I \end{bmatrix}. \tag{8.71}$$

Then the overall closed-loop transfer function of $w \mapsto z$ is given by

$$\begin{aligned} z &= \mathrm{LFT_l}\,(P, K)\, w \\ &= \mathrm{CSD_l}\left(\tilde{G}_1, \mathrm{CSD_r}\,(G_2, K)\right) w \\ &= \mathrm{CSD_l}\left(\tilde{G}_1, \mathrm{CSD_r}\,(\Theta, \Phi)\right) w, \end{aligned} \tag{8.72}$$

where $\mathrm{CSD_r}\,(\Theta, \Phi) = \begin{bmatrix} Y + M\Phi \\ I - N\Phi - X \end{bmatrix}$. Therefore, it yields

$$\begin{aligned} &\mathrm{LFT_l}\,(P, K) \\ &\quad = (\Theta_{11}\Phi + \Theta_{12})\,\tilde{G}_{22} - \tilde{G}_{12} \\ &\quad = \left(\begin{bmatrix} M \\ -N \end{bmatrix}\Phi + \begin{bmatrix} Y \\ I - X \end{bmatrix}\right)\begin{bmatrix} -\tilde{M} & -\tilde{N} \end{bmatrix} + \begin{bmatrix} 0 & 0 \\ \tilde{M} - I & \tilde{N} \end{bmatrix} \\ &\quad = \begin{bmatrix} -(M\Phi + Y)\,\tilde{M} & -(M\Phi + Y)\,\tilde{N} \\ N\Phi\tilde{M} + X\tilde{M} - I & (N\Phi + X)\,\tilde{N} \end{bmatrix} \\ &\quad = \begin{bmatrix} (Y + M\Phi)\,\tilde{M} & (M\Phi + Y)\,\tilde{N} \\ N\left(\tilde{Y} + \Phi\tilde{M}\right) & N\left(\Phi\tilde{N} + \tilde{X}\right) \end{bmatrix}. \end{aligned} \tag{8.73}$$

As expected, the result of (8.73) is identical to the one in (8.60). Then the stabilizing controller determined by (8.70) is given by

$$K = \mathrm{CSD_r}\,(\Pi, \Phi) = (-Y + M\Phi)\,(-X + N\Phi)^{-1}, \quad \forall\Phi \in RH_\infty. \tag{8.74}$$

It can be verified from (8.57) and (8.70) that

$$\tilde{\Pi}(s)\,\Pi(s) \stackrel{s}{=} \left[\begin{array}{cc|cc} A+BF & 0 & B & H \\ HC-BF & A+HC & -B & -H \\ \hline F & F & I & 0 \\ -C & -C & 0 & I \end{array}\right] \tag{8.75}$$

whose minimum realization gives the Bezout identity $\tilde{\Pi}\Pi = I$, which is the same as that defined in (8.51). Therefore, the set of all internally stabilizing controllers is given by $K = \mathrm{CSD_l}\left(\tilde{\Pi}, \Phi\right) = \mathrm{CSD_r}\,(\Pi, \Phi)$, for any $\Phi \in RH_\infty$.

8.5 Stabilization of Special SCC Formulations

The procedure to find two successive coprime factorizations is utilized for stabilization controller synthesis and, later, H_2 problems, in the proposed CSD framework. The above solution procedure shows that the feedback control problem is reduced to the solutions of two less complicated problems which contain the determination of a feedback regulator gain matrix, an observer gain matrix, and two accompanying nonsingular matrices. It can be found that both Methods I and II proceed via two coprime factorizations although the operational sequences are different from each other. In fact as far as result is concerned, using Method I is identical to using Method II for a general $P(s)$ of (8.14). However, for specific control problems [3], it is sometime beneficial to choose one of the two methods to make the solution process more efficient. In this section, the six popular synthesis problems, listed in Table 8.1, will be discussed in the CSD stabilization framework. The formulae for Π (or $\tilde{\Pi}$) are shown and sets of all stabilizing controllers to be characterized. Appropriate selection of methods and corresponding results will be summarized at the end of this section.

In the following, the above specific problems are characterized as examples to show which method, I or II, would be more appropriate and how only one coprime factorization is needed in these cases.

Table 8.1 Special SCC formulations

Formulation of system	Corresponding plant	Formulation of system	Corresponding plant
Disturbance feedforward (DF)	$P_{DF}(s) \overset{s}{=} \left[\begin{array}{c\|cc} A & B_1 & B_2 \\ \hline C_1 & D_{11} & D_{12} \\ C_2 & I & 0 \end{array}\right]$ *with* $(A - B_1C_2)$ *being Hurwitz*	Output estimation (OE)	$P_{OE}(s) \overset{s}{=} \left[\begin{array}{c\|cc} A & B_1 & B_2 \\ \hline C_1 & D_{11} & I \\ C_2 & D_{21} & 0 \end{array}\right]$ *with* $(A - B_2C_1)$ *being Hurwitz*
Full information (FI)	$P_{FI}(s) \overset{s}{=} \left[\begin{array}{c\|cc} A & B_1 & B_2 \\ \hline C_1 & D_{11} & D_{12} \\ \begin{bmatrix} I \\ 0 \end{bmatrix} & \begin{bmatrix} 0 \\ I \end{bmatrix} & \begin{bmatrix} 0 \\ 0 \end{bmatrix} \end{array}\right]$	Full control (FC)	$P_{FC}(s) \overset{s}{=} \left[\begin{array}{c\|cc} A & B_1 & [I \ \ 0] \\ \hline C_1 & D_{11} & [0 \ \ I] \\ C_2 & D_{21} & [0 \ \ 0] \end{array}\right]$
State feedback (SF)	$P_{SF}(s) \overset{s}{=} \left[\begin{array}{c\|cc} A & B_1 & B_2 \\ \hline C_1 & D_{11} & D_{12} \\ I & 0 & 0 \end{array}\right]$	Output injection (OI)	$P_{OI}(s) \overset{s}{=} \left[\begin{array}{c\|cc} A & B_1 & I \\ \hline C_1 & D_{11} & 0 \\ C_2 & D_{21} & 0 \end{array}\right]$

8.5.1 Disturbance Feedforward (DF) Case

At first, DF problem is noted, where the state-space realization is given by

$$P_{DF}(s)=\begin{bmatrix} P_{11}(s) & P_{12}(s) \\ P_{21}(s) & P_{22}(s) \end{bmatrix} \stackrel{s}{=} \left[\begin{array}{c|cc} A & B_1 & B_2 \\ \hline C_1 & D_{11} & D_{12} \\ C_2 & I & 0 \end{array}\right], \tag{8.76}$$

where $A-B_1C_2$ is Hurwitz and $D_{21}=I$. In the so-called DF problem, the framework of $\mathrm{CSD_r} \rightleftarrows \mathrm{CSD_l}$ is adopted in which one just needs to carry out a single coprime factorization for obtaining stabilizing controllers. Note that P_{21} is square and its inverse is stable (i.e., a 2-block problem). Then by the procedure of Step 1, one has, from (8.17) with $D_{21}=I$,

$$\begin{bmatrix} P_{1*,DF}(s) \\ \hdashline P_{2*,DF}(s) \end{bmatrix} \stackrel{s}{=} \left[\begin{array}{c|cc} A & B_1 & B_2 \\ \hline C_1 & D_{11} & D_{12} \\ 0 & 0 & I \\ \hdashline 0 & I & 0 \\ C_2 & 0 & I \end{array}\right]. \tag{8.77}$$

According to the procedure of Step 1, one has $P_{*,\mathrm{DF}} = \begin{bmatrix} P_{1*,\mathrm{DF}} \\ P_{2*,\mathrm{DF}} \end{bmatrix} = \begin{bmatrix} G_{1,\mathrm{DF}} \\ \tilde{G}_{2,\mathrm{DF}} \end{bmatrix} M_{\mathrm{DF}}^{-1}$ from (8.20) with $D_{21}=I$ where

$$G_{1,DF}(s)=\begin{bmatrix} G_{11}(s) & G_{12}(s) \\ 0 & I \end{bmatrix} \stackrel{s}{=} \left[\begin{array}{c|cc} A+B_2F_u & B_2W_{uu} & B_1 \\ \hline C_1+D_{12}F_u & D_{12}W_{uu} & D_{11} \\ 0 & 0 & I \end{array}\right] \in RH_\infty \tag{8.78}$$

and

$$\tilde{G}_{2,DF}(s) \stackrel{s}{=} \left[\begin{array}{c|cc} A+B_2F_u & B_2W_{uu} & B_1 \\ \hline F_u & W_{uu} & 0 \\ C_2 & 0 & I \end{array}\right] \in RH_\infty. \tag{8.79}$$

Furthermore, for the DF problem listed in Table 8.1 with the assumption that $A-B_1C_2$ is Hurwitz, the state-space formula of $\tilde{G}_{2,\mathrm{DF}} = \tilde{\Pi}_{\mathrm{DF}}^{-1}\tilde{\Theta}_{\mathrm{DF}}$ is given by (8.25), with replacements of $D_{21}=I$, $H_y=-B_1$, and $\tilde{W}_{yy}=-I$,

$$\left[\tilde{\Theta}_{DF}(s) \;\vdots\; \tilde{\Pi}_{DF}(s)\right] \overset{s}{=} \left[\begin{array}{c|cc:cc} A - B_1C_2 & 0 & 0 & -B_2 & -B_1 \\ \hline W_{uu}^{-1}F_u & I & 0 & W_{uu}^{-1} & 0 \\ C_2 & 0 & I & 0 & I \end{array}\right], \tag{8.80}$$

where $\tilde{\Theta}_{\mathrm{DF}} = \begin{bmatrix} I & 0 \\ 0 & I \end{bmatrix}$ and $\tilde{\Pi}_{\mathrm{DF}} = \tilde{G}_{2,\mathrm{DF}}^{-1}$. Note that $\tilde{\Theta}_{\mathrm{DF}}$ derived for this DF problem is actually an identity matrix. Then the stabilizing controllers can be generated by $K = \mathrm{CSD}_1\left(\tilde{G}_{2,\mathrm{DF}}^{-1}, \Phi\right) = \mathrm{CSD}_r\left(\tilde{G}_{2,\mathrm{DF}}, \Phi\right),\ \forall \Phi \in RH_\infty$. With such G_1 and $\tilde{\Theta}$, the input/output relation between w and z in can be easily determined from (8.7) as

$$z = \mathrm{CSD}_r\left(\begin{bmatrix} G_{11} & G_{12} \\ 0 & I \end{bmatrix}, \mathrm{CSD}_1\left(\begin{bmatrix} I & 0 \\ 0 & I \end{bmatrix}, \Phi\right)\right) = (G_{12} + G_{11}\Phi)\, w. \tag{8.81}$$

8.5.2 Full Information (FI) Case

With an analogous topology of the DF problems, the SCC plant of FI problems is given by

$$P_{FI}(s) = \begin{bmatrix} P_{11}(s) & P_{12}(s) \\ P_{21}(s) & P_{22}(s) \end{bmatrix} \overset{s}{=} \left[\begin{array}{c|cc} A & B_1 & B_2 \\ \hline C_1 & D_{11} & D_{12} \\ \begin{pmatrix} I \\ 0 \end{pmatrix} & \begin{pmatrix} 0 \\ I \end{pmatrix} & \begin{pmatrix} 0 \\ 0 \end{pmatrix} \end{array}\right] \tag{8.82}$$

Similarly, the stacked matrix is derived by the procedure of Step 1 from (8.17) as

$$\begin{bmatrix} P_{1*,FI}(s) \\ \hdashline P_{2*,FI}(s) \end{bmatrix} \overset{s}{=} \left[\begin{array}{c:cc} A & B_2 & B_1 \\ \hdashline C_1 & D_{12} & D_{11} \\ 0 & 0 & I \\ \hdashline 0 & I & 0 \\ \begin{pmatrix} I \\ 0 \end{pmatrix} & \begin{pmatrix} 0 \\ 0 \end{pmatrix} & \begin{pmatrix} 0 \\ I \end{pmatrix} \end{array}\right], \tag{8.83}$$

and the right coprime factorization $P_{*,\mathrm{FI}} = \begin{bmatrix} P_{1*,\mathrm{FI}} \\ P_{2*,\mathrm{FI}} \end{bmatrix} = \begin{bmatrix} G_{1,\mathrm{FI}} \\ \tilde{G}_{2,\mathrm{FI}} \end{bmatrix} M_{\mathrm{FI}}^{-1}$ is given from (8.20) as

$$G_{1,FI}(s) = \begin{bmatrix} G_{11}(s) & G_{12}(s) \\ 0 & I \end{bmatrix} \overset{s}{=} \left[\begin{array}{c|cc} A + B_2F_u & B_2W_{uu} & B_1 \\ \hline C_1 + D_{12}F_u & D_{12}W_{uu} & D_{11} \\ 0 & 0 & I \end{array}\right] \in RH_\infty \tag{8.84}$$

and

$$\tilde{G}_{2,FI}(s) \stackrel{s}{=} \left[\begin{array}{c|cc} A+B_2F_u & B_2W_{uu} & B_1 \\ \hline F_u & W_{uu} & 0 \\ \begin{pmatrix} I \\ 0 \end{pmatrix} & \begin{pmatrix} 0 \\ 0 \end{pmatrix} & \begin{pmatrix} 0 \\ I \end{pmatrix} \end{array}\right] \in RH_\infty . \tag{8.85}$$

Finally, by applying Step 2, the state-space formula of $\tilde{G}_{2,\mathrm{FI}} = \tilde{\Pi}_{\mathrm{FI}}^{-1}\tilde{\Theta}_{\mathrm{FI}}$ in (8.25) can be realized by

$$\left[\begin{array}{c|ccc} I & -B_2 & B_2F_u & -B_1 \\ \hline 0 & W_{uu}^{-1} & -W_{uu}^{-1}F_u & 0 \\ 0 & 0 & I & 0 \\ 0 & 0 & 0 & I \end{array}\right] \times \left[\begin{array}{c|cc|cc} A+B_2F_u & B_2W_{uu} & B_1 & 0 & \begin{pmatrix} 0 & 0 \end{pmatrix} \\ \hline F_u & W_{uu} & 0 & I & \begin{pmatrix} 0 & 0 \end{pmatrix} \\ \begin{pmatrix} I \\ 0 \end{pmatrix} & \begin{pmatrix} 0 \\ 0 \end{pmatrix} & \begin{pmatrix} 0 \\ I \end{pmatrix} & \begin{pmatrix} 0 \\ 0 \end{pmatrix} & \begin{pmatrix} I & 0 \\ 0 & I \end{pmatrix} \end{array}\right]$$

$$\Rightarrow$$

$$\left[\tilde{\Theta}_{FI}(s) \mid \tilde{\Pi}_{FI}(s)\right] \stackrel{s}{=}$$

$$\left[\begin{array}{c|cc|cc} A+B_2F_u & 0 & 0 & -B_2 & \begin{pmatrix} B_2F_u & -B_1 \end{pmatrix} \\ \hline 0 & I & 0 & W_{uu}^{-1} & \begin{pmatrix} -W_{uu}^{-1}F_u & 0 \end{pmatrix} \\ \begin{pmatrix} I \\ 0 \end{pmatrix} & \begin{pmatrix} 0 \\ 0 \end{pmatrix} & \begin{pmatrix} 0 \\ I \end{pmatrix} & \begin{pmatrix} 0 \\ 0 \end{pmatrix} & \begin{pmatrix} I & 0 \\ 0 & I \end{pmatrix} \end{array}\right]. \tag{8.86}$$

This gives that

$$\left[\tilde{\Theta}_{FI}(s) \mid \tilde{\Pi}_{FI}(s)\right] = \left[\begin{array}{cc|cc} I & 0 & W_{uu}^{-1} & \begin{pmatrix} -W_{uu}^{-1}F_u & 0 \end{pmatrix} \\ \begin{pmatrix} 0 \\ 0 \end{pmatrix} & \begin{pmatrix} 0 \\ I \end{pmatrix} & \tilde{\Pi}_{21}(s) & \tilde{\Pi}_{22}(s) \end{array}\right], \tag{8.87}$$

where the central solution is given by a state feedback gain $\mathrm{CSD}_l\left(\tilde{\Pi}_{\mathrm{FI}}(s), 0\right) = \left[\, F_u \;\, 0 \,\right]$, i.e., $u(t) = Kx(t) = F_u x(t)$, and therefore $z = G_{12}w$.

8.5.3 State Feedback (SF) Case

Consider the SF problems as

$$P_{SF}(s) = \begin{bmatrix} P_{11}(s) & P_{12}(s) \\ P_{21}(s) & P_{22}(s) \end{bmatrix} \stackrel{s}{=} \left[\begin{array}{c|cc} A & B_1 & B_2 \\ \hline C_1 & D_{11} & D_{12} \\ I & 0 & 0 \end{array}\right]. \tag{8.88}$$

The state-space formula of stacked matrix and the right coprime factorization can be calculated from (8.17) and (8.20) with $C_2 = I$ and $D_{21} = 0$; the *lcf* of $\tilde{G}_{2,\mathrm{SF}} = \tilde{\Pi}_{\mathrm{SF}}^{-1}\tilde{\Theta}_{\mathrm{SF}}$ is realized by

$$\begin{array}{l}\left[\begin{array}{c|cc} I & -B_2 & B_2F_u \\ \hline 0 & W_{uu}^{-1} & -W_{uu}^{-1}F_u \\ 0 & 0 & I \end{array}\right]\times\left[\begin{array}{c|cc|cc} A+B_2F_u & BW_{uu} & B_1 & 0 & 0 \\ \hline F_u & W_{uu} & 0 & I & 0 \\ I & 0 & 0 & 0 & I \end{array}\right] \\ \Rightarrow \\ \left[\tilde{\Theta}_{SF}(s) \mid \tilde{\Pi}_{SF}(s)\right]\overset{s}{=}\left[\begin{array}{c|cc|cc} A+B_2F_u & 0 & B_1 & -B_2 & B_2F_u \\ \hline 0 & I & 0 & W_{uu}^{-1} & -W_{uu}^{-1}F_u \\ I & 0 & 0 & 0 & I \end{array}\right].\end{array} \tag{8.89}$$

This gives that

$$\left[\tilde{\Theta}_{SF}(s) \mid \tilde{\Pi}_{SF}(s)\right]=\left[\begin{array}{cc|cc} I & 0 & W_{uu}^{-1} & -W_{uu}^{-1}F_u \\ 0 & \tilde{\Theta}_{22}(s) & \tilde{\Pi}_{21}(s) & \tilde{\Pi}_{22}(s) \end{array}\right]. \tag{8.90}$$

With such G_1 and $\tilde{\Theta}$, the input/output relation between w and z can be easily determined from

$$z = \mathrm{CSD_r}\left(\begin{bmatrix} G_{11} & G_{12} \\ 0 & I \end{bmatrix}, \mathrm{CSD_l}\left(\begin{bmatrix} I & 0 \\ 0 & \tilde{\Theta}_{22} \end{bmatrix}, \Phi\right)\right) = \left(G_{12} + G_{11}\Phi\tilde{\Theta}_{22}\right) w, \tag{8.91}$$

where the central solutions is given by a state feedback gain $\mathrm{CSD_l}\left(\tilde{\Pi}_{\mathrm{SF}}(s), 0\right) = F_u$, i.e., $u(t) = Kx(t) = F_u x(t)$, and therefore $z = G_{12}w$.

8.5.4 Output Estimation (OE) Case

On the other hand, similar to the DF problem, for the OE problem in which $D_{12} = I$, the framework of $\mathrm{CSD_l} \rightleftarrows \mathrm{CSD_r}$ is better to be utilized by Method II, for the same reason. Consider the OE problem of

$$P_{OE}(s)=\begin{bmatrix} P_{11}(s) & P_{12}(s) \\ P_{21}(s) & P_{22}(s) \end{bmatrix}\overset{s}{=}\left[\begin{array}{c|cc} A & B_1 & B_2 \\ \hline C_1 & D_{11} & I \\ C_2 & D_{21} & 0 \end{array}\right], \tag{8.92}$$

where $A - B_2C_1$ is Hurwitz and $D_{12} = I$. Note that P_{12} is square and its inverse is stable (i.e., a 2-block problem). Then one has, from (8.31) with $D_{12} = I$,

$$\left[P_{*1,OE}(s) \mid P_{*2,OE}(s)\right] \overset{s}{=} \left[\begin{array}{c|cc:cc} A & 0 & -B_1 & B_2 & 0 \\ \hline C_1 & I & -D_{11} & I & 0 \\ C_2 & 0 & -D_{21} & 0 & -I \end{array}\right]. \tag{8.93}$$

According to the procedure of Step 1 in Method II, one has $\left[P_{*1,OE} \mid P_{*2,OE}\right] = \tilde{M}_{OE}^{-1}\left[\tilde{G}_{1,OE} \mid G_{2,OE}\right]$ from (8.33) with $D_{21} = I$ where

$$\tilde{G}_{1,OE}(s) = \begin{bmatrix} I & \tilde{G}_{12}(s) \\ 0 & \tilde{G}_{22}(s) \end{bmatrix} \overset{s}{=} \left[\begin{array}{c|cc} A + H_yC_2 & 0 & -(B_1 + H_y) \\ \hline C_1 & I & -D_{11} \\ \tilde{W}_{yy}C_2 & 0 & -\tilde{W}_{yy} \end{array}\right] \in RH_\infty \tag{8.94}$$

and

$$G_{2,OE}(s) \overset{s}{=} \left[\begin{array}{c|cc} A + H_yC_2 & B_2 & -H_y \\ \hline C_1 & I & 0 \\ \tilde{W}_{yy}C_2 & 0 & -\tilde{W}_{yy} \end{array}\right] \in RH_\infty. \tag{8.95}$$

Since $A - B_2C_1$ is Hurwitz, $G_{2,\mathrm{OE}} = \Theta_{\mathrm{OE}}\Pi_{OE}^{-1}$ in Step 2 are given by (8.37) with $D_{12} = I$, $F_u = -C_1$, and $W_{uu} = I$ as

$$\begin{bmatrix} \Theta_{OE}(s) \\ \hdashline \Pi_{OE}(s) \end{bmatrix} \overset{s}{=} \left[\begin{array}{c|cc} A - B_2C_1 & B_2 & H_y\tilde{W}_{yy}^{-1} \\ \hline 0 & I & 0 \\ 0 & 0 & I \\ \hdashline -C_1 & I & 0 \\ C_2 & 0 & -\tilde{W}_{yy}^{-1} \end{array}\right], \tag{8.96}$$

such that $\Theta_{\mathrm{OE}} = \begin{bmatrix} I & 0 \\ 0 & I \end{bmatrix}$ and $\Pi_{\mathrm{OE}} = G_{2,OE}^{-1}$. Then all stabilizing controllers can be generated by $K = \mathrm{CSD_r}(G_{2,OE}^{-1}, \Phi) = \mathrm{CSD_l}(G_{2,\mathrm{OE}}, \Phi)$, $\forall\, \Phi \in RH_\infty$. With such $\tilde{G}_1$ and Θ, the input/output relation between w and z can be easily determined as

$$z = \mathrm{CSD_l}\left(\begin{bmatrix} I & \tilde{G}_{12} \\ 0 & \tilde{G}_{22} \end{bmatrix}, \mathrm{CSD_r}\left(\begin{bmatrix} I & 0 \\ 0 & I \end{bmatrix}, \Phi\right)\right) = \left(\Phi\tilde{G}_{22} - \tilde{G}_{12}\right) w. \tag{8.97}$$

8.5.5 Full Control (FC) Case

Similarly, with an analogous topology of the OE problems, the SCC plant of FC problems is given by

$$P_{FC}(s)=\begin{bmatrix} P_{11}(s) & P_{12}(s) \\ P_{21}(s) & P_{22}(s) \end{bmatrix} \overset{s}{=} \left[\begin{array}{c|cc} A & B_1 & (I \;\; 0) \\ \hline C_1 & D_{11} & (0 \;\; I) \\ C_2 & D_{21} & (0 \;\; 0) \end{array}\right], \tag{8.98}$$

and the left coprime factorization $\left[P_{*1,FC} \mid P_{*2,FC}\right]=\tilde{M}_{FC}^{-1}\left[\tilde{G}_{1,FC} \mid G_{2,FC}\right]$ is given from (8.33) as

$$\tilde{G}_{1,FC}(s)=\begin{bmatrix} I & \tilde{G}_{12}(s) \\ 0 & \tilde{G}_{22}(s) \end{bmatrix} \overset{s}{=} \left[\begin{array}{c|cc} A+H_yC_2 & 0 & -(B_1+H_yD_{21}) \\ \hline C_1 & I & -D_{11} \\ \tilde{W}_{yy}C_2 & 0 & -\tilde{W}_{yy}D_{21} \end{array}\right] \in RH_\infty \tag{8.99}$$

and

$$G_{2,FC}(s) \overset{s}{=} \left[\begin{array}{c|cc} A+H_yC_2 & (I \;\; 0) & -H_y \\ \hline C_1 & (0 \;\; I) & 0 \\ \tilde{W}_{yy}C_2 & (0 \;\; 0) & -\tilde{W}_{yy} \end{array}\right] \in RH_\infty . \tag{8.100}$$

Then the state-space formula of $G_{2,\mathrm{OE}}=\Theta_{\mathrm{OE}}\Pi_{OE}^{-1}$ can be realized by

$$\begin{aligned} &\left[\begin{array}{c|cc} A+H_yC_2 & (I \;\; 0) & -H_y \\ \hline C_1 & (0 \;\; I) & 0 \\ \tilde{W}_{yy}C_2 & (0 \;\; 0) & -\tilde{W}_{yy} \\ \hline \begin{pmatrix} 0 \\ 0 \end{pmatrix} & \begin{pmatrix} I & 0 \\ 0 & I \end{pmatrix} & \begin{pmatrix} 0 \\ 0 \end{pmatrix} \\ 0 & (0 \;\; 0) & I \end{array}\right] \times \left[\begin{array}{c|ccc} I & 0 & 0 & 0 \\ \hline H_yC_2 & I & 0 & -H_y\tilde{W}_{yy}^{-1} \\ -C_1 & 0 & I & 0 \\ C_2 & 0 & 0 & -\tilde{W}_{yy}^{-1} \end{array}\right] \\ &\Rightarrow \\ &\begin{bmatrix} \Theta_{FC}(s) \\ \hline \Pi_{FC}(s) \end{bmatrix} \overset{s}{=} \left[\begin{array}{c|ccc} A+H_yC_2 & I & 0 & 0 \\ \hline 0 & 0 & I & 0 \\ 0 & 0 & 0 & I \\ \hline H_yC_2 & I & 0 & -H_y\tilde{W}_{yy}^{-1} \\ -C_1 & 0 & I & 0 \\ C_2 & 0 & 0 & -\tilde{W}_{yy}^{-1} \end{array}\right]. \end{aligned} \tag{8.101}$$

This gives that

$$\begin{bmatrix} \Theta_{FC}(s) \\ \hdashline \Pi_{FC}(s) \end{bmatrix} = \begin{bmatrix} 0 \quad I & 0 \\ 0 \quad 0 & I \\ \hdashline \Pi_{11}(s) & \begin{pmatrix} -H_y \tilde{W}_{yy}^{-1} \\ 0 \end{pmatrix} \\ \Pi_{21}(s) & -\tilde{W}_{yy}^{-1} \end{bmatrix}, \tag{8.102}$$

where the central solution is given by an output injection gain $\mathrm{CSD_r}\,(\Pi_{FC}(s), 0) = \begin{bmatrix} H_y \\ 0 \end{bmatrix}$, and therefore, $z = -\tilde{G}_{12} w$.

8.5.6 *Output Injection (OI) Case*

Finally, consider the OI problems as

$$P_{OI}(s) = \begin{bmatrix} P_{11}(s) & P_{12}(s) \\ P_{21}(s) & P_{22}(s) \end{bmatrix} \overset{s}{=} \left[\begin{array}{c|cc} A & B_1 & I \\ \hline C_1 & D_{11} & 0 \\ C_2 & D_{21} & 0 \end{array}\right]. \tag{8.103}$$

The state-space formula of stacked matrix and the left coprime factorization can be calculated from (8.33) with $B_2 = I$ and $D_{12} = 0$; the *rcf* of $G_{2,\mathrm{OI}} = \Theta_{\mathrm{OI}} \Pi_{OI}^{-1}$ is realized by

$$\begin{aligned} &\left[\begin{array}{c:cc} A + H_y C_2 & I & -H_y \\ \hdashline C_1 & 0 & 0 \\ \tilde{W}_{yy} C_2 & 0 & -\tilde{W}_{yy} \\ \hdashline 0 & I & 0 \\ 0 & 0 & I \end{array}\right] \times \left[\begin{array}{c:cc} I & 0 & 0 \\ \hdashline H_y C_2 & I & -H_y \tilde{W}_{yy}^{-1} \\ C_2 & 0 & -\tilde{W}_{yy}^{-1} \end{array}\right] \\ \Rightarrow \\ &\begin{bmatrix} \Theta_{OI}(s) \\ \hdashline \Pi_{OI}(s) \end{bmatrix} \overset{s}{=} \left[\begin{array}{c|cc} A + H_y C_2 & W_{uu} & 0 \\ \hline C_1 & 0 & 0 \\ 0 & 0 & I \\ \hdashline H_y C_2 & I & -H_y \tilde{W}_{yy}^{-1} \\ C_2 & 0 & -\tilde{W}_{yy}^{-1} \end{array}\right]. \end{aligned} \tag{8.104}$$

This gives that

$$\begin{bmatrix} \Theta_{OI}(s) \\ \hdashline \Pi_{OI}(s) \end{bmatrix} = \begin{bmatrix} \Theta_{11}(s) & 0 \\ 0 & I \\ \hdashline \Pi_{11}(s) & -H_y \tilde{W}_{yy}^{-1} \\ \Pi_{21}(s) & -\tilde{W}_{yy}^{-1} \end{bmatrix}. \tag{8.105}$$

With such $\tilde{G}_1$ and Θ, the input/output relation between w and z can be easily determined from (8.12) as

$$z = \mathrm{CSD_l}\left(\begin{bmatrix} I & \tilde{G}_{12} \\ 0 & \tilde{G}_{22} \end{bmatrix}, \mathrm{CSD_r}\left(\begin{bmatrix} \Theta_{11} & 0 \\ 0 & I \end{bmatrix}, \Phi\right)\right) = \begin{bmatrix} \Theta_{11}\Phi\tilde{G}_{22} - \tilde{G}_{12} \end{bmatrix} w, \tag{8.106}$$

and the central solutions are given by an output injection gain $\mathrm{CSD_r}(\Pi_{\mathrm{OI}}(s), 0) = H_y$ and therefore $z = -\tilde{G}_{12}w$.

In summary, with respect to the FI and FC issues, one adopts $\mathrm{CSD_r} \rightleftarrows \mathrm{CSD_l}$ and $\mathrm{CSD_l} \rightleftarrows \mathrm{CSD_r}$ for each one, respectively, and consequently, $\tilde{\Theta}(s)$ (or $\Theta(s)$) in the second coprime factorization would always be of a diagonal form. Noted that the central solution is given by $\mathrm{CSD_l}\left(\tilde{\Pi}_{\mathrm{FI}}(s), 0\right) = \begin{bmatrix} F_u & 0 \end{bmatrix}$ and $\mathrm{CSD_r}\left(\Pi_{\mathrm{FC}}(s), 0\right) = \begin{bmatrix} H_y \\ 0 \end{bmatrix}$, respectively. For the state feedback (SF) and output injection (OI) problems, the framework of $\mathrm{CSD_r} \rightleftarrows \mathrm{CSD_l}$ and $\mathrm{CSD_l} \rightleftarrows \mathrm{CSD_r}$ are utilized, respectively, and each central solution is given by $\mathrm{CSD_l}\left(\tilde{\Pi}_{\mathrm{SF}}(s), 0\right) = F_u$ and $\mathrm{CSD_r}(\Pi_{\mathrm{OI}}(s), 0) = H_y$. Explicit state-space solutions for those special problems are summarized in Table 8.2.

The structure of Tables 8.1 and 8.2 clearly shows that FC, OE, and OI are duals of FI, DF, and SF; furthermore, FI and FC are equivalent to DF and OE; SF and OI are the simplified cases of FI and FC. These relationships are depicted in Fig. 8.17.

To clarify the meaning of "equivalent" described in Fig. 8.17, one examines the connection between the problems of DF and FI as an example. From (8.78) and (8.84), one concludes that $G_{1,\mathrm{DF}} = G_{1,\mathrm{FI}}$, and furthermore, from (8.79) and (8.85), it can be verified that

$$\begin{gathered} \left[\begin{array}{c:ccc} I & 0 & 0 & 0 \\ \hdashline 0 & I & 0 & 0 \\ 0 & 0 & C_2 & I \end{array}\right] \times \left[\begin{array}{c:cc} A + B_2 F_u & B_2 W_{uu} & B_1 \\ \hdashline F_u & W_{uu} & 0 \\ \begin{pmatrix} I \\ 0 \end{pmatrix} & \begin{pmatrix} 0 \\ 0 \end{pmatrix} & \begin{pmatrix} 0 \\ I \end{pmatrix} \end{array}\right] \\ \Rightarrow \\ \tilde{G}_{2,DF}(s) = \begin{bmatrix} I & 0 & 0 \\ 0 & C_2 & I \end{bmatrix} \tilde{G}_{2,FI}(s) \stackrel{s}{=} \left[\begin{array}{c|cc} A + B_2 F_u & B_2 W_{uu} & B_1 \\ \hline F_u & W_{uu} & 0 \\ C_2 & 0 & I \end{array}\right]. \end{gathered} \tag{8.107}$$

Table 8.2 Solutions of special SCC formulations

System	Utilized framework	The second *lcf* or the second *rcf*
DF	$\mathrm{CSD_r} \rightleftarrows \mathrm{CSD_l}$	$\left[\tilde{\Theta}_{DF}(s) \;\vdots\; \tilde{\Pi}_{DF}(s)\right] \stackrel{s}{=} \left[\begin{array}{ccccc} A-B_1C_2 & 0 & 0 & -B_2 & -B_1 \\ \hline W_{uu}^{-1}F_u & I & 0 & W_{uu}^{-1} & 0 \\ C_2 & 0 & I & 0 & I \end{array}\right]$
FI	$\mathrm{CSD_r} \rightleftarrows \mathrm{CSD_l}$	$\left[\tilde{\Theta}_{FI}(s) \;\vdots\; \tilde{\Pi}_{FI}(s)\right] \stackrel{s}{=} \left[\begin{array}{cccccc} A+B_2F_u & 0 & 0 & -B_2 & B_2F_u & B_1 \\ \hline 0 & I & 0 & W_{uu}^{-1} & -W_{uu}^{-1}F_u & 0 \\ \begin{pmatrix} I \\ 0 \end{pmatrix} & \begin{pmatrix} 0 \\ 0 \end{pmatrix} & \begin{pmatrix} 0 \\ I \end{pmatrix} & \begin{pmatrix} 0 \\ 0 \end{pmatrix} & \multicolumn{2}{c}{\begin{pmatrix} I & 0 \\ 0 & I \end{pmatrix}} \end{array}\right]$
SF	$\mathrm{CSD_r} \rightleftarrows \mathrm{CSD_l}$	$\left[\tilde{\Theta}_{SF}(s) \;\vdots\; \tilde{\Pi}_{SF}(s)\right] \stackrel{s}{=} \left[\begin{array}{ccccc} A+B_2F_u & 0 & B_1 & -B_2 & B_2F_u \\ \hline 0 & I & 0 & W_{uu}^{-1} & -W_{uu}^{-1}F_u \\ I & 0 & 0 & 0 & I \end{array}\right]$
OE	$\mathrm{CSD_l} \rightleftarrows \mathrm{CSD_r}$	$\begin{bmatrix} \Theta_{OE}(s) \\ \hdashline \Pi_{OE}(s) \end{bmatrix} \stackrel{s}{=} \left[\begin{array}{ccc} A-B_2C_1 & B_2 & H_y\tilde{W}_{yy}^{-1} \\ \hline 0 & I & 0 \\ 0 & 0 & I \\ \hdashline -C_1 & I & 0 \\ C_2 & 0 & -\tilde{W}_{yy}^{-1} \end{array}\right]$
FC	$\mathrm{CSD_l} \rightleftarrows \mathrm{CSD_r}$	$\begin{bmatrix} \Theta_{FC}(s) \\ \hdashline \Pi_{FC}(s) \end{bmatrix} \stackrel{s}{=} \left[\begin{array}{cccc} A+H_yC_2 & I & 0 & 0 \\ \hline 0 & 0 & I & 0 \\ 0 & 0 & 0 & I \\ \hdashline H_yC_2 & I & 0 & -H_y\tilde{W}_{yy}^{-1} \\ -C_1 & 0 & I & 0 \\ C_2 & 0 & 0 & -\tilde{W}_{yy}^{-1} \end{array}\right]$
OI	$\mathrm{CSD_l} \rightleftarrows \mathrm{CSD_r}$	$\begin{bmatrix} \Theta_{OI}(s) \\ \hdashline \Pi_{OI}(s) \end{bmatrix} \stackrel{s}{=} \left[\begin{array}{ccc} A+H_yC_2 & W_{uu} & 0 \\ \hline C_1 & 0 & 0 \\ 0 & 0 & I \\ \hdashline H_yC_2 & I & -H_y\tilde{W}_{yy}^{-1} \\ C_2 & 0 & -\tilde{W}_{yy}^{-1} \end{array}\right]$

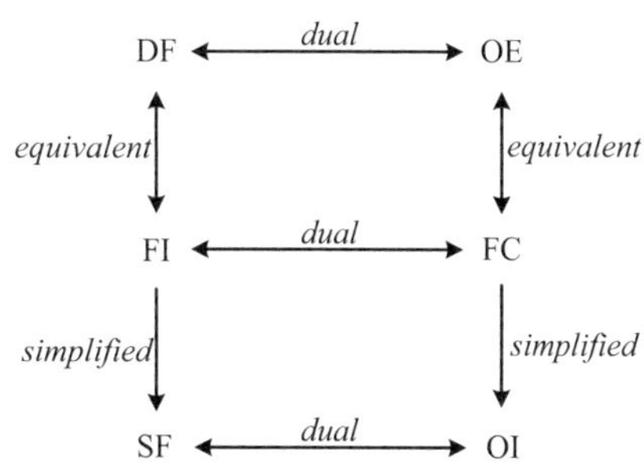

Fig. 8.17 Relationships between each special case

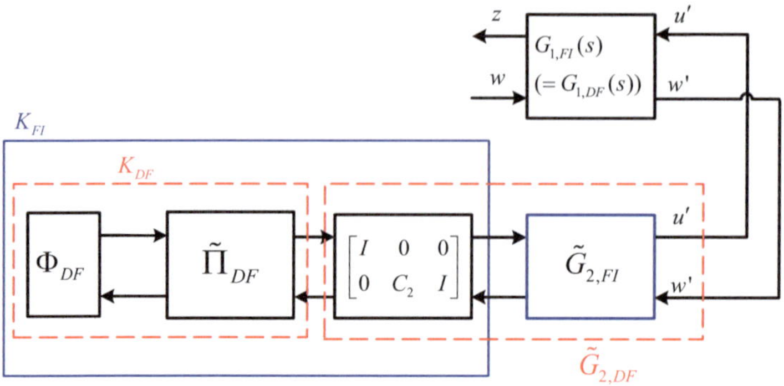

Fig. 8.18 Relationship between FI and DF problem

Figure 8.18 illustrates the topologies of FI and DF problems. Finally, the I/O relationship can be realized by

$$\begin{aligned} z &= \mathrm{LFT}_l\left(P_{\mathrm{FI}}, K_{\mathrm{FI}}\right) w \\ &= \mathrm{CSD}_r\left(G_{1,\mathrm{FI}}, \mathrm{CSD}_l\left(\tilde{G}_{2,\mathrm{FI}}, K_{\mathrm{FI}}\right)\right) w \\ &= \mathrm{CSD}_r\left(G_{1,\mathrm{FI}}, \mathrm{CSD}_l(\tilde{G}_{2,\mathrm{FI}}, \mathrm{CSD}_l\left(\tilde{\Pi}_{\mathrm{DF}}\begin{bmatrix} I & 0 & 0 \\ 0 & C_2 & I \end{bmatrix}, \Phi_{\mathrm{DF}}\right)\right)) w \\ &= \mathrm{CSD}_r\left(G_{1,\mathrm{FI}}, \mathrm{CSD}_l(\tilde{G}_{2,\mathrm{DF}}, \mathrm{CSD}_l\left(\tilde{\Pi}_{\mathrm{DF}}, \Phi_{\mathrm{DF}}\right)\right)) w \\ &= \mathrm{CSD}_r\left(G_{1,\mathrm{FI}}, \mathrm{CSD}_l\left(I, \Phi_{\mathrm{DF}}\right)\right) w \\ &= \mathrm{CSD}_r\left(G_{1,\mathrm{DF}}, \Phi_{\mathrm{DF}}\right) w. \end{aligned} \tag{8.108}$$

Similarly, problem OE is equivalent to FC. From (8.94) and (8.99), one has $\tilde{G}_{1,\mathrm{OE}}(s) = \tilde{G}_{1,\mathrm{FC}}(s)$. From (8.95) and (8.100), one can find that

$$\begin{aligned} &\left[\begin{array}{c:ccc} A+H_yC_2 & I & 0 & -H_y \\ \hdashline C_1 & 0 & I & 0 \\ \tilde{W}_{yy}C_2 & 0 & 0 & -\tilde{W}_{yy} \end{array}\right] \times \left[\begin{array}{c:cc} I & 0 & 0 \\ \hdashline 0 & B_2 & 0 \\ 0 & I & 0 \\ 0 & 0 & I \end{array}\right] \\ &\Rightarrow \\ &G_{2,OE}(s) = G_{2,FC}(s)\begin{bmatrix} B_2 & 0 \\ I & 0 \\ 0 & I \end{bmatrix} \overset{s}{=} \left[\begin{array}{c|cc} A+H_yC_2 & B_2 & -H_y \\ \hline C_1 & I & 0 \\ \tilde{W}_{yy}C_2 & 0 & -\tilde{W}_{yy} \end{array}\right]. \end{aligned} \tag{8.109}$$

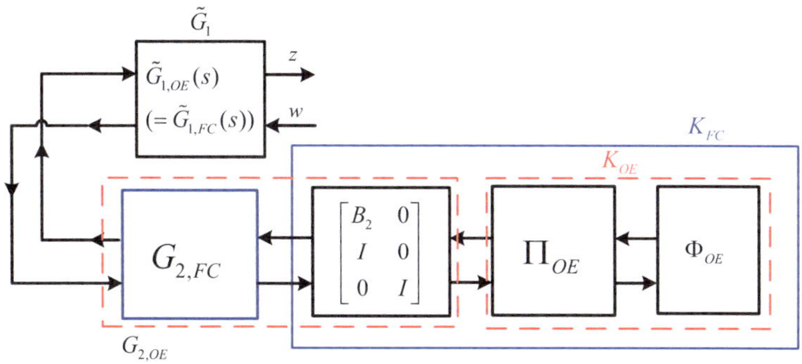

Fig. 8.19 Relationship between FC and OE problem

Figure 8.19 illustrates the topologies of OE and FC problems. Finally, the I/O relationship can be realized by

$$
\begin{aligned}
z &= \mathrm{LFT_l}\left(P_{\mathrm{FC}}, K_{\mathrm{FC}}\right) w \\
&= \mathrm{CSD_l}\left(\tilde{G}_{1,\mathrm{FC}}, \mathrm{CSD_r}\left(G_{2,\mathrm{FC}}, K_{\mathrm{FC}}\right)\right) w \\
&= \mathrm{CSD_l}\left(\tilde{G}_{1,\mathrm{FC}}, \mathrm{CSD_r}\left(G_{2,\mathrm{FC}}, \mathrm{CSD_r}\left(\begin{bmatrix} B_2 & 0 \\ I & 0 \\ 0 & I \end{bmatrix} \Pi_{\mathrm{OE}}, \Phi_{\mathrm{OE}}\right)\right)\right) w \\
&= \mathrm{CSD_l}\left(\tilde{G}_{1,\mathrm{FC}}, \mathrm{CSD_r}\left(G_{2,\mathrm{OE}}, \mathrm{CSD_r}\left(\Pi_{\mathrm{OE}}, \Phi_{\mathrm{OE}}\right)\right)\right) w \\
&= \mathrm{CSD_l}\left(\tilde{G}_{1,\mathrm{FC}}, \mathrm{CSD_r}\left(I, \Phi_{\mathrm{OE}}\right)\right) w \\
&= \mathrm{CSD_l}\left(G_{1,\mathrm{OE}}, \Phi_{\mathrm{OE}}\right) w.
\end{aligned}
\tag{8.110}
$$

8.6 Optimal H_2 Controller

For analytic design of control systems, system performance is often measured by means of norm of the closed-loop system from the exogenous signals to the regulated variables. One of performance indices used in a linear control system is the H_2 norm, which is the root-mean-square of the energy of its (unit) impulse response. The impulse response of a linear time-invariant system G is denoted by $g(t)$. The squared H_2 norm of G may be computed as

$$
\|G\|_2^2 = \int_{-\infty}^{\infty} \mathrm{trace}\left[g^{\mathrm{T}}(t) g(t)\right] \mathrm{d}t. \tag{8.111}
$$

Due to different mathematical tools applied, the H_2 control problem can be solved at various levels of generality. In this section, we extend the proposed CSD approach to the optimal H_2 controller synthesis and derive the solution in the state-space form. The benefit of the state-space solution lies in its computational efficiency. The four parameters H_y, F_u, W_{uu}, and $\tilde{W}_{yy}$ in the controller generator of (8.27) or (8.39), the remaining design freedom from stabilization, are designated here in order to minimize the H_2 norm of $G = \mathrm{LFT_l}(P,K)$. Of course, these parameters have to satisfy that F_u and H_y make $A + B_2F_u$ and $A + H_yC_2$ Hurwitz, respectively, and W_{uu}, $\tilde{W}_{yy}$ are invertible.

Let the SCC plant in (8.14) with $D_{11} = 0$ satisfy the following assumptions:

(I) $\mathrm{rank}\begin{bmatrix} A - j\omega I & B_2 \\ C_1 & D_{12} \end{bmatrix} = m_2 + n$ and $\mathrm{rank}\, D_{12} = m_2$.

(II) $\mathrm{rank}\begin{bmatrix} A - j\omega I & B_1 \\ C_2 & D_{21} \end{bmatrix} = q_2 + n$ and $\mathrm{rank}\, D_{21} = q_2$.

Note that for some special cases as depicted in Table 8.1, (I) and (II) would not always be satisfied. Particular factorizations are therefore required in the second step, as shown in Table 8.2.

The H_2 optimization problem requires finding a stabilizing controller K such that the cost function $\|\mathrm{LFT_l}(P,K)\|_2^2$ is minimized. The solution approaches using CSD framework are presented below.

8.6.1 Method I: Using a Right CSD Associated with a Left One

Considering the H_2 norm of the closed-loop transfer function from w to z in Fig. 8.5, one has, from (8.1) and (8.7),

$$\|\mathrm{LFT_l}\,(P,K)\|_2^2 = \left\| \mathrm{CSD_r}\left(\begin{bmatrix} G_{11} & G_{12} \\ 0 & I \end{bmatrix}, \mathrm{CSD_l}\left(\begin{bmatrix} I & \tilde{\Theta}_{12} \\ 0 & \tilde{\Theta}_{22} \end{bmatrix}, \Phi \right) \right) \right\|_2^2 = \|G_{12} + G_{11}\Psi\|_2^2, \tag{8.112}$$

where

$$\Psi = \mathrm{CSD_l}\left(\begin{bmatrix} I & \tilde{\Theta}_{12} \\ 0 & \tilde{\Theta}_{22} \end{bmatrix}, \Phi \right) = \Phi\tilde{\Theta}_{22} - \tilde{\Theta}_{12}. \tag{8.113}$$

The H_2 solution is summarized in the following lemma. The detailed construction of such an optimal controller is given in the proof of the lemma.

Lemma 8.2 *The H_2 norm of (8.112) is minimized with $\Phi = 0$, if G_{11} is inner and $\tilde{\Theta}_{22}$ co-inner. Furthermore, the H_2 optimal controller is obtained as*

$$K_2(s) = CSD_l(\tilde{\Pi}, 0) \stackrel{s}{=} \left[\begin{array}{c|c} A + H_yC_2 + B_2F_u & H_y \\ \hline -F_u & 0 \end{array} \right], \tag{8.114}$$

where

$$F_u = -\left(D_{12}^{\mathrm{T}} D_{12}\right)^{-1}\left(B_2^{\mathrm{T}} X + D_{12}^{\mathrm{T}} C_1\right), \tag{8.115}$$

$$H_y = -\left(Y C_2^T + B_1 D_{21}^T\right)\left(D_{21} D_{21}^T\right)^{-1}, \tag{8.116}$$

in which X and Y satisfy, respectively,

$$\begin{aligned}&\left(A - B_2\left(D_{12}^{\mathrm{T}} D_{12}\right)^{-1} D_{12}^{\mathrm{T}} C_1\right)^{\mathrm{T}} X + X\left(A - B_2(D_{12}^{\mathrm{T}} D_{12})^{-1} D_{12}^{\mathrm{T}} C_1\right)\\ &\quad - X B_2\left(D_{12}^{\mathrm{T}} D_{12}\right)^{-1} B_2^{\mathrm{T}} X + C_1^{\mathrm{T}}\left(I - D_{12}(D_{12}^{\mathrm{T}} D_{12})^{-1} D_{12}^{\mathrm{T}}\right) C_1 = 0,\end{aligned} \tag{8.117}$$

and

$$\begin{aligned}&\left(A - B_1 D_{21}^{\mathrm{T}}\left(D_{21} D_{21}^{\mathrm{T}}\right)^{-1} C_2\right) Y + Y\left(A - B_1 D_{21}^{\mathrm{T}}(D_{21} D_{21}^{\mathrm{T}})^{-1} C_2\right)^{\mathrm{T}}\\ &\quad - Y C_2^{\mathrm{T}}\left(D_{21} D_{21}^{\mathrm{T}}\right)^{-1} C_2 Y + B_1\left(I - D_{21}^{\mathrm{T}}(D_{21} D_{21}^{\mathrm{T}})^{-1} D_{21}\right) B_1^{\mathrm{T}} = 0.\end{aligned} \tag{8.118}$$

The minimized H_2 norm is given by

$$\begin{aligned}\min_{\text{stabilizing } K} \|\mathrm{LFT}_l(P, K)\|_2^2 &= \|G_{12}\|_2^2 + \left\|\tilde{\Theta}_{12}\right\|_2^2\\ &= \mathrm{trace}\left(B_1^{\mathrm{T}} X B_1\right) + \mathrm{trace}\left(\left(D_{12}^{\mathrm{T}} D_{12}\right)^{\frac{1}{2}} F_u Y F_u^{\mathrm{T}}\left(D_{12}^{\mathrm{T}} D_{12}\right)^{\frac{1}{2}}\right).\end{aligned} \tag{8.119}$$

Proof From (8.21), one has $G_{11}(s) \stackrel{s}{=} \left[\begin{array}{c|c} A + B_2 F_u & B_2 W_{uu} \\ \hline C_1 + D_{12} F_u & D_{12} W_{uu} \end{array}\right]$ and $G_{12}(s) \stackrel{s}{=} \left[\begin{array}{c|c} A + B_2 F_u & B_1 \\ \hline C_1 + D_{12} F_u & 0 \end{array}\right]$. First, the following shows how to make G_{11} inner. It can be derived that

$$G_{11}^{\sim}(s) G_{11}(s) \stackrel{s}{=} \left[\begin{array}{cc|c} A + B_2 F_u & 0 & B_2 W_{uu} \\ -(F_u^T D_{12}^T + C_1^T)(C_1 + D_{12} F_u) & -(A + B_2 F_u)^T & -(F_u^T D_{12}^T + C_1^T) D_{12} W_{uu} \\ \hline W_{uu}^T D_{12}^T (C_1 + D_{12} F_u) & W_{uu}^T B_2^T & W_{uu}^T D_{12}^T D_{12} W_{uu} \end{array}\right] \tag{8.120}$$

Applying the state similarity transformation $\begin{bmatrix} I & 0 \\ -X & 0 \end{bmatrix}$ on the left and $\begin{bmatrix} I & 0 \\ -X & 0 \end{bmatrix}^{-1} = \begin{bmatrix} I & 0 \\ X & 0 \end{bmatrix}$ on the right will yield

$$\left[\begin{array}{cc|c} \begin{array}{c} A+B_2F_u \\ -(C_1+D_{12}F_u)^T(C_1+D_{12}F_u) \\ -(A+B_2F_u)^TX-X(A+B_2F_u) \end{array} & \begin{array}{c} 0 \\ -(A+B_2F_u)^T \end{array} & \begin{array}{c} B_2W_{uu} \\ -(C_1+D_{12}F_u)^TD_{12}W_{uu}-XB_2W_{uu} \end{array} \\ \hline W_{uu}^TD_{12}^T(C_1+D_{12}F_u)+W_{uu}^TB_2^TX & W_{uu}^TB_2^T & W_{uu}^TD_{12}^TD_{12}W_{uu} \end{array}\right]. \tag{8.121}$$

Let $W_{uu} = (D_{12}^TD_{12})^{-0.5}$ and $F_u = -(D_{12}^TD_{12})^{-1}(B_2^TX + D_{12}^TC_1)$, where X satisfies

$$(A + B_2F_u)^{\mathrm{T}}X + X(A + B_2F_u) + \left(F_u^{\mathrm{T}}D_{12}^{\mathrm{T}} + C_1^{\mathrm{T}}\right)(C_1 + D_{12}F_u) = 0.$$

Then the ARE of (8.117) is derived by substituting such defined F_u into the above equation. This gives $G_{11}^{\sim}G_{11} = I$. Note that X is the observability gramian of G_{12} and the realization of $G_{11}^{\sim}G_{12}$ is given by, with the similarity transformation used in the state-space model,

$$\begin{aligned} G_{11}^{\sim}(s)G_{12}(s) &\overset{s}{=} \left[\begin{array}{cc|c} A+B_2F_u & 0 & B_1 \\ -(F_u^TD_{12}^T+C_1^T)(C_1+D_{12}F_u) & -(A+B_2F_u)^T & 0 \\ \hline W_{uu}^TD_{12}^T(C_1+D_{12}F_u) & W_{uu}^TB_2^T & 0 \end{array}\right] \\ &\overset{s}{=} \left[\begin{array}{c|c} -(A+B_2F_u)^T & -XB_1 \\ \hline W_{uu}^TB_2^T & 0 \end{array}\right]. \end{aligned} \tag{8.122}$$

Therefore, one concludes that for an inner G_{11} in (8.112), $G_{11}^{\sim}G_{12}$ is always an anti-stable function in $RH_2{}^{\perp}$. The H_2 norm of (8.112) can then be rewritten by

$$\begin{aligned} \min_{\Psi\in RH_\infty} \|G_{12}+G_{11}\Psi\|_2^2 &= \min_{\Psi\in RH_\infty} \|G_{11}^{\sim}G_{12}+\Psi\|_2^2 = \min_{\Psi\in RH_\infty} \left(\|G_{11}^{\sim}G_{12}\|_2^2+\|\Psi\|_2^2\right) \\ &= \|G_{11}^{\sim}G_{12}\|_2^2 + \min_{\Psi\in RH_\infty} \left(\|\Psi\|_2^2\right). \end{aligned} \tag{8.123}$$

Thus the optimal 2-norm of the overall transfer function is equivalent to minimize the 2-norm of the stable transfer function Ψ.

Similarly, for a co-inner $\tilde{\Theta}_{22}$, one has

$$\begin{aligned} \min_{\Phi\in RH_\infty} \|\Psi\|_2^2 &= \min_{\Phi\in RH_\infty} \left\|\tilde{\Theta}_{12} - \Phi\tilde{\Theta}_{22}\right\|_2^2 = \min_{\Phi\in RH_\infty} \left\|\tilde{\Theta}_{12}\tilde{\Theta}_{22}^{\sim} - \Phi\right\|_2^2 \\ &= \min_{\Phi\in RH_\infty} \left(\left\|\tilde{\Theta}_{12}\tilde{\Theta}_{22}^{\sim}\right\|_2^2 + \|\Phi\|_2^2\right), \end{aligned} \tag{8.124}$$

where $\tilde{W}_{yy} = \left(D_{21}D_{21}^{\mathrm{T}}\right)^{-0.5}$, $H_y = -(YC_2^T + B_1D_{21}^T)(D_{21}D_{21}^T)^{-1}$, and Y satisfies

$$\left(B_1 + H_yD_{21}\right)\left(B_1 + H_yD_{21}\right)^{\mathrm{T}} + \left(A + H_yC_2\right)Y + Y\left(A + H_yC_2\right)^{\mathrm{T}} = 0, \tag{8.125}$$

and the ARE of (8.118) is given by substituting H_y into above equation. Obviously, the optimal solution is given by $\Phi = 0$ and then

$$\begin{aligned}\min_{\text{stabilizing } K} \|\mathrm{LFT_l}(P,K)\|_2^2 &= \|G_{12}\|_2^2 + \left\|\tilde{\Theta}_{12}\right\|_2^2 \\ &= \mathrm{trace}\left(B_1^{\mathrm{T}} X B_1\right) + \mathrm{trace}\left(\left(D_{12}^{\mathrm{T}} D_{12}\right)^{\frac{1}{2}} F_u Y F_u^{\mathrm{T}} \left(D_{12}^{\mathrm{T}} D_{12}\right)^{\frac{1}{2}}\right). \end{aligned} \tag{8.126}$$

■

The dual approach can also be derived as in the following.

8.6.2 Method II: Using a Left CSD Associated with a Right One

Dually for a left CSD associated with a right one, one has, from (8.1) and (8.12),

$$\begin{aligned}&\|\mathrm{LFT_l}(P,K)\|_2^2 \\ &= \left\|\mathrm{CSD_l}\left(\begin{bmatrix} I & \tilde{G}_{12}(s) \\ 0 & \tilde{G}_{22}(s)\end{bmatrix}, \mathrm{CSD_r}\left(\begin{bmatrix} \Theta_{11} & \Theta_{12} \\ 0 & I \end{bmatrix}, \Phi\right)\right)\right\|^2 = \left\|\left(\tilde{G}_{12} - \tilde{\Psi}\tilde{G}_{22}\right)\right\|_2^2,\end{aligned} \tag{8.127}$$

where

$$\tilde{\Psi} = \mathrm{CSD_r}\left(\begin{bmatrix} \Theta_{11} & \Theta_{12} \\ 0 & I \end{bmatrix}, \Phi\right) = \Theta_{11}\Phi + \Theta_{12}. \tag{8.128}$$

The optimal H_2 controller is given by Lemma 8.3; proof of the dual part is skipped. It is interesting but not surprising to find that the two resultant controllers are identical. The optimal controller is actually unique.

Lemma 8.3 *The H_2 norm of* (8.127) *is minimized with a co-inner $\tilde{G}_{22}$ and inner Θ_{11}, where $\Phi = 0$, the minimized H_2 norm, is given by*

$$\begin{aligned}\min_{\text{stabilizing } K} \|\mathrm{LFT_l}(P,K)\|_2^2 &= \left\|\tilde{G}_{12}\right\|_2^2 + \|\Theta_{12}\|_2^2 \\ &= \mathrm{trace}\left(C_1 Y C_1^{\mathrm{T}}\right) + \mathrm{trace}\left(\left(D_{21} D_{21}^{\mathrm{T}}\right)^{\frac{1}{2}} H_y^{\mathrm{T}} X H_y \left(D_{21} D_{21}^{\mathrm{T}}\right)^{\frac{1}{2}}\right). \end{aligned} \tag{8.129}$$

The H_2 optimal controller can be obtained as

$$K_2(s) = CSD_r(\Pi, 0) \stackrel{s}{=} \left[\begin{array}{c|c} A + H_y C_2 + B_2 F_u & H_y \\ \hline -F_u & 0 \end{array}\right], \tag{8.130}$$

where $\Pi(s)$ *is obtained from* (8.39) *and other elements are identical to the ones in* (8.114). ■

It should be reminded that for the H_∞ synthesis, the numerators of the coprime factorizations in (8.6) are devoted to make $G_1(s)$ J-lossless and $\tilde{\Theta}(s)$ dual J-lossless, and the dual work of (8.11) is to have a dual J-lossless matrix $\tilde{G}_2(s)$ and a J-lossless matrix $\Theta(s)$, as will be discussed in Chap. 9.

8.7 Example of the Output Feedback H_2 Optimal Control Problem

Consider a standard control problem shown in Fig. 8.20 where the realization of $G(s) \overset{s}{=} \left[\begin{array}{c|c} A & B \\ \hline C & 0 \end{array}\right]$ is minimum with $\dim(A) = n$ and the weighting matrices U, V, Q, and R are symmetric with $U \geq 0$, $Q \geq 0$, $V > 0$, and $R > 0$. The H_2 control problem in Fig. 8.20, which is essentially the same as the one shown in Fig. 8.11 but inclusive of the weighting function, is to find a stabilizing feedback controller K which minimizes the 2-norm of the closed-loop transfer function matrix from $\begin{bmatrix} w_1 \\ w_2 \end{bmatrix}$ to $\begin{bmatrix} z_1 \\ z_2 \end{bmatrix}$, i.e.,

$$\min_K \| \mathrm{LFT}_l\,(P\ \ K) \|_2$$

where

$$P(s) = \begin{bmatrix} P_{11}(s) & P_{12}(s) \\ P_{21}(s) & P_{22}(s) \end{bmatrix} \overset{s}{=} \left[\begin{array}{c|cc} A & B_1 & B_2 \\ \hline C_1 & 0 & D_{12} \\ C_2 & D_{21} & 0 \end{array}\right] = \left[\begin{array}{c|cc} A & \begin{pmatrix} BU^{\frac{1}{2}} & 0 \end{pmatrix} & B \\ \hline \begin{pmatrix} Q^{\frac{1}{2}}C \\ 0 \end{pmatrix} & \begin{pmatrix} 0 & 0 \\ 0 & 0 \end{pmatrix} & \begin{pmatrix} 0 \\ R^{\frac{1}{2}} \end{pmatrix} \\ C & \begin{pmatrix} 0 & V^{\frac{1}{2}} \end{pmatrix} & 0 \end{array}\right].$$

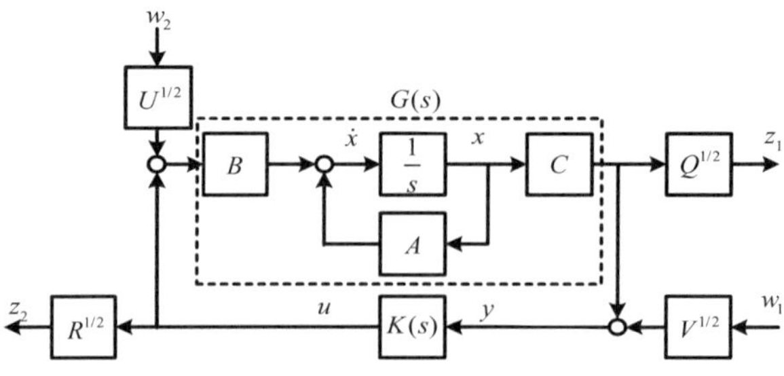

Fig. 8.20 H_2 control problem

It can be found in this example that $D_{12}^T C_1 = 0$ and $B_1 D_{21}^T = 0$. Suppose that $U \geq 0$ and $Q \geq 0$ are chosen such that $P_{12}(s) \stackrel{s}{=} \left[\begin{array}{c|c} A & B \\ \hline Q^{\frac{1}{2}}C & 0 \\ 0 & R^{\frac{1}{2}} \end{array}\right]$ and $P_{21}(s) \stackrel{s}{=} \left[\begin{array}{c|cc} A & BU^{\frac{1}{2}} & 0 \\ \hline C & 0 & V^{\frac{1}{2}} \end{array}\right]$ are of full column rank and of full row rank on the imaginary axis, respectively.

By applying a right CSD associated with a left one in Sect. 8.3, one has, by assumption of Sect. 8.5 that from Lemma 8.2,

$$W_{uu} = \left(D_{12}^{\mathrm{T}} D_{12}\right)^{-\frac{1}{2}} = R^{-\frac{1}{2}}, \quad \tilde{W}_{yy} = \left(D_{21} D_{21}^{\mathrm{T}}\right)^{-\frac{1}{2}} = V^{-\frac{1}{2}}$$

$$F_u = -R^{-1} B^{\mathrm{T}} X, \quad H_y = -Y C^{\mathrm{T}} V^{-1}$$

where

$$X = \mathrm{Ric}\,(H_X) = \mathrm{Ric}\begin{bmatrix} A & -BR^{-1}B^{\mathrm{T}} \\ -C^{\mathrm{T}}QC & -A^{\mathrm{T}} \end{bmatrix}, \tag{8.131}$$

$$Y = \mathrm{Ric}\,(H_Y) = \mathrm{Ric}\begin{bmatrix} A^{\mathrm{T}} & -C^{\mathrm{T}}V^{-1}C \\ -BUB^{\mathrm{T}} & -A \end{bmatrix}, \tag{8.132}$$

such that G_{11} is inner and $\tilde{\Theta}_{22}$ is co-inner. Note that all eigenvalues of $A + BF_u = A - BR^{-1}B^{\mathrm{T}}X$ and $A + H_yC = A - YC^{\mathrm{T}}V^{-1}C$ are in the open left-half plane. Thus one has

$$G_{11}(s) \stackrel{s}{=} \left[\begin{array}{c|c} A - BR^{-1}B^T X & BR^{-\frac{1}{2}} \\ \hline \begin{pmatrix} Q^{\frac{1}{2}}C \\ R^{\frac{1}{2}}F_u \end{pmatrix} & \begin{pmatrix} 0 \\ I \end{pmatrix} \end{array}\right], \; G_{12}(s) \stackrel{s}{=} \left[\begin{array}{c|c} A - BR^{-1}B^T X & BU^{\frac{1}{2}} \quad 0 \\ \hline \begin{pmatrix} Q^{\frac{1}{2}}C \\ R^{\frac{1}{2}}F_u \end{pmatrix} & \begin{pmatrix} 0 & 0 \\ 0 & 0 \end{pmatrix} \end{array}\right], \tag{8.133}$$

and

$$\tilde{\Theta}(s) = \begin{bmatrix} I & \tilde{\Theta}_{12}(s) \\ 0 & \tilde{\Theta}_{22}(s) \end{bmatrix} \stackrel{s}{=} \left[\begin{array}{c|cc} A + H_yC & 0 & \begin{pmatrix} BU^{\frac{1}{2}} & H_yV^{\frac{1}{2}} \end{pmatrix} \\ \hline R^{\frac{1}{2}}F_u & I & \begin{pmatrix} 0 & 0 \end{pmatrix} \\ V^{-\frac{1}{2}}C & 0 & \begin{pmatrix} 0 & I \end{pmatrix} \end{array}\right] \in RH_\infty. \tag{8.134}$$

Furthermore, it can be verified that

$$G_{11}^{\sim}(s)G_{12}(s) \stackrel{s}{=} \left[\begin{array}{c|cc} -(A - BR^{-1}B^T X)^T & -XBR^{-\frac{1}{2}} & 0 \\ \hline (BR^{-\frac{1}{2}})^T & 0 & 0 \end{array}\right] \in RH_2^{\perp}, \tag{8.135}$$

$$\tilde{\Theta}_{12}(s)\tilde{\Theta}_{22}^{\sim}(s) \stackrel{s}{=} \left[\begin{array}{c|c} -\left(A - YC^T V^{-1} C\right)^T & -(V^{-\frac{1}{2}}C)^T \\ \hline R^{\frac{1}{2}} F_u Y & 0 \end{array}\right] \in RH_2^{\perp}. \tag{8.136}$$

Then one has, from (8.27),

$$\tilde{\Pi}(s) = \begin{bmatrix} \tilde{\Pi}_{11}(s) & \tilde{\Pi}_{12}(s) \\ \tilde{\Pi}_{21}(s) & \tilde{\Pi}_{22}(s) \end{bmatrix} \stackrel{s}{=} \left[\begin{array}{c|cc} A + H_y C & -B & H_y \\ \hline R^{\frac{1}{2}} F_u & R^{\frac{1}{2}} & 0 \\ V^{-\frac{1}{2}} C & 0 & V^{-\frac{1}{2}} \end{array}\right] \in GH_\infty \tag{8.137}$$

and from (8.114), the optimal controller by letting $\Phi = 0$ is given by

$$K_{opt}(s) \stackrel{s}{=} \left[\begin{array}{c|c} A + BF_u + H_y C & H_y \\ \hline -F_u & 0 \end{array}\right]. \tag{8.138}$$

It shows that the optimal controller is an observer-based type where F_u and H_y are solved, such that G_{11} is inner and $\tilde{\Theta}_{22}$ is co-inner.

8.7.1 A Numerical Example

Consider the motion of an antenna discussed in the book [12], which can be described by the state differential equation with defining proper state variables:

$$\dot{x}(t) = \begin{bmatrix} 0 & 1 \\ 0 & -\alpha \end{bmatrix} x(t) + \begin{bmatrix} 0 \\ \kappa \end{bmatrix} u(t) + \begin{bmatrix} 0 \\ \gamma \end{bmatrix} \tau_d(t) = Ax(t) + Bu(t) + \begin{bmatrix} 0 \\ \gamma \end{bmatrix} \tau_d(t),$$

where τ_d denotes the disturbing torque. Furthermore, one assumes that the observed variable is given by

$$\eta(t) = \begin{bmatrix} 1 & 0 \end{bmatrix} x(t) + v_m(t) = Cx(t) + v_m(t),$$

in which $v_m(t)$ denotes the white noise with constant scalar intensity V_m. The simplified block diagram of the control system is depicted in Fig. 8.21. Then, an optimal H_2 observer-based control synthesis problem is proposed in Fig. 8.22. In this example, the purpose of the control scheme is to minimize the criterion

$$\int_0^\infty \left[Ru^2(t) + Qy^2(t)\right] dt = \int_0^\infty \left[z_1^2(t) + z_2^2(t)\right] dt.$$

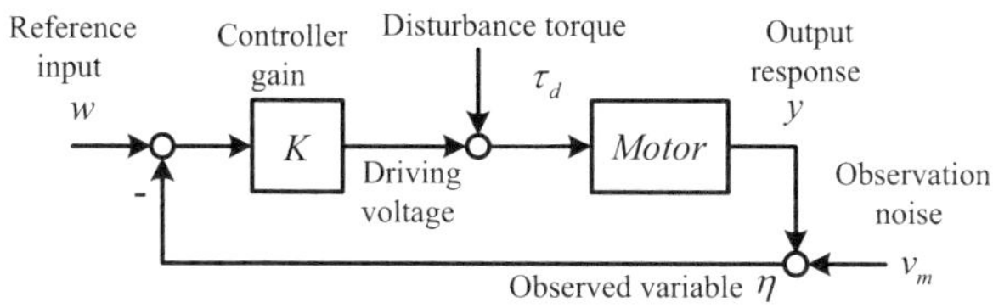

Fig. 8.21 Block diagram of system

Fig. 8.22 Block diagram of the observer-based controller

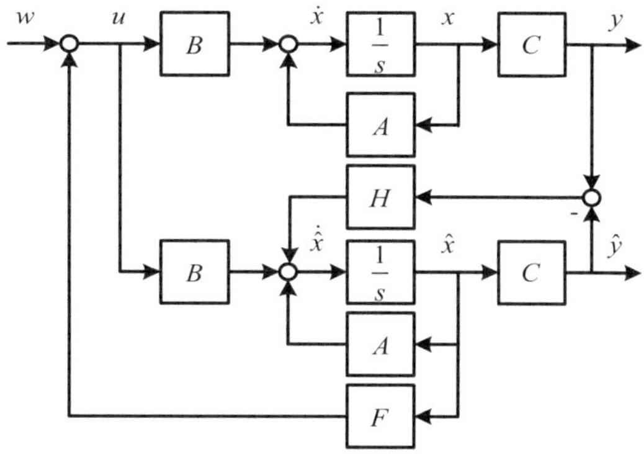

With the specified yardstick, one can rewrite Fig. 8.22 into Fig. 8.23, in which U, Q, R, and V are the weighting functions. The corresponding LFT and CSD representations are illustrated in Fig. 8.1, where

$$P(s) \stackrel{s}{=} \left[\begin{array}{c|cc} A & \begin{pmatrix} BU^{\frac{1}{2}} & 0 \end{pmatrix} & B \\ \hline \begin{pmatrix} Q^{\frac{1}{2}}C \\ 0 \end{pmatrix} & \begin{pmatrix} 0 & 0 \\ 0 & 0 \end{pmatrix} & \begin{pmatrix} 0 \\ R^{\frac{1}{2}} \end{pmatrix} \\ C & \begin{pmatrix} 0 & V^{\frac{1}{2}} \end{pmatrix} & 0 \end{array}\right] = \left[\begin{array}{cc|ccc} 0 & 1 & 0 & 0 & 0 \\ 0 & -4.6 & 0.787U^{\frac{1}{2}} & 0 & 0.787 \\ \hline Q^{\frac{1}{2}} & 0 & 0 & 0 & 0 \\ 0 & 0 & 0 & 0 & R^{\frac{1}{2}} \\ 1 & 0 & 0 & V^{\frac{1}{2}} & 0 \end{array}\right].$$

One now considers the following numerical values [12]: $\kappa = 0.787$ rad/(Vs^2), $\alpha = 4.6\ \text{s}^{-1}$, $\gamma = 0.1\ \text{kg}^{-1}\text{m}^{-2}$, $V_d = 10\ \text{N}^2\text{m}^2\text{s}$, and $V_m = 10^{-7}\ \text{rad}^2\text{s}$. Furthermore, one has $U = 0.4018$, $Q = 1$, $R = 0.00002$, $V = 10^{-7}$; then by computing the CSD form derived before, the optimal controller and observer gain could be obtained by Lemma 8.2 such as

$$W_{uu} = R^{-\frac{1}{2}} = (0.0002)^{-\frac{1}{2}}, \quad \tilde{W}_{yy} = \left(10^{-7}\right)^{-\frac{1}{2}},$$

$$X = \text{Ric}\begin{bmatrix} A & -BR^{-1}B^{\mathrm{T}} \\ -C^{\mathrm{T}}QC & -A^{\mathrm{T}} \end{bmatrix} = \begin{bmatrix} 0.1098 & 0.0059 \\ 0.0059 & 0.0005 \end{bmatrix},$$

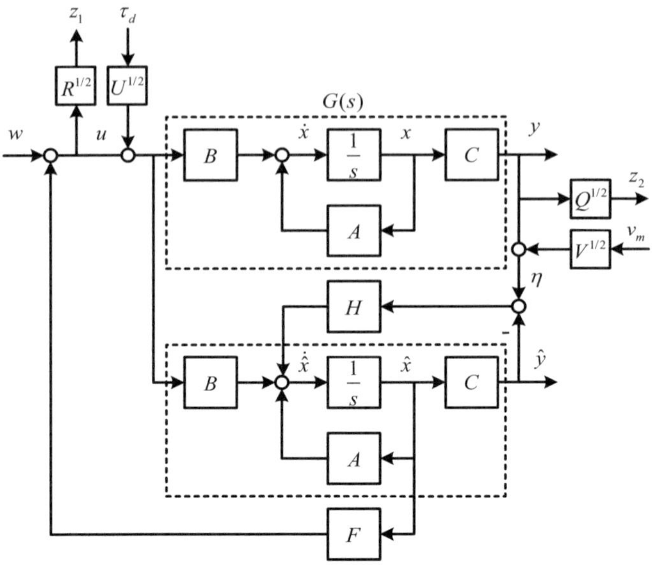

Fig. 8.23 Block diagram with the weighting function

$$Y = \text{Ric}\begin{bmatrix} A^{\mathrm{T}} & -C^{\mathrm{T}}V^{-1}C \\ -BUB^{\mathrm{T}} & -A \end{bmatrix} = \begin{bmatrix} 4.0357\cdot 10^{-6} & 8.1436\cdot 10^{-5} \\ 8.1436\cdot 10^{-5} & 3.6611\cdot 10^{-3} \end{bmatrix},$$

then

$$F = -R^{-1}B^{\mathrm{T}}X = -\begin{bmatrix} 223.6068 & 18.6992 \end{bmatrix}$$

$$H = -YC^{\mathrm{T}}V^{-1} = -\begin{bmatrix} 40.3573 \\ 814.3574 \end{bmatrix}.$$

The optimal controller is given by

$$K_2(s) \stackrel{s}{=} \left[\begin{array}{c|c} A+BF+HC & -H \\ \hline F & 0 \end{array}\right] = \left[\begin{array}{cc|c} -40.3573 & 1 & 40.3573 \\ -990.3359 & -19.3162 & 814.3574 \\ \hline -223.6068 & -223.6068 & 0 \end{array}\right].$$

The optimal 2-norm of the closed-loop system can be obtained by Lemma 8.2 in (8.119) as

$$\min_{\text{stabilizing } K} \|\mathrm{LFT}_l(P,K)\|_2^2 = \mathrm{trace}\left(B_1^{\mathrm{T}}XB_1\right) + \mathrm{trace}\left(F_u Y F_u^{\mathrm{T}}\right) = 9.0779\cdot 10^{-5},$$

or from the dual part, one can also verify the same result by (8.129) such as

$$\min_{\text{stabilizing } K} \|\text{LFT}_l(P, K)\|_2^2$$

$$= \text{trace}\left(C_1 Y C_1^{\mathrm{T}}\right) + \text{trace}\left(\left(D_{21} D_{21}^{\mathrm{T}}\right)^{\frac{1}{2}} H_y^{\mathrm{T}} X H_y \left(D_{21} D_{21}^{\mathrm{T}}\right)^{\frac{1}{2}}\right) = 9.0779 \cdot 10^{-5}.$$

Those two calculation results above reveal again that these two topologies are in fact the same in essence. Note that all numerical solutions are identical to the ones in the reference [12].

8.8 Example of LQR Controller

In Sect. 7.4, one first discussed the LQR problem via the coprime factorization. Consider this problem again as depicted in Fig. 8.24. This problem involves a stabilizing state feedback $u = F_u x + W_{uu} u'$ such that

$$\min_{u=Fx} \left\| Q^{1/2} y \right\|_2^2 + \left\| R^{1/2} u \right\|_2^2, \tag{8.139}$$

where $Q = Q^{\mathrm{T}} \geq 0$ and $R = R^{\mathrm{T}} > 0$ subject to

$$\begin{aligned} \dot{x}(t) &= Ax(t) + Bu(t), \quad x(0) = x_o \\ y(t) &= Cx(t). \end{aligned} \tag{8.140}$$

Now consider the problem in Fig. 8.24 in the form of the H_2 optimal control problem. That is to find a stabilizing feedback gain which minimizes the 2-norm of the closed-loop transfer function from w to $\begin{bmatrix} y_w \\ u_w \end{bmatrix}$, i.e.,

$$\min_K \|\text{LFT}_l(P, K)\|_2,$$

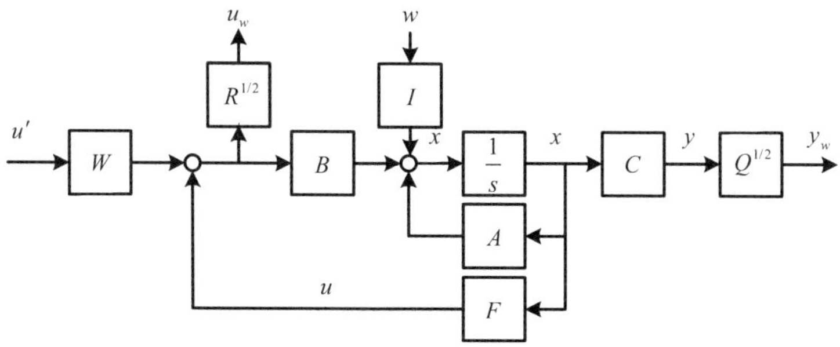

Fig. 8.24 Feedback system with weighting functions

where

$$P(s)=\begin{bmatrix} P_{11}(s) & P_{12}(s) \\ P_{21}(s) & P_{22}(s) \end{bmatrix} \overset{s}{=} \left[\begin{array}{c|cc} A & B_1 & B_2 \\ \hline C_1 & 0 & D_{12} \\ C_2 & 0 & 0 \end{array}\right] = \left[\begin{array}{c|cc} A & I & B \\ \hline \begin{pmatrix} Q^{1/2}C \\ 0 \end{pmatrix} & \begin{pmatrix} 0 \\ 0 \end{pmatrix} & \begin{pmatrix} 0 \\ R^{1/2} \end{pmatrix} \\ I & 0 & 0 \end{array}\right].$$

It can be found here that $D_{11}=0$, $D_{12}^T C_1=0$, $D_{12}^T D_{12}=R>0$, and $B_1 D_{21}^T=0$. Suppose that $G(s)\overset{s}{=}\left[\begin{array}{c|c} A & B \\ \hline C & 0 \end{array}\right]$ is a minimum realization and then $\begin{bmatrix} G(s) \\ R^{\frac{1}{2}} \end{bmatrix} \overset{s}{=} \left[\begin{array}{c|c} A & B \\ \hline C & 0 \\ 0 & R^{\frac{1}{2}} \end{array}\right]$ for $R>0$ is the full column rank in the imaginary axis. Suppose that Q is given such that $P_{12}(s)\overset{s}{=}\left[\begin{array}{c|c} A & B \\ \hline Q^{\frac{1}{2}}C & 0 \\ 0 & R^{\frac{1}{2}} \end{array}\right]$ is of full column rank on the imaginary axis.

For a right CSD associated with a left CSD, one has, from (8.21) and (8.22),

$$\begin{bmatrix} G_{11}(s) & G_{12}(s) \\ 0 & I \end{bmatrix} \overset{s}{=} \left[\begin{array}{c|cc} A+BF_u & BW_{uu} & I \\ \hline \begin{pmatrix} Q^{1/2}C \\ R^{1/2}F_u \end{pmatrix} & \begin{pmatrix} 0 \\ R^{1/2}W_{uu} \end{pmatrix} & \begin{pmatrix} 0 \\ 0 \end{pmatrix} \\ 0 & 0 & I \end{array}\right]$$

and

$$\begin{bmatrix} G_{11}(s) & G_{12}(s) \\ 0 & I \end{bmatrix} \overset{s}{=} \left[\begin{array}{c|cc} A+BF_u & BW_{uu} & I \\ \hline \begin{pmatrix} Q^{1/2}C \\ R^{1/2}F_u \end{pmatrix} & \begin{pmatrix} 0 \\ R^{1/2}W_{uu} \end{pmatrix} & \begin{pmatrix} 0 \\ 0 \end{pmatrix} \\ 0 & 0 & I \end{array}\right] \text{ and} \tag{8.141}$$

$$\tilde{G}_2(s) \overset{s}{=} \left[\begin{array}{c|cc} A+BF_u & BW_{uu} & I \\ \hline F_u & W_{uu} & 0 \\ I & 0 & 0 \end{array}\right],$$

Since $P_{12}(s)$ is of full column rank on the imaginary axis, one has by Lemma 8.2 that $W_{uu}=\left(D_{12}^{\mathrm{T}}D_{12}\right)^{-\frac{1}{2}}=R^{-\frac{1}{2}}$, $F_u=-R^{-1}B^{\mathrm{T}}X$ where

$$X=\mathrm{Ric}\begin{bmatrix} A & -BR^{-1}B^{\mathrm{T}} \\ -C^{\mathrm{T}}QC & -A^{\mathrm{T}} \end{bmatrix}, \tag{8.142}$$

and then $G_{11}(s) \stackrel{s}{=} \left[\begin{array}{c|c} A+BF & BR^{-\frac{1}{2}} \\ \hline \begin{pmatrix} Q^{1/2}C \\ R^{1/2}F_u \end{pmatrix} & \begin{pmatrix} 0 \\ I \end{pmatrix} \end{array}\right]$ is inner.

Since $D_{21} = 0$ and $C_2 = I$ in this example, this is in fact an SF problem as listed in Table 8.1. Then the state-space realization of the *lcf* $\tilde{G}_2 = \tilde{\Pi}^{-1}\tilde{\Theta}$ can be found as

$$\left[\begin{array}{c:cc} I & -B & BF_u \\ \hdashline 0 & R^{\frac{1}{2}} & -R^{\frac{1}{2}}F_u \\ 0 & 0 & I \end{array}\right] \times \left[\begin{array}{c:cc:cc} A+BF_u & BW_{uu} & I & 0 & 0 \\ \hdashline F_u & W_{uu} & 0 & I & 0 \\ I & 0 & 0 & 0 & I \end{array}\right]$$
$$\Rightarrow \tag{8.143}$$
$$\left[\tilde{\Theta}(s) \;\vdots\; \tilde{\Pi}(s)\right] \stackrel{s}{=} \left[\begin{array}{c|cc:cc} A+BF_u & 0 & I & -B & BF_u \\ \hline 0 & I & 0 & R^{\frac{1}{2}} & -R^{\frac{1}{2}}F_u \\ I & 0 & 0 & 0 & I \end{array}\right] \in RH_\infty$$

where

$$\tilde{\Theta}(s) = \begin{bmatrix} I & 0 \\ 0 & \tilde{\Theta}_{22}(s) \end{bmatrix} \stackrel{s}{=} \left[\begin{array}{c|cc} A+BF_u & 0 & I \\ \hline 0 & I & 0 \\ I & 0 & 0 \end{array}\right] \tag{8.144}$$

and

$$\tilde{\Pi}(s) = \begin{bmatrix} R^{\frac{1}{2}} & -R^{\frac{1}{2}}F_u \\ \tilde{\Pi}_{21}(s) & \tilde{\Pi}_{22}(s) \end{bmatrix} \stackrel{s}{=} \left[\begin{array}{c|cc} A+BF_u & -B & BF_u \\ \hline 0 & R^{\frac{1}{2}} & -R^{\frac{1}{2}}F_u \\ I & 0 & I \end{array}\right]. \tag{8.145}$$

The optimal H_2 controller is then determined by $K_2 = \mathrm{CSD}_l\left(\tilde{\Pi}, 0\right) = F_u$ with

$$\min_{\text{stabilizing } K} \|\mathrm{LFT}_l\,(P, K)\,\|_2 = \|G_{12}\|_2 = \mathrm{trace}(X) \tag{8.146}$$

for $B_1 = I$.

8.9 More Numerical Examples

To demonstrate the validity and potential effectiveness of the theoretical results as well as state-space formulae presented above, one considers a model reference control problem shown in Fig. 8.25, which was originally discussed in the book [22].

Fig. 8.25 Model reference control

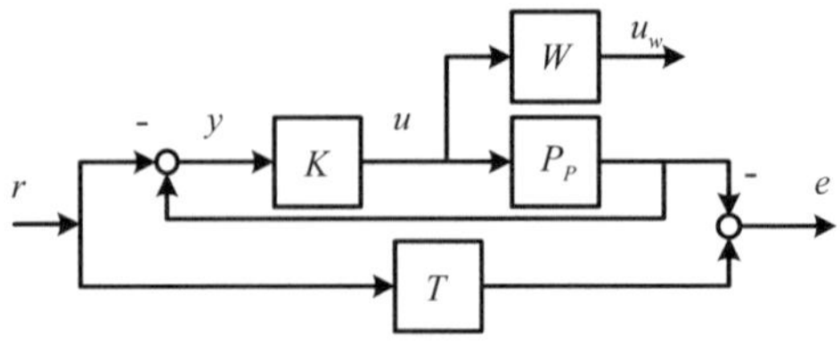

Fig. 8.26 LFT framework for the model reference control

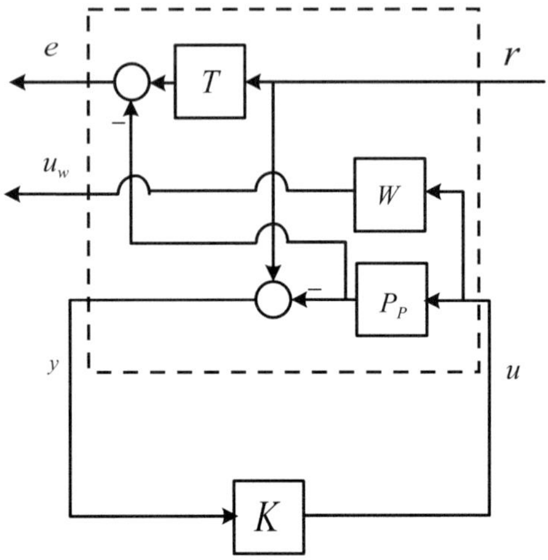

The target of this problem is to minimize the weighted control effort u_w and the model tracking error e through minimizing the H_2 norm of the transfer function matrix from r to $\begin{bmatrix} e \\ u_w \end{bmatrix}$ over all stabilizing controllers, i.e.,

$$\min\left(\int_0^{\infty} \|u_w\|^2 \mathrm{d}t + \int_0^{\infty} \|\mathrm{e}\|^2 \mathrm{d}t\right).$$

It is assumed that $P_P(s) \stackrel{s}{=} \left[\begin{array}{c|c} A_P & B_P \\ \hline C_P & D_P \end{array}\right]$ is a minimal realization, $T(s) \stackrel{s}{=} \left[\begin{array}{c|c} A_T & B_T \\ \hline C_T & D_T \end{array}\right] \in RH_\infty$, and $W(s) \stackrel{s}{=} \left[\begin{array}{c|c} A_W & B_W \\ \hline C_W & D_W \end{array}\right] \in RH_\infty$.

Figure 8.26 describes explicitly the input/output signals of SCC with regard to this particular design problem, of which a state-space block diagram is depicted in Fig. 8.27. By taking out the integrators and controller, the interconnection matrix $P(s)$ in the SCC is actually the "open-loop" system from all "input" signals $\begin{bmatrix} x_T \\ x_p \\ x_W \\ w \\ u \end{bmatrix}$

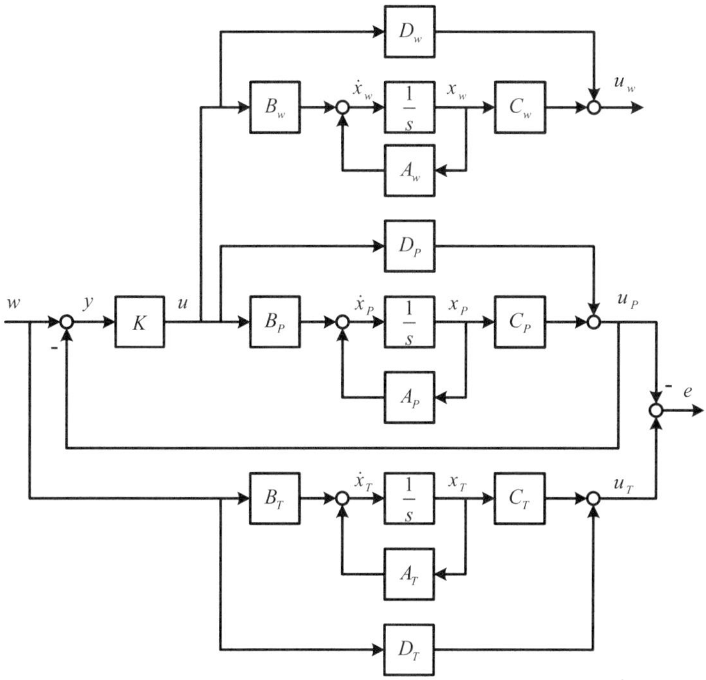

Fig. 8.27 State-space representation

to the "output" $\begin{bmatrix} \dot{x}_T \\ \dot{x}_p \\ \dot{x}_w \\ e \\ u_w \\ y \end{bmatrix}$. The state-space form of P can then be directly obtained from Fig. 8.27 as

$$P(s) = \begin{bmatrix} P_{11}(s) & P_{12}(s) \\ P_{21}(s) & P_{22}(s) \end{bmatrix} \overset{s}{=} \left[\begin{array}{c|cc} A & B_1 & B_2 \\ \hline C_1 & D_{11} & D_{12} \\ C_2 & D_{21} & D_{22} \end{array}\right]$$

$$= \left[\begin{array}{ccc|c:c} A_T & 0 & 0 & B_T & 0 \\ 0 & A_P & 0 & 0 & B_P \\ 0 & 0 & A_w & 0 & B_w \\ \hline C_T & -C_P & 0 & D_T & -D_P \\ 0 & 0 & C_w & 0 & D_w \\ \hdashline 0 & -C_P & 0 & I & -D_P \end{array}\right].$$

As an illustration example, the following textbook-like data are assumed for this design. Let

$$P_P(s)=\frac{10(s+2)}{(s+1)^3}\overset{s}{=}\left[\begin{array}{ccc|c}0&1&0&0\\0&0&1&0\\-1&-3&-3&1\\\hline 20&10&0&0\end{array}\right],\quad T(s)=\frac{4}{s^2+2s+4}\overset{s}{=}\left[\begin{array}{cc|c}0&1&0\\-4&-2&1\\\hline 4&0&0\end{array}\right]$$

$$\text{and } W(s)=\frac{0.1(s+1)}{(s+10)}\overset{s}{=}\left[\begin{array}{c|c}-10&1\\\hline -0.9&0.1\end{array}\right].$$

Then the state-space representation of P is determined as

$$P(s)\overset{s}{=}\left[\begin{array}{c|cc}A&B_1&B_2\\\hline C_1&D_{11}&D_{12}\\C_2&D_{21}&D_{22}\end{array}\right]=\left[\begin{array}{cccccc|c:c}0&1&0&0&0&0&0&0\\-4&-2&0&0&0&0&1&0\\0&0&0&1&0&0&0&0\\0&0&0&0&1&0&0&0\\0&0&-1&-3&-3&0&0&1\\0&0&0&0&0&-10&0&1\\\hline 4&0&-20&-10&0&0&0&0\\0&0&0&0&0&-0.9&0&0.1\\\hdashline 0&0&-20&-10&0&0&1&0\end{array}\right].$$

It can be verified that (A,B_2) is stabilizable and (C_2,A) detectable with $D_{11}=0$, $D_{22}=0$ and the system satisfies the following assumptions:

1. $\text{rank}\begin{bmatrix}A-j\omega I & B_2\\ C_1 & D_{12}\end{bmatrix}=1+1$ and $\text{rank}\, D_{12}=1$.
2. $\text{rank}\begin{bmatrix}A-j\omega I & B_1\\ C_2 & D_{21}\end{bmatrix}=1+1$ and $\text{rank}\, D_{21}=1$.

To determine the H_2 optimal solution, the four parameters H_y, F_u, W_{uu}, and $\tilde{W}_{yy}$ in (8.27) with (8.39) can be selected as, according to Lemma 8.2,

$$W_{uu}=\left(D_{12}^{\mathrm{T}}D_{12}\right)^{-\frac{1}{2}}=10,\quad \tilde{W}_{yy}=\left(D_{21}D_{21}^{\mathrm{T}}\right)^{-\frac{1}{2}}=1,$$

$$F_u=\left[38.6975\ 4.5182\ -199.1704\ -123.494\ -12.1515\ 1.7038\right],$$

$$H_y=\left[0\ -1\ 0\ 0\ 0\ 0\right]^{\mathrm{T}}.$$

Therefore, the optimal controller was found to yield a closed-loop H_2 performance with the SCC plant P. All stabilizing controllers can be characterized by (8.27) as $K=\text{CSD}_l\left(\tilde{\Pi},\Phi\right)$, $\forall\Phi\in RH_\infty$ or by (8.39) as $K=\text{CSD}_r(\Pi,\Phi)$, $\forall\,\Phi\in RH_\infty$

where

$$\tilde{\Pi}(s) \overset{s}{=} \left[\begin{array}{cccccc|cc} 0 & 1 & 0 & 0 & 0 & 0 & 0 & 0 \\ -4 & -2 & 20 & 10 & 0 & 0 & 0 & -1 \\ 0 & 0 & 0 & 1 & 0 & 0 & 0 & 0 \\ 0 & 0 & 0 & 0 & 1 & 0 & 0 & 0 \\ 0 & 0 & -1 & -3 & -3 & 0 & -1 & 0 \\ 0 & 0 & 0 & 0 & 0 & -10 & -1 & 0 \\ \hline 3.87 & 0.45 & -19.92 & -12.35 & -1.215 & 0.17 & 0.1 & 0 \\ 0 & 0 & 20 & 10 & 0 & 0 & 0 & -1 \end{array}\right],$$

$$\Pi(s) \overset{s}{=} \left[\begin{array}{cccccc|cc} 0 & 1 & 0 & 0 & 0 & 0 & 0 & 0 \\ -4 & -2 & 0 & 0 & 0 & 0 & 0 & -1 \\ 0 & 0 & 0 & 1 & 0 & 0 & 0 & 0 \\ 0 & 0 & 0 & 0 & 1 & 0 & 0 & 0 \\ 38.7 & 4.52 & -200.17 & -126.49 & -15.15 & 1.7 & 10 & 0 \\ 38.7 & 4.52 & -199.17 & -123.49 & -12.15 & -8.3 & 10 & 0 \\ \hline 38.7 & 4.52 & -199.17 & -123.49 & -12.15 & 1.7 & 10 & 0 \\ 0 & 0 & -20 & -10 & 0 & 0 & 0 & -1 \end{array}\right].$$

Note that it can be verified that $\tilde{\Pi}\Pi = I$, which in turn shows $\mathrm{CSD_r}(\Pi, \Phi) = \mathrm{CSD_l}(\tilde{\Pi}, \Phi)$.

Consequently, the optimal H_2 controller $K(s)$ is given from (8.114) or (8.130) as

$$K_{opt}(s) = CSD_r(\Pi, 0) = CSD_l(\tilde{\Pi}, 0) \overset{s}{=} \left[\begin{array}{c|c} A + H_y C_2 + B_2 F_u & H_y \\ \hline -F_u & 0 \end{array}\right].$$

This gives

$$K_{\mathrm{opt}}(s) = \frac{4.5s^5 + 97.4s^4 + 652.2s^3 + 1417.1s^2 + 1244.8s + 387}{s^6 + 25.4s^5 + 323.8s^4 + 2054.4s^3 + 5082.4s^2 + 4292.5s + 260.5}.$$

The optimal 2-norm of the closed-loop system can be obtained from (8.119) as

$$\min_{\text{stabilizing } K} \|\mathrm{LFT_l}(P, K)\|_2^2 = \mathrm{trace}\left(B_1^{\mathrm{T}} X B_1\right) + \mathrm{trace}\left(\left(D_{12}^{\mathrm{T}} D_{12}\right)^{\frac{1}{2}} F_u Y F_u^{\mathrm{T}} \left(D_{12}^{\mathrm{T}} D_{12}\right)^{\frac{1}{2}}\right) = 0.013,$$

or from the dual part, one can also verify the same result by (8.129) such as

$$\min_{\text{stabilizing } K} \|\mathrm{LFT}_l(P, K)\|_2^2$$

$$= \mathrm{trace}\left(C_1 Y C_1^{\mathrm{T}}\right) + \mathrm{trace}\left(\left(D_{21} D_{21}^{\mathrm{T}}\right)^{\frac{1}{2}} H_y^{\mathrm{T}} X H_y \left(D_{21} D_{21}^{\mathrm{T}}\right)^{\frac{1}{2}}\right) = 0.013.$$

The two calculation results above reconfirm that these two approaches are in fact the same in essence. All the calculations and results can be verified by using the function *h2syn* in MATLAB® Robust Control Toolbox.

8.10 Summary

This chapter has proposed a unified approach to describe and synthesize the stabilizing controllers and the H_2 optimal controller by finding two coupled CSD matrices. Note that the selection of weighting function is not the major issue concerned in this book. The obtained results reveal an interesting feature in that the original output feedback problem can be simplified to the solutions of two less complicated subproblems. It is found that using the proposed approach admits separate computations of estimator and regulator gains. In fact, this result is similar to the "separation principle" in linear control systems theory. The feedback control gain and the observer gain are found, respectively, in the subproblems to satisfy a specific cost function. Notice that on the basis of the proposed CSD method, the solution of specific control problems can be easily solved in an explicit form. The explicit formulae obtained from the coupled CSD method are beneficial for analyzing the closed-loop characteristics in various control problems.

References

1. Boulet B, Francis BA (1998) Consistency of open-loop experimental frequency-response date with coprime factor plant models. IEEE Trans Autom Control 43:1680–1691
2. Chen J (1997) Frequency-domain tests for validation of linear fractional uncertain models. IEEE Trans Autom Control 42:748–760
3. Doyle JC, Glover K, Khargonekar PP, Francis BA (1989) State-space solutions to standard H_2 and H_∞ control problems. IEEE Trans Autom Control 34:831–847
4. Evans WR (1948) Graphical analysis of control systems. Trans AIEE 67:547–551
5. Francis BA, Zames G (1987) On H_∞-optimal sensitivity theory for SISO feedback systems. IEEE Trans Autom Control 29:9–16
6. Glover K, Doyle JC (1988) State-space formulae for all stabilizing controllers that satisfy an H_∞-norm bounded and relations to risk sensitivity. Syst Control Lett 11:167–172
7. Green M, Limebeer DJN (1995) Linear robust control. Prentice Hall, Englewood Cliffs
8. Kimura H (1987) Directional interpolation approach to H_∞ optimization and robust stabilization. IEEE Trans Autom Control 32:1085–1093

9. Kimura H (1995) Chain-scattering representation, J-lossless factorization and H_∞ control. J Math Syst Estim Control 5:203–255
10. Kimura H (1997) Chain-scattering approach to H_∞ control. Birkhäuser, Boston
11. Kimura H, Okunishi F (1995) Chain-scattering approach to control system design. In: Isidori A (ed) Trends in control: an European perspective. Springer, Berlin
12. Kwakernaak H, Sivan R (1972) Linear optimal control systems. Wiley, New York
13. MacFarlane AGJ, Postlethwaite I (1977) Characteristic frequency functions and characteristic gain functions. Int J Control 26:265–278
14. Nyquist H (1932) Regeneration theory. Bell Syst Tech J 11:126–147
15. Rosenbrock HH (1974) Computer aided control system design. Academic, New York
16. Safonov MG, Athans M (1977) Gain and phase margin for multiloop LQG regulators. IEEE Trans Autom Control 22:173–178
17. Smith RS, Doyle JC (1992) Model validation: a connection between robust control and identification. IEEE Trans Autom Control 37:942–952
18. Tsai MC, Tsai CS, Sun YY (1993) On discrete-time H_∞ control: a J-lossless coprime factorization approach. IEEE Trans Autom Control 38:1143–1147
19. Zames G (1981) Feedback and optimal sensitivity: model reference transformation, multiplicative seminorms, and approximated inverse. IEEE Trans Autom Control 26:301–320
20. Zames G, Francis BA (1983) Feedback, minimax sensitivity, and optimal robustness. IEEE Trans Autom Control 28:585–601
21. Zhou K, Doyle JC, Glover K (1995) Robust and optimal control. Prentice Hall, Upper Saddle River
22. Zhou K, Doyle JC (1998) Essentials of robust control. Prentice Hall, Upper Saddle River

Chapter 9
A CSD Approach to H-Infinity Controller Synthesis

H_∞ optimal control, which minimizes the H_∞-norm of a closed-loop system, has been developed in the last 30 years and been applied in various domains. The original H_∞ optimal control problem involves an equivalent model matching problem, which can be transformed into a four-block distance problem. By applying spectral factorizations, the four-block distance problem can be reduced to a Nehari problem, and Hankel norm approximation can be considered [2–4, 6]. Operator theory approach is very mathematics involved, and numerical solution procedures are difficult to be developed for general form problems. Notable progress was made in finding suboptimal solutions of a general control synthesis problem by solving two algebraic Riccati equations (AREs) [2, 3, 10, 13]. However, even with such solution procedures, questions of "why?" and "how?" often arise from students and engineers who want to understand and use them. An alternative development based on the framework of J-lossless coprime factorizations was proposed by Green in which the solutions can be characterized in terms of transfer function matrices [5]. A similar framework based on a single chain scattering description (CSD) was initially proposed by Kimura [10]. As described in Chap. 8, the general four-block problem can be solved by augmenting with some fictitious signals. Furthermore, since the transformation from LFT to CSD does not guarantee stability of the resulting CSD matrix, the J-lossless factorization with an outer matrix cannot be found directly by the coprime factorization-based method. In this book, the proposed H_∞ CSD solution framework involves constructing two coupled (right and left) CSD matrices by solving two J-lossless coprime factorizations and is fairly straightforward. The method is generally valid and does not need to introduce any fictitious signals for matrix augmentation. Based on Green's approach of J-lossless coprime factorizations, the proposed CSDs framework is significantly different from Kimura's approach.

Many applications using the CSD and J-lossless factorization approaches can be found in various control problems, such as the influence of the weighting function adjustment [1], the state-delayed problem [8], the nonlinear systems

M.-C. Tsai and D.-W. Gu, *Robust and Optimal Control: A Two-port Framework Approach*, Advances in Industrial Control, DOI 10.1007/978-1-4471-6257-5_9,
© Springer-Verlag London 2014

control problem [9], and simultaneous stabilization problem [11]. This chapter aims at developing a CSD framework to solve H_∞ suboptimal control synthesis problem. As in Chap. 8, the standard control configuration (SCC) description is first formulated into a coupled chain scattering-matrix description where graphic representations are utilized to interpret the matrix description. Specific right and left CSDs are to be constructed, in the state-space form and at the transfer function level, to characterize the H_∞ solutions. This approach provides a comprehensive understanding for control engineers. Illustration examples are given to show the solution procedures where the general H_∞ solutions are derived from finding two coupled CSDs (one right and one left) as well as two successive coprime factorizations with J-lossless numerators.

9.1 H_∞ Control Problem

Consider the general feedback control framework discussed in Chap. 8. For a given SCC plant $P = \begin{bmatrix} P_{11} & P_{12} \\ P_{21} & P_{22} \end{bmatrix}$ and a prespecified $\gamma > 0$, the general H_∞ suboptimal control problem is to find a stabilizing controller, denoted by K_∞, such that the closed-loop transfer function from w to z satisfies

$$\|\mathrm{LFT_l}(P, K_\infty)\|_\infty < \gamma \tag{9.1}$$

or, equivalently,

$$\left\|\mathrm{LFT_l}\left(P_\gamma, K_\infty\right)\right\|_\infty < 1, \tag{9.2}$$

where

$$P_\gamma = \begin{bmatrix} \frac{1}{\gamma}P_{11} & \frac{1}{\gamma}P_{12} \\ P_{21} & P_{22} \end{bmatrix} \quad \text{or} \quad P_\gamma = \begin{bmatrix} \frac{1}{\gamma}P_{11} & P_{12} \\ \frac{1}{\gamma}P_{21} & P_{22} \end{bmatrix}. \tag{9.3}$$

Similar to the deduction from Figs. 8.1 to 8.4, one has, as depicted in Fig. 9.1,

$$z = \mathrm{LFT_l}\left(P_\gamma, K_\infty\right) w = \mathrm{CSD_r}\left(\begin{bmatrix} G_{11} & G_{12} \\ G_{21} & G_{22} \end{bmatrix}, \mathrm{CSD_l}\left(\begin{bmatrix} \tilde{\Theta}_{11} & \tilde{\Theta}_{12} \\ \tilde{\Theta}_{21} & \tilde{\Theta}_{22} \end{bmatrix}, \Phi\right)\right) w. \tag{9.4}$$

As mentioned earlier in Chap. 8, it is not so straightforward to determine the overall stability of the feedback system here from Fig. 9.1. However, for the particular case when G_1 is J-lossless and $\tilde{\Theta}$ dual J-lossless, one can ensure the stability of the interconnection system for any $\Phi \in BH_\infty$ by the small gain theory.

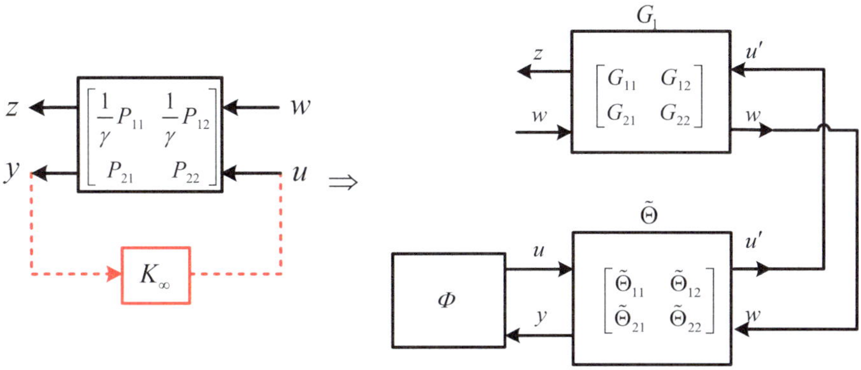

Fig. 9.1 Equivalent framework

Dually, for a system formulated in terms of a left CSD matrix $\tilde{G}_1 = \begin{bmatrix} \tilde{G}_{11} & \tilde{G}_{12} \\ \tilde{G}_{21} & \tilde{G}_{22} \end{bmatrix} \in RH_\infty$ associated with a right CSD matrix $\Theta = \begin{bmatrix} \Theta_{11} & \Theta_{12} \\ \Theta_{21} & \Theta_{22} \end{bmatrix} \in RH_\infty$ as illustrated in Fig. 8.7, the closed-loop transfer function from w to z is given by

$$z = \text{LFT}_l\left(P_\gamma, K_\infty\right) w = \text{CSD}_l\left(\begin{bmatrix} \tilde{G}_{11} & \tilde{G}_{12} \\ \tilde{G}_{21} & \tilde{G}_{22} \end{bmatrix}, \text{CSD}_l\left(\begin{bmatrix} \Theta_{11} & \Theta_{12} \\ \Theta_{21} & \Theta_{22} \end{bmatrix}, \Phi\right)\right) w. \tag{9.5}$$

Apparently, one can also ensure the stability of the interconnection system for $\tilde{G}_1$ dual J-lossless, ΘJ-lossless, and $\Phi \in BH_\infty$ by the small gain theory. The rest part of this chapter is devoted to show how to construct the required right and left CSD matrices and then the H_∞ solutions of (9.2) from a given SCC plant P_γ. The approach is similar to that of the H_2 optimal control case as shown in the previous chapter. It will show that the H_∞ control problem of (9.2) is reduced to two solutions which are linked to J-lossless coprime factorizations. State-space formulae of the solution procedure are provided in the next section which contains the determination of a feedback gain F, an observer gain H, and two accompanying nonsingular matrices (W and $\tilde{W}$).

9.1.1 Method I: $\text{CSD}_r \rightleftarrows \text{CSD}_l$ Right CSD Coupled with Left CSD

In summary, a set of the H_∞ controllers satisfying (9.2) can be constructed via three solution steps as illustrated in Fig. 9.2. Obviously, this CSD method relies on solving two J-lossless coprime factorizations. The following summarizes the key components of the whole framework.

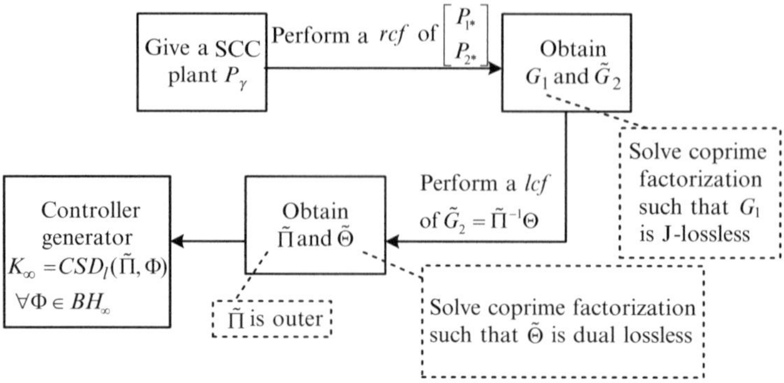

Fig. 9.2 Flowchart of Method I

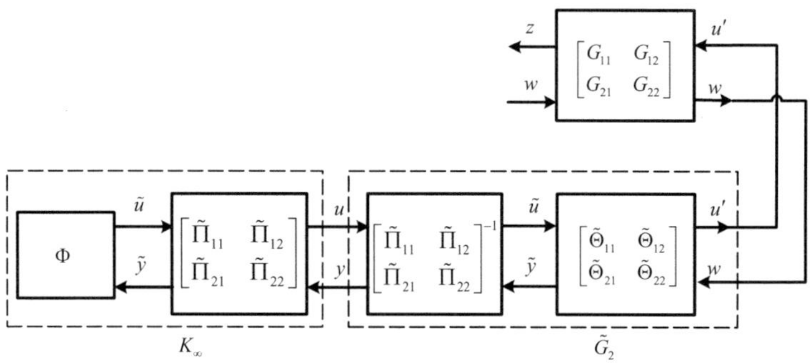

Fig. 9.3 $K_\infty = \mathrm{CSD_l}\left(\tilde{\Pi}, \Phi\right)$

Step 1: Find a right coprime factorization of $\begin{bmatrix} P_{1*} \\ P_{2*} \end{bmatrix} = \begin{bmatrix} G_1 \\ \tilde{G}_2 \end{bmatrix} M_\infty^{-1}$ over RH_∞ such that G_1 is J-lossless where $\begin{bmatrix} P_{1*} \\ \hdashline P_{2*} \end{bmatrix} = \begin{bmatrix} \frac{1}{\gamma}P_{12} & \frac{1}{\gamma}P_{11} \\ 0 & I \\ \hdashline I & 0 \\ P_{22} & P_{21} \end{bmatrix}$.

Step 2: Find a left coprime factorization of $\tilde{G}_2 = \tilde{\Pi}^{-1}\tilde{\Theta}$ over RH_∞ such that $\tilde{\Pi} \in GH_\infty$ and $\tilde{\Theta}$ is dual J-lossless.

Step 3: The H_∞ controller satisfying (9.2) is generated by (Fig. 9.3)

$$K_\infty = \mathrm{CSD_l}\left(\tilde{\Pi}, \Phi\right) = -\left(\tilde{\Pi}_{11} - \Phi\tilde{\Pi}_{21}\right)^{-1}\left(\tilde{\Pi}_{12} - \Phi\tilde{\Pi}_{22}\right), \quad \text{where} \quad \forall \Phi \in BH_\infty. \tag{9.6}$$

Finally, the input/output relationship could be realized by

$$\begin{aligned} z &= \mathrm{LFT}_l\left(P_\gamma, K_\infty\right) w \\ &= \mathrm{CSD_r}\left(P_{1*}, \mathrm{CSD_l}\left(P_{2*}, K_\infty\right)\right) w \\ &= \mathrm{CSD_r}\left(G_1, \mathrm{CSD_l}\left(\tilde{G}_2, K_\infty\right)\right) w \\ &= \mathrm{CSD_r}\left(G_1, \mathrm{CSD_l}\left(\mathrm{CSD_l}\left(\tilde{\Theta}, \tilde{\Pi}^{-1}\right), K_\infty\right)\right) w \\ &= \mathrm{CSD_r}\left(G_1, \mathrm{CSD_l}\left(\mathrm{CSD_l}\left(\tilde{\Theta}, \tilde{\Pi}^{-1}\right), \mathrm{CSD_l}\left(\tilde{\Pi}, \Phi\right)\right)\right) w \\ &= \mathrm{CSD_r}\left(G_1, \mathrm{CSD_l}\left(\tilde{\Theta}, \Phi\right)\right) w. \end{aligned} \tag{9.7}$$

Theorem 9.1 *For a given SCC plant P_γ, there exists an internally stabilizing controller K_∞ such that $\|LFT_l(P_\gamma,K_\infty)\|_\infty < 1$ if there exists an rcf of $\begin{bmatrix} P_{1*} \\ P_{2*} \end{bmatrix} = \begin{bmatrix} G_1 \\ \tilde{G}_2 \end{bmatrix} M_\infty^{-1}$ such that G_1 is J-lossless and then an lcf of $\tilde{G}_2 = \tilde{\Pi}^{-1}\tilde{\Theta}$ such that $\tilde{\Theta}$ is dual J-lossless and $\tilde{\Pi}$ outer. Then, all proper real rational stabilizing controllers K_∞ satisfying $\|LFT_l(P_\gamma,K_\infty)\|_\infty < 1$ are given by $K_\infty = \mathrm{CSD_l}\left(\tilde{\Pi}, \Phi\right)$, $\forall \Phi \in BH_\infty$.*

9.1.2 Method II: $CSD_l \rightleftarrows CSD_r$ Left CSD Coupled with Right CSD

A dual scheme of Method I in which a left CSD is coupled with a right CSD can also be developed. The procedure of Method II is similar to that of Method I. The detail is summarized as follows.

Recall from Chap. 5 where a row stacked transfer matrix from the SCC plant P_γ is defined by $\tilde{P}_* = \left[P_{*1} \;\vdots\; P_{*2} \right] = \left[\begin{array}{cc:cc} I & -\frac{1}{\gamma}P_{11} & P_{12} & 0 \\ 0 & -\frac{1}{\gamma}P_{21} & P_{22} & -I \end{array}\right]$. Then the LFT formulation of the H_∞ control problem can be expressed by a coupled CSD representation. A particular *lcf* of the stacked transfer matrix, denoted by $\tilde{P}_* = \left[P_{*1} \;\vdots\; P_{*2} \right] = \tilde{M}_\infty^{-1}\left[\tilde{G}_1 \;\vdots\; G_2 \right]$, is to be found such that the numerator $\tilde{G}_1$ is dual J-lossless, and a particular *rcf* of the right part numerator, denoted by $G_2 = \Theta\Pi^{-1}$, is found such that its numerator Θ is J-lossless. Then the H_∞ controllers can be generated by the denominator Π. Figure 9.4 shows a graphical flowchart for the proposed solution process. The key concept of each step is described below (Fig. 9.4).

Step 1: Find a left coprime factorization of $\tilde{P}_* = \left[P_{*1} \;\vdots\; P_{*2} \right] = \tilde{M}_\infty^{-1}\left[\tilde{G}_1 \;\vdots\; G_2 \right]$ over RH_∞ such that the left part numerator $\tilde{G}_1$ is dual J-lossless (Fig. 9.5).

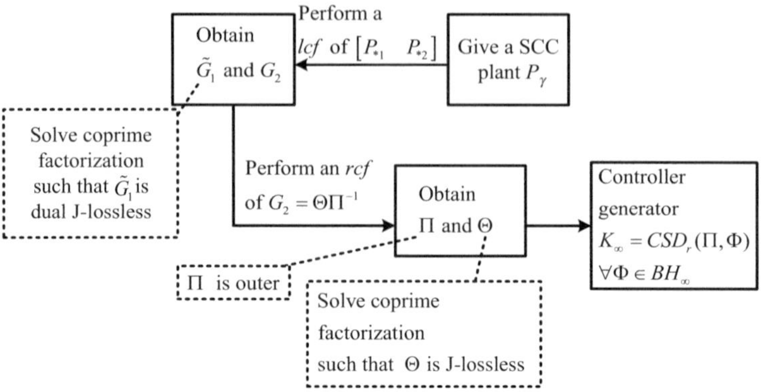

Fig. 9.4 Flowchart of Method II

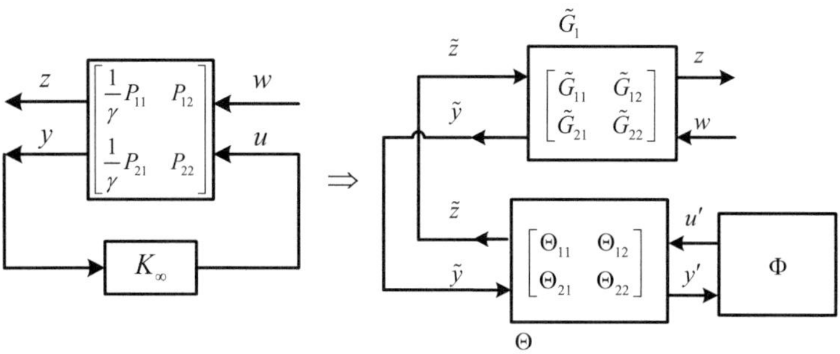

Fig. 9.5 Framework of Method II

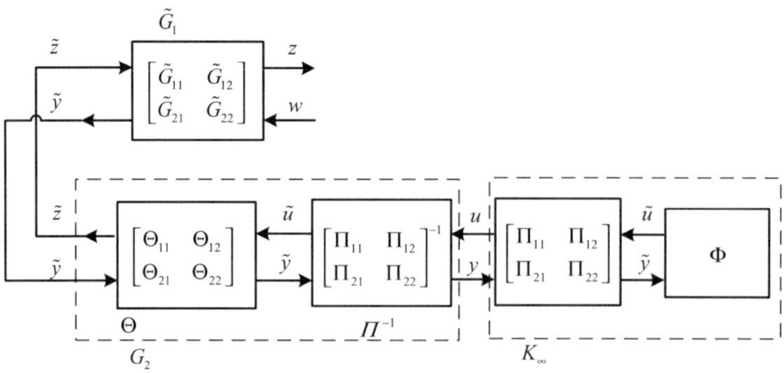

Fig. 9.6 $K_\infty = \mathrm{CSD_r}(\prod, \Phi)$

Step 2: Find a right coprime factorization of $G_2 = \Theta\Pi^{-1}$ over RH_∞ such that $\Pi \in GH_\infty$ and Θ is J-lossless.

Step 3: The H_∞ controller set is generated by the denominator Π as (Fig. 9.6)

$$K_\infty = \mathrm{CSD_r}(\Pi, \Phi) = (\Pi_{11}\Phi + \Pi_{12})(\Pi_{21}\Phi + \Pi_{22})^{-1}, \quad \text{where} \quad \forall\Phi \in BH_\infty. \tag{9.8}$$

Table 9.1 Three rupled CSD formulations ($CSD_r \rightleftarrows CSD_l$ case)

	Objective: find a K such that	Requirements on coupled CSD matrices
Stabilization problem	$LFT_l(P,K)$ is stable	$G_1 = \begin{bmatrix} G_{11} & G_{12} \\ 0 & I \end{bmatrix} \in RH_\infty$, $\tilde{\Theta} = \begin{bmatrix} I & \tilde{\Theta}_{12} \\ 0 & \tilde{\Theta}_{22} \end{bmatrix} \in RH_\infty$
H_2 control problem	$\min \lVert LFT_l(P,K) \rVert_2$ over stabilizing K	(i) $G_1 = \begin{bmatrix} G_{11} & G_{12} \\ 0 & I \end{bmatrix} \in RH_\infty$, $\tilde{\Theta} = \begin{bmatrix} I & \tilde{\Theta}_{12} \\ 0 & \tilde{\Theta}_{22} \end{bmatrix} \in RH_\infty$ (ii) G_{11} inner and $\tilde{\Theta}_{22}$ co-inner
H_∞ control problem	$\lVert LFT_l(P_\gamma,K) \rVert_\infty < 1$ over stabilizing K	(i) G_1 J-lossless (ii) $\tilde{\Theta}$ dual J-lossless

Finally, the input/output relationship could be realized by

$$
\begin{aligned}
z &= LFT_l\left(P_\gamma, K_\infty\right) w \\
&= CSD_l\left(P_{*1}, CSD_r\left(P_{*2}, K_\infty\right)\right) w \\
&= CSD_l\left(\tilde{G}_1, CSD_r\left(G_2, K_\infty\right)\right) w \\
&= CSD_l\left(\tilde{G}_1, CSD_r\left(CSD_r\left(\Theta, \Pi^{-1}\right), K_\infty\right)\right) w \\
&= CSD_l\left(\tilde{G}_1, CSD_r\left(CSD_r\left(\Theta, \Pi^{-1}\right), CSD_r\left(\Pi, \Phi\right)\right)\right) w \\
&= CSD_l\left(\tilde{G}_1, CSD_r\left(\Theta, \Phi\right)\right) w.
\end{aligned} \tag{9.9}
$$

Theorem 9.2 *For a given SCC plant P_γ, there exists an internally stabilizing controller K_∞ such that $\lVert LFT_l(P_\gamma,K_\infty)\rVert_\infty < 1$ if there exists an lcf of $\left[P_{*1} \mid P_{*2}\right] = \tilde{M}_\infty^{-1}\left[\tilde{G}_1 \mid G_2\right]$ such that $\tilde{G}_1$ is dual J-lossless and then an rcf of $G_2 = \Theta\Pi^{-1}$ such that Θ is J-lossless and Π outer. Then, all proper real rational stabilizing controllers K_∞ satisfying $\lVert LFT_l(P_\gamma,K_\infty)\rVert_\infty < 1$ (or $\left\lVert CSD_r\left(\tilde{G}_1, CSD_l\left(\Theta, \Phi\right)\right)\right\rVert_\infty < 1$) are given by $K_\infty = CSD_r(\Pi,\Phi)$, $\forall\, \Phi \in BH_\infty$.*

In the following sections, some specific H_∞ control problems will be discussed to demonstrate the important issues of H_∞ synthesis, after a section on state-space formulae. It will be seen that both Methods II and I aim to solve two J-lossless coprime factorizations although the operational sequences are different from each other. In fact, using Method I is identical to using Method II, for a general four-block problem. However, for specific control problems (e.g., the two-block problems), it is necessary to choose a proper method to make the solution process more efficient. In the two-block problems with P_{21} invertible, Method I is more appealing to solve the problems since the numerator of the second J-lossless coprime factorization of Method I can be an identity matrix. On the other hand, Method II is more suitable for two-block problems of P_{12} invertible for the similar reason.

Overall, solutions to the stabilizing H_2/H_∞ control problems are established by means of a unified procedure within the CSD framework with respect to finding corresponding coprime factorizations as summarized in Table 9.1.

In the CSD framework for controller synthesis, the procedure to find two successive coprime factorizations is unified for solving stabilization, H_2, and H_∞ (sub)optimal problems. The difference is that, as mentioned above, the numerators of the coprime factorizations in the problems of finding all stabilization solutions have to be in triangular forms while in the H_2 or H_∞ problems, the numerators (or part of them) are related to inner/J-lossless matrices.

9.2 State-Space Formulae of H_∞ Controllers

Similar to the H_2 control problem defined in (8.14), let

$$P_\gamma(s) \stackrel{s}{=} \left[\begin{array}{c|c} A_P & B_P \\ \hline C_P & D_P \end{array}\right] = \left[\begin{array}{c|cc} A & B_1 & B_2 \\ \hline C_1 & D_{11} & D_{12} \\ C_2 & D_{21} & 0 \end{array}\right] \tag{9.10}$$

be a $(q_1+q_2)\times(m_1+m_2)$ transfer function matrix with $\dim(A)=n$, and the following assumptions are satisfied.

(a) (A,B_2) is stabilizable and (C_2,A) is detectable.

(b) rank $\begin{bmatrix} A-j\omega I & B_2 \\ C_1 & D_{12} \end{bmatrix} = m_2+n$ and rank $D_{12}=m_2$.

(c) rank $\begin{bmatrix} A-j\omega I & B_1 \\ C_2 & D_{21} \end{bmatrix} = q_2+n$ and rank $D_{21}=q_2$.

Notice that Assumption (a) ensures that there are right and left coprime factorizations of P_γ, and Assumptions (b) and (c) are necessary, respectively, to find two successive J-lossless coprime factorizations in solving the H_∞ suboptimal control problem.

Many steps in the H_∞ suboptimal controller synthesis are similar to those in Chap. 8, and symbols/variables are deliberately kept the same. To avoid excessively repetitive description, derivation of the state-space formulae below will be simplified. Readers are advised to consult relevant parts in previous chapters if a doubt arises.

9.2.1 *Method I: $CSD_r \rightleftarrows CSD_l$*

Step 1: Find the particular *rcf* such that G_1 is J-lossless.

For a proper rational matrix of (9.10), the state-space realization of $\left[\begin{array}{c} P_{1\cdot}\ \ s \\ \hdashline P_{2\cdot}\ \ s \end{array}\right]$ is given, by (8.17), as

$$\begin{bmatrix} \dot{x} \\ \hline z \\ y \\ \hdashline w \\ u \end{bmatrix} = \left[\begin{array}{c|cc} A & B_1 & B_2 \\ \hline C_1 & D_{11} & D_{12} \\ C_2 & D_{21} & 0 \\ \hdashline 0 & I & 0 \\ 0 & 0 & I \end{array}\right] \begin{bmatrix} x \\ w \\ u \end{bmatrix} \Rightarrow \begin{bmatrix} \dot{x} \\ \hline z \\ w \\ u \\ y \end{bmatrix} = \left[\begin{array}{c|cc} A & B_2 & B_1 \\ \hline C_1 & D_{12} & D_{11} \\ 0 & 0 & I \\ \hdashline 0 & I & 0 \\ C_2 & 0 & D_{21} \end{array}\right] \begin{bmatrix} I & 0 & 0 \\ F_{u1} & W_{uu} & 0 \\ F_w & W_{wu} & W_{ww} \end{bmatrix} \begin{bmatrix} x \\ u' \\ w' \end{bmatrix}. \tag{9.11}$$

To construct the H_∞ controllers following the procedure of a right CSD associated with a left CSD, by (8.19), the particular right coprime factorization is given

$$G_1(s) \stackrel{s}{=} \left[\begin{array}{c|cc} A + B_2F_{u1} + B_1F_w & B_2W_{uu} + B_1W_{wu} & B_1W_{ww} \\ \hline C_1 + D_{12}F_{u1} + D_{11}F_w & D_{12}W_{uu} + D_{11}W_{wu} & D_{11}W_{ww} \\ F_w & W_{wu} & W_{ww} \end{array}\right], \tag{9.12}$$

$$\tilde{G}_2(s) \stackrel{s}{=} \left[\begin{array}{c|cc} A + B_2F_{u1} + B_1F_w & B_2W_{uu} + B_1W_{wu} & B_1W_{ww} \\ \hline F_{u1} & W_{uu} & 0 \\ C_2 + D_{21}F_w & D_{21}W_{wu} & D_{21}W_{ww} \end{array}\right] = \left[\begin{array}{c|c} A_{\tilde{G}_2} & B_{\tilde{G}_2} \\ \hline C_{\tilde{G}_2} & D_{\tilde{G}_2} \end{array}\right], \tag{9.13}$$

$$M_\infty(s) \stackrel{s}{=} \left[\begin{array}{c|cc} A + B_2F_{u1} + B_1F_w & B_2W_{uu} + B_1W_{wu} & B_1W_{ww} \\ \hline F_{u1} & W_{uu} & 0 \\ F_w & W_{wu} & W_{ww} \end{array}\right], \tag{9.14}$$

where $F_I = \begin{bmatrix} F_{u1} \\ F_w \end{bmatrix}$ and $W_I = \begin{bmatrix} W_{uu} & 0 \\ W_{wu} & W_{ww} \end{bmatrix}$ are found such that $G_1 \in RH_\infty$ is J-lossless, i.e., $G_1 JG_1 = J$ and $G_1^*(s)JG_1(s) \le J,\ \forall\, \mathrm{Re}(s) \ge 0$. It can be verified that G_1 defined by (9.12) is J-lossless if

$$W_I = \begin{bmatrix} \left[D_{12}^{\mathrm{T}}(I - D_{11}D_{11}^{\mathrm{T}})^{-1}D_{12}\right]^{-\frac{1}{2}} & 0 \\ (I - D_{11}^{\mathrm{T}}D_{11})^{-1}D_{11}^{\mathrm{T}}D_{12}\left[D_{12}^{\mathrm{T}}(I - D_{11}D_{11}^{\mathrm{T}})^{-1}D_{12}\right]^{-\frac{1}{2}} & (I - D_{11}^{\mathrm{T}}D_{11})^{-\frac{1}{2}} \end{bmatrix}, \tag{9.15}$$

where $I - D_{11}^T D_{11} > 0$ and

$$\begin{aligned} F_I &= -R_I^{-1}\left(B^{\mathrm{T}}X + D_{1*}^{\mathrm{T}}C_1\right) \\ X &= \mathrm{Ric}\,(H_X) \ge 0 \end{aligned} \tag{9.16}$$

$$H_X = \begin{bmatrix} A - BR_I^{-1}D_{1*}^{\mathrm{T}}C_1 & -BR_I^{-1}B^{\mathrm{T}} \\ -C_1^{\mathrm{T}}\left(I - D_{1*}R_I^{-1}D_{1*}^{\mathrm{T}}\right)C_1 & -\left(A - BR_I^{-1}D_{1*}^{\mathrm{T}}C_1\right)^{\mathrm{T}} \end{bmatrix} \in \mathrm{dom}\,(\mathrm{Ric}), \tag{9.17}$$

$$R_I = \begin{bmatrix} D_{12}^{\mathrm{T}} D_{12} & D_{12}^{\mathrm{T}} D_{11} \\ D_{11}^{\mathrm{T}} D_{12} & -\left(I - D_{11}^{\mathrm{T}} D_{11}\right) \end{bmatrix}, \tag{9.18}$$

$$D_{1*} = \begin{bmatrix} D_{12} & D_{11} \end{bmatrix}, \quad B = \begin{bmatrix} B_2 & B_1 \end{bmatrix}. \tag{9.19}$$

Step 2: Find the particular *lcf* such that $\tilde{\Theta}$ is dual J-lossless.

Since $\tilde{G}_2 \in RH_\infty$ is a stable transfer matrix defined by (9.13), there exists an *lcf* of $\tilde{G}_2 = \tilde{\Pi}^{-1}\tilde{\Theta}$ such that $\tilde{\Pi} \in GH_\infty$ and $\tilde{\Theta} \in RH_\infty$ is dual J-lossless by Assumption (a). Similar to the derivation of state-space formulae in Step 1, the coprime factors $\begin{bmatrix} \tilde{\Theta}(s) & \tilde{\Pi}(s) \end{bmatrix}$ can be constructed by

$$\left[\begin{array}{c|cc} I & H_u & H_{y1} \\ \hline 0 & \tilde{W}_{uu} & 0 \\ 0 & \tilde{W}_{yu} & \tilde{W}_{yy} \end{array}\right] \times \left[\begin{array}{c|cc|cc} A + B_2 F_{u1} + B_1 F_w & B_2 W_{uu} + B_1 W_{wu} & B_1 W_{ww} & 0 & 0 \\ \hline F_{u1} & W_{uu} & 0 & I & 0 \\ C_2 + D_{21} F_w & D_{21} W_{wu} & D_{21} W_{ww} & 0 & I \end{array}\right] \tag{9.20}$$

while $\tilde{\Theta}(s)$ is given by

$$\tilde{\Theta}(s) \overset{s}{=} \left[\begin{array}{c|cc} A + B_2 F_{u1} + B_1 F_w + H_u F_{u1} + H_{y1}\ C_2 + D_{21} F_w & B_2 W_{uu} + B_1 W_{wu} + H_u W_{uu} + H_{y1} D_{21} W_{wu} & B_1 W_{ww} + H_{y1} D_{21} W_{ww} \\ \hline \tilde{W}_{uu} F_{u1} & \tilde{W}_{uu} W_{uu} & 0 \\ \tilde{W}_{yu} F_{u1} + \tilde{W}_{yy}\ C_2 + D_{21} F_w & \tilde{W}_{yu} W_{uu} + \tilde{W}_{yy} D_{21} W_{wu} & \tilde{W}_{yy} D_{21} W_{ww} \end{array}\right]. \tag{9.21}$$

For $\tilde{\Theta}$ to be dual J-lossless, the nonsingular matrix $\tilde{W}_I = \begin{bmatrix} \tilde{W}_{uu} & 0 \\ \tilde{W}_{yu} & \tilde{W}_{yy} \end{bmatrix} \in R_{(m_2+q_2)\times(m_2+q_2)}$ should satisfy

$$\tilde{W}_I D_{\tilde{G}_2} J_{m_2,m_1} D_{\tilde{G}_2}^{\mathrm{T}} \tilde{W}_I^{\mathrm{T}} = J_{m_2,q_2}. \tag{9.22}$$

Let

$$H_Z = \begin{bmatrix} \left(A_{\tilde{G}_2} - B_{\tilde{G}_2} J_{m_2,m_1} D_{\tilde{G}_2}^{\mathrm{T}} \tilde{R}_I^{-1} C_{\tilde{G}_2}\right)^{\mathrm{T}} & C_{\tilde{G}_2}^{\mathrm{T}} \tilde{R}_I^{-1} C_{\tilde{G}_2} \\ B_{\tilde{G}_2}\left(J_{m_2,m_1} - J_{m_2,m_1} D_{\tilde{G}_2}^{\mathrm{T}} \tilde{R}_I^{-1} D_{\tilde{G}_2} J_{m_2,m_1}\right) B_{\tilde{G}_2}^{\mathrm{T}} & -\left(A_{\tilde{G}_2} - B_{\tilde{G}_2} J_{m_2,m_1} D_{\tilde{G}_2}^{\mathrm{T}} \tilde{R}_I^{-1} C_{\tilde{G}_2}\right) \end{bmatrix}, \tag{9.23}$$

where

$$\tilde{R}_I = D_{\tilde{G}_2} J_{m_2,m_1} D_{\tilde{G}_2}^{\mathrm{T}}. \tag{9.24}$$

Assume that $H_Z \in \mathrm{dom(Ric)}$ with $Z = \mathrm{Ric}(H_Z) \geq 0$, then define

$$H_I = \begin{bmatrix} H_u & H_{y1} \end{bmatrix} = \left(ZC_{\tilde{G}_2}^{\mathrm{T}} - B_{\tilde{G}_2} J_{m_2,m_1} D_{\tilde{G}_2}^{\mathrm{T}} \right) \tilde{R}^{-1} \tag{9.25}$$

and

$$\begin{bmatrix} \tilde{W}_{uu} & 0 \\ \tilde{W}_{yu} & \tilde{W}_{yy} \end{bmatrix} = \begin{bmatrix} \left[D_{12}^{\mathrm{T}} \left(I - D_{11} D_{11}^{\mathrm{T}} \right)^{-1} D_{12} \right]^{\frac{1}{2}} & 0 \\ -\left[D_{21} \left(I - D_{11}^{\mathrm{T}} D_{11} \right)^{-1} D_{21}^{\mathrm{T}} \right]^{-\frac{1}{2}} & \left[D_{21} \left(I - D_{11}^{\mathrm{T}} D_{11} \right)^{-1} D_{21}^{\mathrm{T}} \right]^{-\frac{1}{2}} \\ \times D_{21} \left(I - D_{11}^{\mathrm{T}} D_{11} \right)^{-1} D_{11}^{\mathrm{T}} D_{12} & \end{bmatrix}. \tag{9.26}$$

It can be verified that such defined H_I and $\tilde{W}_I$ lead to a dual J-lossless $\tilde{\Theta}$ and $\tilde{\Pi} \in GH_\infty$. As can be seen above, the first coprime factorization with J-lossless G_1 determines the free parameters F_I and W_I, and the second coprime factorization with J-lossless $\tilde{\Theta}$ resolves for the other two free parameters H_I and $\tilde{W}_I$. The above solution process shows an interesting feature in that the general output feedback H_∞ control problem is reduced to the solutions of two less complicated problems which contain the determination of a feedback control gain matrix F_I, an observer gain matrix H_I, and two accompanying nonsingular matrices (W_I and $\tilde{W}_I$).

Step 3: Find the H_∞ (suboptimal) controllers.

For the H_∞ control problem, if there exists a particular *rcf* of P_{1*} such that G_1 is J-lossless and there also exists an *lcf* of $\tilde{G}_2$ such that $\tilde{\Theta}$ is dual J-lossless, then one has by the small gain theorem

$$\left\| \mathrm{LFT}_{\mathrm{l}} \left(P_\gamma, K_\infty \right) \right\|_\infty = \left\| \mathrm{CSD}_{\mathrm{r}} \left(G_1, \mathrm{CSD}_{\mathrm{l}} \left(\tilde{\Theta}, \Phi \right) \right) \right\|_\infty < 1 \tag{9.27}$$

where $K_\infty = \mathrm{CSD}_{\mathrm{l}} \left(\tilde{\Pi}, \Phi \right)$, for any $\Phi \in BH_\infty$.

This has shown that the CSD coprime factorization approach enables the designer to formulate the general feedback control problem into a pair of simple two-port network connections.

To characterize the H_∞ controllers in terms of state-space realizations, the left CSD characterization can be equivalently transformed into its LFT form, denoted by $\tilde{\Pi}_P$, such as

$$K_\infty = \mathrm{CSD}_{\mathrm{l}} \left(\tilde{\Pi}, \Phi \right) = \mathrm{LFT}_{\mathrm{l}} \left(\tilde{\Pi}_P, \Phi \right), \quad \forall \Phi \in BH_\infty \tag{9.28}$$

where, from (9.20),

$$\tilde{\Pi}(s) \stackrel{s}{=} \left[\begin{array}{c|cc} A + B_1 F_w + B_2 F_{u1} + H_u F_{u1} + H_{y1} \; C_2 + D_{21} F_w & H_u & H_{y1} \\ \hline \tilde{W}_{uu} F_{u1} & \tilde{W}_{uu} & 0 \\ \tilde{W}_{yu} F_{u1} + \tilde{W}_{yy} \; C_2 + D_{21} F_w & \tilde{W}_{yu} & \tilde{W}_{yy} \end{array} \right] \tag{9.29}$$

and, by (5.123),

$$\tilde{\Pi}_P(s) \overset{s}{=} \left[\begin{array}{c|cc} A + B_1F_w + B_2F_{u1} + H_{y1}\ \ C_2 + D_{21}F_w & H_{y1} & H_u\tilde{W}_{uu}^{-1} \\ \hline -F_{u1} & 0 & \tilde{W}_{uu}^{-1} \\ \tilde{W}_{yy}\ \ C_2 + D_{21}F_w & \tilde{W}_{yy} & \tilde{W}_{yu}\tilde{W}_{uu}^{-1} \end{array}\right]. \tag{9.30}$$

Thus, the central controller is given by

$$K_0(s) \overset{s}{=} \left[\begin{array}{c|c} A + B_1F_w + B_2F_{u1} + H_{y1}\ \ C_2 + D_{21}F_w & H_{y1} \\ \hline -F_{u1} & 0 \end{array}\right]. \tag{9.31}$$

9.2.2 Method II: $CSD_l \rightleftarrows CSD_r$

Step 1: Find a particular *lcf* such that $\tilde{G}_1$ is dual J-lossless.

For the row stacked matrix of (9.10) inducted by (8.9), the left coprime factors of $\tilde{P}_* = P_{*1} \mid P_{*2} = \tilde{M}_\infty^{-1}\left[\tilde{G}_1 \mid G_2\right]$ can be constructed from (8.33) as

$$\begin{gathered}\left[\begin{array}{c|cc} I & H_z & H_{y2} \\ \hline 0 & \tilde{W}_{zz} & 0 \\ 0 & \tilde{W}_{yz} & \tilde{W}_{yy} \end{array}\right] \times \left[\begin{array}{c|cccc|cc} A & 0 & -B_1 & B_2 & 0 & 0 & 0 \\ \hline C_1 & I & -D_{11} & D_{12} & 0 & I & 0 \\ C_2 & 0 & -D_{21} & 0 & -I & 0 & I \end{array}\right] \\ \Rightarrow \left[\tilde{G}_1(s)\ \ G_2(s) \mid \tilde{M}_\infty(s)\right]. \end{gathered} \tag{9.32}$$

This gives

$$\tilde{G}_1(s) \overset{s}{=} \left[\begin{array}{c|cc} A + H_zC_1 + H_{y2}C_2 & H_z & -B_1 - H_zD_{11} - H_{y2}D_{21} \\ \hline \tilde{W}_{zz}C_1 & \tilde{W}_{zz} & -\tilde{W}_{zz}D_{11} \\ \tilde{W}_{yz}C_1 + \tilde{W}_{yy}C_2 & \tilde{W}_{yz} & -\tilde{W}_{yz}D_{11} - \tilde{W}_{yy}D_{21} \end{array}\right], \tag{9.33}$$

$$G_2(s) \overset{s}{=} \left[\begin{array}{c|cc} A + H_zC_1 + H_{y2}C_2 & B_2 + H_zD_{12} & -H_{y2} \\ \hline \tilde{W}_{zz}C_1 & \tilde{W}_{zz}D_{12} & 0 \\ \tilde{W}_{yz}C_1 + \tilde{W}_{yy}C_2 & \tilde{W}_{yz}D_{12} & -\tilde{W}_{yy} \end{array}\right] := \left[\begin{array}{c|c} A_{G_2} & B_{G_2} \\ \hline C_{G_2} & D_{G_2} \end{array}\right], \tag{9.34}$$

where $H_{II} = \left[H_z\ \ H_{y2}\right]$ and $\tilde{W}_{II} = \begin{bmatrix} \tilde{W}_{zz} & 0 \\ \tilde{W}_{yz} & \tilde{W}_{yy} \end{bmatrix}$ are two free parameters to be found such that $\tilde{G}_1 = \begin{bmatrix} \tilde{G}_{11} & \tilde{G}_{12} \\ \tilde{G}_{21} & \tilde{G}_{22} \end{bmatrix}$ is dual J-lossless. Then a dual J-lossless $\tilde{G}_1$ can be constructed if $I - D_{11}D_{11}^T > 0$ and $H_Y \in \mathrm{dom(Ric)}$ where

$$\tilde{W}_{II} = \begin{bmatrix} \left(I - D_{11}D_{11}^{\mathrm{T}}\right)^{-\frac{1}{2}} & 0 \\ \left[D_{21}\left(I - D_{11}^{\mathrm{T}}D_{11}\right)^{-1}D_{21}^{\mathrm{T}}\right]^{-\frac{1}{2}} D_{21}D_{11}^{\mathrm{T}}\left(I - D_{11}D_{11}^{\mathrm{T}}\right)^{-1} & \left[D_{21}\left(I - D_{11}^{\mathrm{T}}D_{11}\right)^{-1}D_{21}^{\mathrm{T}}\right]^{-\frac{1}{2}} \end{bmatrix} \tag{9.35}$$

$$H_{II} = -\left(YC^{\mathrm{T}} + B_1D_{*1}^{\mathrm{T}}\right)\tilde{R}_{II}^{-1} \tag{9.36}$$

$$Y = \mathrm{Ric}\left(H_Y\right) \geq 0 \tag{9.37}$$

$$H_Y = \begin{bmatrix} \left(A - B_1D_{*1}^{\mathrm{T}}\tilde{R}_{II}^{-1}C\right)^{\mathrm{T}} & -C^{\mathrm{T}}\tilde{R}_{II}^{-1}C \\ -B_1\left(I - D_{*1}^{\mathrm{T}}\tilde{R}_{II}^{-1}D_{*1}\right)B_1^{\mathrm{T}} & -\left(A - B_1D_{*1}^{\mathrm{T}}\tilde{R}_{II}^{-1}C\right)^{\mathrm{T}} \end{bmatrix} \tag{9.38}$$

$$\tilde{R}_{II} = \begin{bmatrix} -\left(I - D_{11}D_{11}^{\mathrm{T}}\right) & D_{11}D_{21}^{\mathrm{T}} \\ D_{21}D_{11}^{\mathrm{T}} & D_{21}D_{21}^{\mathrm{T}} \end{bmatrix} \tag{9.39}$$

$$D_{*1} = \begin{bmatrix} D_{11} \\ D_{21} \end{bmatrix}, \quad C = \begin{bmatrix} C_1 & C_2 \end{bmatrix}. \tag{9.40}$$

Step 2: Find a particular *rcf* such that Θ is J-lossless.

Similarly, the assumptions, an *rcf* of $G_2 = \Theta\Pi^{-1}$, can be constructed by

$$\left[\begin{array}{c|cc} A + H_zC_1 + H_{y2}C_2 & B_2 + H_zD_{12} & -H_{y2} \\ \hline \tilde{W}_{zz}C_1 & \tilde{W}_{zz}D_{12} & 0 \\ \tilde{W}_{yz}C_1 + \tilde{W}_{yy}C_2 & \tilde{W}_{yz}D_{12} & -\tilde{W}_{yy} \\ \hline 0 & I & 0 \\ 0 & 0 & I \end{array}\right] \times \left[\begin{array}{c|cc} I & 0 & 0 \\ \hline F_{u2} & W_{uu} & 0 \\ F_y & W_{yu} & W_{yy} \end{array}\right] \Rightarrow \left[\begin{array}{c} \Theta(s) \\ \hline \Pi(s) \end{array}\right] \tag{9.41}$$

Then the required coprime factor $\Theta(s)$ is given by

$$\begin{bmatrix} \Theta_{11}(s) & \Theta_{12}(s) \\ \Theta_{21}(s) & \Theta_{22}(s) \end{bmatrix} \stackrel{s}{=} \left[\begin{array}{c|cc} \begin{array}{c} A + H_zC_1 + H_{y2}C_2 \\ +(B_2 + H_zD_{12})F_u - H_{y2}F_y \end{array} & \begin{array}{c} (B_2 + H_zD_{12})W_{uu} \\ -H_{y2}W_{yu} \end{array} & -H_{y2}W_{yy} \\ \hline \tilde{W}_{zz}C_1 + \tilde{W}_{zz}D_{12}F_{u2} & \tilde{W}_{zz}D_{12}W_{uu} & 0 \\ \tilde{W}_{yz}C_1 + \tilde{W}_{yy}C_2 + \tilde{W}_{yz}D_{12}F_{u2} - \tilde{W}_{yy}F_y & \tilde{W}_{yz}D_{12}W_{uu} - \tilde{W}_{yy}W_{yu} & -\tilde{W}_{yy}W_{yy} \end{array}\right]. \tag{9.42}$$

Furthermore there exists a particular *rcf* such that $\Theta = \begin{bmatrix} \Theta_{11} & \Theta_{12} \\ \Theta_{21} & \Theta_{22} \end{bmatrix} \in RH_\infty$ is J-lossless if

1. There exists a nonsingular matrix $W_{II} = \begin{bmatrix} W_{uu} & 0 \\ W_{yu} & W_{yy} \end{bmatrix} \in R_{(m_2+q_2)\times(m_2+q_2)}$ such that

$$W_{II}^{\mathrm{T}} D_{G_2}^{\mathrm{T}} J_{q_1,q_2} D_{G_2} W_{II} = J_{m_2,q_2} \tag{9.43}$$

2. $H_V \in \mathrm{dom(Ric)}$ with $V = \mathrm{Ric}(H_V) \geq 0$ where

$$H_V = \begin{bmatrix} \left(A_{G_2} - B_{G_2} R_{II}^{-1} D_{G_2}^{\mathrm{T}} J_{q_1,q_2} C_{G_2}\right) & -B_{G_2} R_{II}^{-1} B_{G_2}^{\mathrm{T}} \\ -C_{G_2}^{\mathrm{T}} \left(J_{q_1,q_2} - J_{q_1,q_2} D_{G_2} R_{II}^{-1} D_{G_2}^{\mathrm{T}} J_{q_1,q_2}\right) C_{G_2} & -\left(A_{G_2} - B_{G_2} R_{II}^{-1} D_{G_2}^{\mathrm{T}} J_{q_1,q_2} C_{G_2}\right)^{\mathrm{T}} \end{bmatrix}. \tag{9.44}$$

$$R_{II} = D_{G_2}^{\mathrm{T}} J_{q_1,q_2} D_{G_2}. \tag{9.45}$$

Then one has

$$F_{II} = \begin{bmatrix} F_{u2} \\ F_y \end{bmatrix} = -R_{II}^{-1}\left(B_{G_2}^{\mathrm{T}} V + D_{G_2}^{\mathrm{T}} J_{q_1,q_2} C_{G_2}\right) \quad \text{and}$$

$$W_{II} = \begin{bmatrix} \left[D_{12}^{\mathrm{T}}\left(I - D_{11}D_{11}^{\mathrm{T}}\right)^{-1} D_{12}\right]^{-\frac{1}{2}} & 0 \\ D_{21}D_{11}^{\mathrm{T}}\left(I - D_{11}D_{11}^{\mathrm{T}}\right)^{-1} D_{12}\left[D_{12}^{\mathrm{T}}\left(I - D_{11}D_{11}^{\mathrm{T}}\right)^{-1} D_{12}\right]^{-\frac{1}{2}} & \left[D_{21}\left(I - D_{11}^{\mathrm{T}}D_{11}\right)^{-1} D_{21}^{\mathrm{T}}\right]^{\frac{1}{2}} \end{bmatrix}. \tag{9.46}$$

Step 3: Find the H_∞ (suboptimal) controllers.

Similarly, to characterize the H_∞ controllers in terms of state-space realizations, the right CSD characterization can be equivalently transformed into its LFT form such as

$$K_\infty = \mathrm{CSD_r}\left(\Pi, \Phi\right) = \mathrm{LFT_l}\left(\Pi_P, \Phi\right), \quad \forall \Phi \in BH_\infty \tag{9.47}$$

where, from (9.41),

$$\begin{bmatrix} \Pi_{11}(s) & \Pi_{12}(s) \\ \Pi_{21}(s) & \Pi_{22}(s) \end{bmatrix} \overset{s}{=} \left[\begin{array}{c|cc} \begin{array}{c} A + H_z C_1 + H_{y2} C_2 \\ +(B_2 + H_z D_{12}) F_{u2} \\ -H_{y2} F_y \end{array} & \begin{array}{c} (B_2 + H_z D_{12}) W_{uu} \\ -H_{y2} W_{yu} \end{array} & -H_{y2} W_{yy} \\ \hline F_{u2} & W_{uu} & 0 \\ F_y & W_{yu} & W_{yy} \end{array}\right] \tag{9.48}$$

and

$$\Pi_P(s) \stackrel{s}{=} \left[\begin{array}{c|cc} A+H_zC_1+H_{y2}C_2+(B_2+H_zD_{12})F_{u2} & -H_{y2} & (B_2+H_zD_{12})W_{uu} \\ \hline F_{u2} & 0 & W_{uu} \\ -W_{yy}^{-1}F_y & W_{yy}^{-1} & -W_{yy}^{-1}W_{yu} \end{array}\right]. \tag{9.49}$$

Thus, the central controller is given by

$$K_0(s) \stackrel{s}{=} \left[\begin{array}{c|c} A+H_zC_1+H_{y2}C_2+(B_2+H_zD_{12})F_{u2} & -H_{y2} \\ \hline F_{u2} & 0 \end{array}\right]. \tag{9.50}$$

9.3 H_∞ Solution of Special SCC Formulations

As described in Sect. 9.1 (or 9.2), two successive coprime factorizations are sought for H_∞ controller synthesis in the proposed CSD framework. Instead of characterizing upper triangular matrices in the stabilizing controller or H_2 optimal synthesis, the factorization in the H_∞ synthesis proceeds to find a J-lossless (or dual J-lossless) matrix factor. Similar to Sect. 8.5, for specific control problems [2], it is beneficial to choose a better method to make the solution process more efficient. In this section, the six popular synthesis problems, listed in Table 8.1, will be discussed in the CSD framework. The formulae for Π (or $\tilde{\Pi}$) are presented and H_∞ controllers to be characterized.

9.3.1 Disturbance Feedforward (DF) Problem

For the disturbance feedforward (DF) problem described in Table 8.1, the framework of $\mathrm{CSD_r} \rightleftarrows \mathrm{CSD_l}$ is utilized, and the *rcf* of $\begin{bmatrix} P_{1*,\mathrm{DF}} \\ P_{2*,\mathrm{DF}} \end{bmatrix} = \begin{bmatrix} G_{1,\mathrm{DF}} \\ \tilde{G}_{2,\mathrm{DF}} \end{bmatrix} M_{\mathrm{DF}}^{-1}$ can be obtained from (9.11) as

$$\left[\begin{array}{c|cc} A & B_2 & B_1 \\ \hline C_1 & D_{12} & D_{11} \\ 0 & 0 & I \\ \hline 0 & I & 0 \\ C_2 & 0 & I \\ \hline 0 & I & 0 \\ 0 & 0 & I \end{array}\right] \times \left[\begin{array}{c|cc} I & 0 & 0 \\ \hline F_{u1} & W_{uu} & 0 \\ F_w & W_{wu} & W_{ww} \end{array}\right] \Rightarrow \left[\begin{array}{c} G_{1,DF}(s) \\ \hline \tilde{G}_{2,DF}(s) \\ \hline M_{DF}(s) \end{array}\right] \tag{9.51}$$

where

$$G_{1,DF}(s) \overset{s}{=} \left[\begin{array}{c|cc} A+B_2F_{u1}+B_1F_w & B_2W_{uu}+B_1W_{wu} & B_1W_{ww} \\ \hline C_1+D_{12}F_{u1}+D_{11}F_w & D_{12}W_{uu}+D_{11}W_{wu} & D_{11}W_{ww} \\ F_w & W_{wu} & W_{ww} \end{array}\right] \in RH_\infty, \tag{9.52}$$

$$\tilde{G}_{2,DF}(s) \overset{s}{=} \left[\begin{array}{c|cc} A+B_2F_{u1}+B_1F_w & B_2W_{uu}+B_1W_{wu} & B_1W_{ww} \\ \hline F_{u1} & W_{uu} & 0 \\ C_2+F_w & W_{wu} & W_{ww} \end{array}\right] \in RH_\infty. \tag{9.53}$$

For H_∞ controllers, $F = \begin{bmatrix} F_{u1} \\ F_w \end{bmatrix}$ and $W = \begin{bmatrix} W_{uu} & 0 \\ W_{wu} & W_{ww} \end{bmatrix}$ should be found such that $G_{1,\mathrm{DF}}(s)$ is J-lossless. Furthermore, the particular *lcf* of $\tilde{G}_{2,\mathrm{DF}} = \tilde{\Pi}_{\mathrm{DF}}^{-1}\tilde{\Theta}_{\mathrm{DF}}$ can be found as

$$\left[\begin{array}{c|cc} I & -B_2 & -B_1 \\ \hline 0 & W_{uu}^{-1} & 0 \\ 0 & -W_{ww}^{-1}W_{wu}W_{uu}^{-1} & W_{ww}^{-1} \end{array}\right] \times \left[\begin{array}{c|cc|cc} A+B_2F_{u1}+B_1F_w & B_2W_{uu}+B_1W_{wu} & B_1W_{ww} & 0 & 0 \\ \hline F_{u1} & W_{uu} & 0 & I & 0 \\ C_2+F_w & W_{wu} & W_{ww} & 0 & I \end{array}\right]$$

$$\Rightarrow$$

$$\left[\tilde{\Theta}_{DF}(s) \;\vdots\; \tilde{\Pi}_{DF}(s)\right] \overset{s}{=}$$

$$\left[\begin{array}{c|cc|cc} A-B_1C_2 & 0 & 0 & -B_2 & -B_1 \\ \hline W_{uu}^{-1}F_{u1} & I & 0 & W_{uu}^{-1} & 0 \\ -W_{ww}^{-1}W_{wu}W_{uu}^{-1}F_{u1}+W_{ww}^{-1}(C_2+F_w) & 0 & I & -W_{ww}^{-1}W_{wu}W_{uu}^{-1} & W_{ww}^{-1} \end{array}\right] \tag{9.54}$$

where

$$\tilde{\Pi}_{DF}(s) \overset{s}{=} \left[\begin{array}{c|cc} A-B_1C_2 & -B_2 & -B_1 \\ \hline W_{uu}^{-1}F_{u1} & W_{uu}^{-1} & 0 \\ -W_{ww}^{-1}W_{wu}W_{uu}^{-1}F_{u1}+W_{ww}^{-1}(C_2+F_w) & -W_{ww}^{-1}W_{wu}W_{uu}^{-1} & W_{ww}^{-1} \end{array}\right] \in RH_\infty. \tag{9.55}$$

Since $A - B_1C_2$ is Hurwitz according to the assumption, $\tilde{\Theta}_{\mathrm{DF}}$ derived from this problem is an identity matrix so that $\tilde{\Pi}_{\mathrm{DF}} = \tilde{G}_{2,\mathrm{DF}}^{-1}$. The transfer function from w to z is given by

$$z = \mathrm{LFT}_l\,(P_{\mathrm{DF}}, K_{\mathrm{DF}})\,w = \mathrm{CSD}_r\,(G_{1,\mathrm{DF}}, \Phi_{\mathrm{DF}})\,w. \tag{9.56}$$

If F and W are found such that $G_{1,\mathrm{DF}}(s)$ is a J-lossless function, then the H_∞ solutions are given by $K_{\mathrm{DF}} = \mathrm{CSD}_l\left(\tilde{\Pi}_{\mathrm{DF}}, \Phi_{\mathrm{DF}}\right)$ for any $\Phi_{\mathrm{DF}} \in BH_\infty$.

9.3.2 Full Information (FI) Problem

Similarly, for the FI problem depicted in Table 8.1, the *rcf* of $P_{*,\mathrm{FI}} = \begin{bmatrix} P_{1*,\mathrm{FI}} \\ P_{2*,\mathrm{FI}} \end{bmatrix} = \begin{bmatrix} G_{1,\mathrm{FI}} \\ \tilde{G}_{2,\mathrm{FI}} \end{bmatrix} M_{\mathrm{FI}}^{-1}$ is given by (9.11) as

$$\left[\begin{array}{c:cc} A & B_2 & B_1 \\ \hdashline C_1 & D_{12} & D_{11} \\ 0 & 0 & I \\ \hdashline 0 & I & 0 \\ \begin{pmatrix} I \\ 0 \end{pmatrix} & \begin{pmatrix} 0 \\ 0 \end{pmatrix} & \begin{pmatrix} 0 \\ I \end{pmatrix} \\ \hdashline 0 & I & 0 \\ 0 & 0 & I \end{array}\right] \times \left[\begin{array}{c:cc} I & 0 & 0 \\ \hdashline F_{u1} & W_{uu} & 0 \\ F_w & W_{wu} & W_{ww} \end{array}\right] \Rightarrow \left[\begin{array}{c} G_{1,FI}(s) \\ \hdashline \tilde{G}_{2,FI}(s) \\ \hdashline M_{FI}(s) \end{array}\right] \tag{9.57}$$

where

$$G_{1,FI}(s) \overset{s}{=} \left[\begin{array}{c|cc} A + B_2F_{u1} + B_1F_w & B_2W_{uu} + B_1W_{wu} & B_1W_{ww} \\ \hline C_1 + D_{12}F_{u1} + D_{11}F_w & D_{12}W_{uu} + D_{11}W_{wu} & D_{11}W_{ww} \\ F_w & W_{wu} & W_{ww} \end{array}\right] \in RH_\infty, \tag{9.58}$$

$$\tilde{G}_{2,FI}(s) \overset{s}{=} \left[\begin{array}{c|cc} A + B_2F_{u1} + B_1F_w & B_2W_{uu} + B_1W_{wu} & B_1W_{ww} \\ \hline F_{u1} & W_{uu} & 0 \\ \begin{pmatrix} I \\ F_w \end{pmatrix} & \begin{pmatrix} 0 \\ W_{wu} \end{pmatrix} & \begin{pmatrix} 0 \\ W_{ww} \end{pmatrix} \end{array}\right] \in RH_\infty. \tag{9.59}$$

It can be seen that $P_{1*,\mathrm{DF}} = P_{1*,\mathrm{FI}}$, then one concludes $G_{1,\mathrm{DF}} = G_{1,\mathrm{FI}}$. The particular *lcf* of $\tilde{G}_{2,\mathrm{FI}} = \tilde{\Pi}_{\mathrm{FI}}^{-1}\tilde{\Theta}_{\mathrm{FI}}$ can be construed as

$$\left[\begin{array}{c:cc} I & -B_2 & B_1F_w + B_2F_{u1} \quad -B_1 \\ \hdashline 0 & W_{uu}^{-1} & \begin{pmatrix} -W_{uu}^{-1}F_{u1} & 0 \end{pmatrix} \\ \begin{pmatrix} 0 \\ 0 \end{pmatrix} & \begin{pmatrix} 0 \\ -W_{ww}^{-1}W_{wu}W_{uu}^{-1} \end{pmatrix} & \begin{pmatrix} I & 0 \\ 0 & W_{ww}^{-1} \end{pmatrix} \end{array}\right] \times \left[\begin{array}{c:cc:cc} A + B_2F_{u1} + B_1F_w & B_2W_{uu} + B_1W_{wu} & B_1W_{ww} & 0 & \begin{pmatrix} 0 & 0 \end{pmatrix} \\ \hdashline F_{u1} & W_{uu} & 0 & I & \begin{pmatrix} 0 & 0 \end{pmatrix} \\ \begin{pmatrix} I \\ F_w \end{pmatrix} & \begin{pmatrix} 0 \\ W_{wu} \end{pmatrix} & \begin{pmatrix} 0 \\ W_{ww} \end{pmatrix} & \begin{pmatrix} 0 \\ 0 \end{pmatrix} & \begin{pmatrix} I & 0 \\ 0 & I \end{pmatrix} \end{array}\right]$$

$\Rightarrow$

$$\left[\tilde{\Theta}_{FI}(s) \;\vdots\; \tilde{\Pi}_{FI}(s)\right] \overset{s}{=} \left[\begin{array}{c|cc:ccc} A + B_2F_{u1} + B_1F_w & 0 & 0 & -B_2 & B_1F_w + B_2F_{u1} & -B_1 \\ \hline 0 & I & 0 & W_{uu}^{-1} & -W_{uu}^{-1}F_{u1} & 0 \\ I & 0 & 0 & 0 & I & 0 \\ -W_{ww}^{-1}W_{wu}W_{uu}^{-1}F_{u1} + W_{ww}^{-1}F_w & 0 & I & -W_{ww}^{-1}W_{wu}W_{uu}^{-1} & 0 & W_{ww}^{-1} \end{array}\right] \tag{9.60}$$

where

$$\tilde{\Theta}_{FI}(s) = \begin{bmatrix} I & 0 \\ \begin{pmatrix} 0 \\ 0 \end{pmatrix} & \begin{pmatrix} 0 \\ I \end{pmatrix} \end{bmatrix} \overset{s}{=} \left[\begin{array}{c|cc} A + B_2F_{u1} + B_1F_w & 0 & 0 \\ \hline 0 & I & 0 \\ \begin{pmatrix} I \\ -W_{ww}^{-1}W_{wu}W_{uu}^{-1}F_{u1} + W_{ww}^{-1}F_w \end{pmatrix} & \begin{pmatrix} 0 \\ 0 \end{pmatrix} & \begin{pmatrix} 0 \\ I \end{pmatrix} \end{array}\right], \tag{9.61}$$

$$\begin{aligned} \tilde{\Pi}_{FI}(s) &= \begin{bmatrix} W_{uu}^{-1} & -W_{uu}^{-1}F_{u1} \quad 0 \\ \tilde{\Pi}_{21}(s) & \tilde{\Pi}_{22}(s) \end{bmatrix} \\ &\overset{s}{=} \left[\begin{array}{c|ccc} A + B_2F_{u1} + B_1F_w & -B_2 & B_1F_w + B_2F_{u1} & -B_1 \\ \hline 0 & W_{uu}^{-1} & W_{uu}^{-1}F_{u1} & 0 \\ I & 0 & I & 0 \\ -W_{ww}^{-1}W_{wu}W_{uu}^{-1}F_{u1} + W_{ww}^{-1}F_w & -W_{ww}^{-1}W_{wu}W_{uu}^{-1} & 0 & W_{ww}^{-1} \end{array}\right]. \end{aligned} \tag{9.62}$$

Note that $\tilde{\Theta}_{\mathrm{FI}}$ derived from this FI problem is a constant matrix. For $\Phi_{\mathrm{FI}} = 0$, one has $\mathrm{CSD_l}\left(\tilde{\Theta}_{\mathrm{FI}}, 0\right) = -\tilde{\Theta}_{11}^{-1}\tilde{\Theta}_{12} = 0$ since $\tilde{\Theta}_{12} = 0$. Then the closed-loop transfer function resulting from the center solution w is given by

$$\begin{aligned} z &= \mathrm{LFT_l}(P_{\mathrm{FI}}, K_{\mathrm{FI}})\, w \\ &= \mathrm{CSD_r}\big(G_{1,\mathrm{FI}}, \mathrm{CSD_l}\left(\tilde{G}_{2,\mathrm{FI}}, K_{\mathrm{FI}}\right)\big)\, w \\ &= \mathrm{CSD_r}\big(G_{1,\mathrm{FI}}, \mathrm{CSD_l}\left(\tilde{\Theta}_{\mathrm{FI}}, 0\right)\big)\, w \\ &= \mathrm{CSD_r}(G_{1,\mathrm{FI}}, 0)\, w \end{aligned} \tag{9.63}$$

where $K_{\mathrm{FI}} = \mathrm{CSD_l}\big(\tilde{\Pi}_{\mathrm{FI}}, 0\big)$. This H_∞ solution appears in a state feedback given by

$$\begin{aligned} u &= K_{\mathrm{FI}}\, y_{\mathrm{FI}} = \mathrm{CSD_l}\left(\tilde{\Pi}_{\mathrm{FI}}, 0\right) y_{\mathrm{FI}} \\ &= -\tilde{\Pi}_{11}^{-1}\tilde{\Pi}_{12} y_{FI} = \begin{bmatrix} F_{u1} & 0 \end{bmatrix} \begin{bmatrix} x \\ w \end{bmatrix} = F_{u1}x. \end{aligned} \tag{9.64}$$

Furthermore, it can also be verified that $\tilde{G}_{2,\mathrm{DF}} = \begin{bmatrix} I & 0 & 0 & 0 \\ 0 & I & 0 & 0 \\ 0 & 0 & C_2 & I \end{bmatrix} \tilde{G}_{2,\mathrm{FI}}$ and $G_{1,\mathrm{DF}} = G_{1,\mathrm{FI}}$ as depicted in Chap. 8. Therefore one then has

$$\begin{aligned} z &= \mathrm{LFT_l}\left(P_{\mathrm{FI}}, K_{\mathrm{FI}}\right) w \\ &= \mathrm{CSD_r}\left(G_{1,\mathrm{FI}}, \mathrm{CSD_l}\left(\tilde{G}_{2,\mathrm{FI}}, K_{\mathrm{FI}}\right)\right) w \\ &= \mathrm{CSD_r}\left(G_{1,\mathrm{FI}}, \mathrm{CSD_l}\left(\tilde{G}_{2,\mathrm{FI}}, \mathrm{CSD_l}\left(\tilde{\Pi}_{\mathrm{DF}} \begin{bmatrix} I & 0 & 0 & 0 \\ 0 & I & 0 & 0 \\ 0 & 0 & C_2 & I \end{bmatrix}, \Phi_{\mathrm{DF}}\right)\right)\right) w \\ &= \mathrm{CSD_r}\left(G_{1,\mathrm{FI}}, \mathrm{CSD_l}\left(\tilde{G}_{2,\mathrm{DF}}, \mathrm{CSD_l}\left(\tilde{\Pi}_{\mathrm{DF}}, \Phi_{\mathrm{DF}}\right)\right)\right) w \\ &= \mathrm{CSD_r}\left(G_{1,\mathrm{FI}}, \mathrm{CSD_l}\left(I, \Phi_{\mathrm{DF}}\right)\right) w \\ &= \mathrm{CSD_r}\left(G_{1,\mathrm{DF}}, \Phi_{\mathrm{DF}}\right) w. \end{aligned} \tag{9.65}$$

9.3.3 State Feedback (SF) Problem

Thirdly, one considers the state feedback (SF) problems. The state-space formula of the stacked matrix and its right coprime factorization can be calculated from (9.12) and (9.13) with $C_2 = I$ and $D_{21} = 0$; the *lcf* of $\tilde{G}_{2,\mathrm{SF}} = \tilde{\Pi}_{\mathrm{SF}}^{-1}\tilde{\Theta}_{\mathrm{SF}}$ is realized by

$$\left[\begin{array}{c:cc} I & -B_2 - B_1W_{wu}W_{uu}^{-1} & B_2F_{u1} + B_1W_{wu}W_{uu}^{-1}F_{u1} \\ \hdashline 0 & W_{uu}^{-1} & -W_{uu}^{-1}F_{u1} \\ 0 & 0 & \alpha \end{array}\right] \times \left[\begin{array}{c:cc:cc} A + B_2F_{u1} + B_1F_w & B_2W_{uu} + B_1W_{wu} & B_1W_{ww} & 0 & 0 \\ \hdashline F_{u1} & W_{uu} & 0 & I & 0 \\ I & 0 & 0 & 0 & I \end{array}\right]$$

$\Rightarrow$

$$\left[\tilde{\Theta}_{SF}(s) \;\vdots\; \tilde{\Pi}_{SF}(s)\right] \overset{s}{=} \left[\begin{array}{c|cc:cc} A + B_2F_{u1} + B_1F_w & 0 & B_1W_{ww} & -B_2 - B_1W_{wu}W_{uu}^{-1} & B_2F_{u1} + B_1W_{wu}W_{uu}^{-1}F_{u1} \\ \hline 0 & I & 0 & W_{uu}^{-1} & -W_{uu}^{-1}F_{u1} \\ \alpha & 0 & 0 & 0 & \alpha \end{array}\right] \tag{9.66}$$

where

$$\tilde{\Theta}_{SF}(s) \overset{s}{=} \left[\begin{array}{c|cc} A + B_2F_{u1} + B_1F_w & 0 & B_1W_{ww} \\ \hline 0 & I & 0 \\ \alpha & 0 & 0 \end{array}\right] \overset{s}{=} \begin{bmatrix} I & 0 \\ 0 & \tilde{\Theta}_{22}(s) \end{bmatrix}, \tag{9.67}$$

and

$$\begin{aligned} \tilde{\Pi}_{SF}(s) &\overset{s}{=} \left[\begin{array}{c|cc} A + B_2F_{u1} + B_1F_w & -B_2 - B_1W_{wu}W_{uu}^{-1} & B_2F_{u1} + B_1W_{wu}W_{uu}^{-1}F_{u1} \\ \hline 0 & W_{uu}^{-1} & -W_{uu}^{-1}F_{u1} \\ \alpha & 0 & \alpha \end{array}\right] \\ &\overset{s}{=} \begin{bmatrix} W_{uu}^{-1} & -W_{uu}^{-1}F_{u1} \\ \tilde{\Pi}_{21}(s) & \tilde{\Pi}_{22}(s) \end{bmatrix}. \end{aligned} \tag{9.68}$$

The central solutions are given by a state feedback gain $\mathrm{CSD_l}\left(\tilde{\Pi}_{\mathrm{SF}}(s), 0\right) = F_{u1}$, i.e., $u(t) = Kx(t) = F_{u1}x(t)$. Note that with defining $\left\|\tilde{\Theta}_{22}\right\|_\infty = \alpha$, the closed loop of $\mathrm{CSD_l}\left(\tilde{\Theta}, \Phi\right)$ belongs to BH_∞ if and only if $\alpha\|\Phi\|_\infty < 1$ since $\left\|\mathrm{CSD_l}\left(\tilde{\Theta}, \Phi\right)\right\|_\infty = \left\|\Phi\tilde{\Theta}_{22}\right\|_\infty$.

9.3.4 Output Estimation (OE) Problem

As depicted in Chap. 8, the output estimation (OE) problem is dual to the disturbance feedforward (DF) problem. In this case, the framework of $\mathrm{CSD_l} \rightleftarrows \mathrm{CSD_r}$ is utilized, and the left coprime factors can be constructed as

$$\left[\begin{array}{c:cc} I & H_z & H_{y2} \\ \hdashline 0 & \tilde{W}_{zz} & 0 \\ 0 & \tilde{W}_{yz} & \tilde{W}_{yy} \end{array}\right] \times \left[\begin{array}{c:cc:cc:cc} A & 0 & -B_1 & B_2 & 0 & 0 & 0 \\ \hdashline C_1 & I & -D_{11} & I & 0 & I & 0 \\ C_2 & 0 & -D_{21} & 0 & -I & 0 & I \end{array}\right]$$

$$\Rightarrow$$

$$\left[\tilde{G}_1(s) \;\vdots\; G_2(s) \;\vdots\; \tilde{M}_*(s)\right] \tag{9.69}$$

$$\stackrel{s}{=} \left[\begin{array}{c|cc:cc:cc} A + H_z C_1 + H_{y2} C_2 & H_z & -B_1 - H_{y2} D_{21} & B_2 + H_z & -H_{y2} & H_z & H_{y2} \\ \hline \tilde{W}_{zz} C_1 & \tilde{W}_{zz} & -\tilde{W}_{zz} D_{11} & \tilde{W}_{zz} & 0 & \tilde{W}_{zz} & 0 \\ \tilde{W}_{yz} C_1 + \tilde{W}_{yy} C_2 & \tilde{W}_{yz} & -\tilde{W}_{yy} D_{21} & \tilde{W}_{yz} & -\tilde{W}_{yy} & 0 & \tilde{W}_{yy} \end{array}\right].$$

Then realizations of $\tilde{G}_{1,\mathrm{OE}}$ and $G_{2,\mathrm{OE}}$ can be found as

$$\tilde{G}_{1,OE}(s) \stackrel{s}{=} \left[\begin{array}{c|cc} A + H_z C_1 + H_{y2} C_2 & H_z & -B_1 - H_{y2} D_{21} \\ \hline \tilde{W}_{zz} C_1 & \tilde{W}_{zz} & -\tilde{W}_{zz} D_{11} \\ \tilde{W}_{yz} C_1 + \tilde{W}_{yy} C_2 & \tilde{W}_{yz} & -\tilde{W}_{yy} D_{21} \end{array}\right] \tag{9.70}$$

$$G_{2,OE}(s) \stackrel{s}{=} \left[\begin{array}{c|cc} A + H_z C_1 + H_{y2} C_2 & B_2 + H_z & -H_{y2} \\ \hline \tilde{W}_{zz} C_1 & \tilde{W}_{zz} & 0 \\ \tilde{W}_{yz} C_1 + \tilde{W}_{yy} C_2 & \tilde{W}_{yz} & -\tilde{W}_{yy} \end{array}\right]. \tag{9.71}$$

Furthermore, the right coprime factors of $G_{2,\mathrm{OE}} = \Theta_{\mathrm{OE}} \Pi_{OE}^{-1}$ can be found as

$$\left[\begin{array}{c:cc} A + H_z C_1 + H_{y2} C_2 & B_2 + H_z & -H_{y2} \\ \hdashline \tilde{W}_{zz} C_1 & \tilde{W}_{zz} & 0 \\ \tilde{W}_{yz} C_1 + \tilde{W}_{yy} C_2 & \tilde{W}_{yz} & -\tilde{W}_{yy} \\ \hdashline 0 & I & 0 \\ 0 & 0 & I \end{array}\right] \times \left[\begin{array}{c:cc} I & 0 & 0 \\ \hdashline -C_1 & \tilde{W}_{zz}^{-1} & 0 \\ C_2 & \tilde{W}_{yy}^{-1} \tilde{W}_{yz} \tilde{W}_{zz}^{-1} & -\tilde{W}_{yy}^{-1} \end{array}\right] \Rightarrow \left[\begin{array}{c} \Theta_{OE}(s) \\ \hdashline \Pi_{OE}(s) \end{array}\right] \tag{9.72}$$

where

$$\Theta_{OE}(s) \stackrel{s}{=} \left[\begin{array}{c|cc} A - B_2 C_1 & B_2 \tilde{W}_{zz}^{-1} + H_z \tilde{W}_{zz}^{-1} - H_{y2} \tilde{W}_{zz}^{-1} \tilde{W}_{yz} \tilde{W}_{yy}^{-1} & H_{y2} \tilde{W}_{yy}^{-1} \\ \hline 0 & I & 0 \\ 0 & 0 & I \end{array}\right] \tag{9.73}$$

$$\Pi_{OE}(s) \stackrel{s}{=} \left[\begin{array}{c|cc} A - B_2 C_1 & B_2 \tilde{W}_{zz}^{-1} + H_z \tilde{W}_{zz}^{-1} - H_{y2} \tilde{W}_{zz}^{-1} \tilde{W}_{yz} \tilde{W}_{yy}^{-1} & H_{y2} \tilde{W}_{yy}^{-1} \\ \hline -C_1 & \tilde{W}_{zz}^{-1} & 0 \\ C_2 & \tilde{W}_{yy}^{-1} \tilde{W}_{yz} \tilde{W}_{zz}^{-1} & -\tilde{W}_{yy}^{-1} \end{array}\right]. \tag{9.74}$$

Note that Θ_{OE} is in fact an identity matrix so that $\Pi_{\text{OE}} = G_{2,OE}^{-1}$. The closed-loop transfer function from w to z is given by

$$\begin{aligned} z &= \text{LFT}_1\left(P_{\text{OE}}, K_{\text{OE}}\right) w \\ &= \text{CSD}_1\left(\tilde{G}_{1,\text{OE}}, \text{CSD}_r\left(G_{2,\text{OE}}, \text{CSD}_r\left(\Pi_{\text{OE}}, \Phi_{\text{OE}}\right)\right)\right) w \\ &= \text{CSD}_1\left(\tilde{G}_{1,\text{OE}}, \text{CSD}_r\left(\Theta_{\text{OE}}\Pi_{\text{OE}}^{-1}, \text{CSD}_r\left(\Pi_{\text{OE}}, \Phi_{\text{OE}}\right)\right)\right) w \\ &= \text{CSD}_1\left(\tilde{G}_{1,\text{OE}}, \Phi_{\text{OE}}\right) w. \end{aligned} \tag{9.75}$$

The central solution ($\Phi_{\text{OE}} = 0$) gives

$$\text{CSD}_r\left(\Theta_{\text{OE}}, 0\right) = \Theta_{12}\Theta_{22}^{-1} = 0 \quad (\text{since } \Theta_{12} = 0) \tag{9.76}$$

and therefore $z = \text{CSD}_1\left(\tilde{G}_{1,\text{OE}}, 0\right) w$. This concludes that the H_∞ solution of this OE problem is given by $\text{CSD}_r(\Pi_{\text{OE}},\Phi_{\text{OE}})$ for any $\Phi_{\text{OE}} \in BH_\infty$, if both $H = [H_z \; H_{y2}]$ and $\tilde{W} = \begin{bmatrix} \tilde{W}_{zz} & 0 \\ \tilde{W}_{yz} & \tilde{W}_{yy} \end{bmatrix}$ are found such that $\tilde{G}_{1,\text{OE}}(s)$ is dual J-lossless.

9.3.5 Full Control (FC) Problem

For the full control (FC) problem depicted in Chap. 8, let

$$\left[P_{*1,FC}(s) \;\vdots\; P_{*2,FC}(s)\right] \stackrel{s}{=} \left[\begin{array}{c|cc:ccc} A & 0 & -B_1 & I & 0 & 0 \\ \hline C_1 & I & 0 & 0 & I & 0 \\ C_2 & 0 & -D_{21} & 0 & 0 & -I \end{array}\right]. \tag{9.77}$$

Then a lcf of $\left[P_{*1,FC} \;\vdots\; P_{*2,FC}\right] = \tilde{M}_{FC}^{-1}\left[\tilde{G}_{1,FC} \;\vdots\; G_{2,FC}\right]$ is given by

$$\begin{aligned} &\left[\begin{array}{c:cc} I & H_z & H_{y2} \\ \hdashline 0 & \tilde{W}_{zz} & 0 \\ 0 & \tilde{W}_{yz} & \tilde{W}_{yy} \end{array}\right] \times \left[\begin{array}{c:cc:ccc:cc} A & 0 & -B_1 & I & 0 & 0 & 0 & 0 \\ \hdashline C_1 & I & 0 & 0 & I & 0 & I & 0 \\ C_2 & 0 & -D_{21} & 0 & 0 & -I & 0 & I \end{array}\right] \\ &\Rightarrow \left[\tilde{G}_{1,FC}(s) \;\vdots\; G_{2,FC}(s) \;\vdots\; \tilde{M}_{FC}(s)\right] \end{aligned} \tag{9.78}$$

where

$$\tilde{G}_{1,FC}(s) \stackrel{s}{=} \left[\begin{array}{c|cc} A + H_z C_1 + H_{y2} C_2 & H_{y2} & -B_1 - H_{y2} D_{21} \\ \hline \tilde{W}_{zz} C_1 & \tilde{W}_{zz} & 0 \\ \tilde{W}_{yz} C_1 + \tilde{W}_{yy} C_2 & \tilde{W}_{yz} & -\tilde{W}_{yy} D_{21} \end{array}\right], \tag{9.79}$$

$$G_{2,FC}(s) \stackrel{s}{=} \left[\begin{array}{c|ccc} A+H_zC_1+H_{y2}C_2 & I & H_z & -H_{y2} \\ \hline \tilde{W}_{zz}C_1 & 0 & \tilde{W}_{zz} & 0 \\ \tilde{W}_{yz}C_1+\tilde{W}_{yy}C_2 & 0 & \tilde{W}_{yz} & -\tilde{W}_{yy} \end{array}\right]. \tag{9.80}$$

It can be seen from (9.79) and (9.70) that $\tilde{G}_{1,\mathrm{OE}}(s) = \tilde{G}_{1,\mathrm{FC}}(s)$.

Furthermore, the right coprime factors of $G_{2,\mathrm{FC}} = \Theta_{\mathrm{FC}}\Pi^{-1}{}_{\mathrm{FC}}$ can be found as

$$\left[\begin{array}{c|ccc} A+H_zC_1+H_{y2}C_2 & I & H_z & -H_{y2} \\ \hline \tilde{W}_{zz}C_1 & 0 & \tilde{W}_{zz} & 0 \\ \tilde{W}_{yz}C_1+\tilde{W}_{yy}C_2 & 0 & \tilde{W}_{yz} & -\tilde{W}_{yy} \\ \hline \begin{pmatrix} 0 \\ 0 \end{pmatrix} & \begin{pmatrix} I & 0 \\ 0 & I \end{pmatrix} & & \begin{pmatrix} 0 \\ 0 \end{pmatrix} \\ 0 & 0 & 0 & I \end{array}\right] \times \left[\begin{array}{c|ccc} I & 0 & 0 & 0 \\ \hline H_zC_1+H_{y2}C_2 & I & 0 & -H_{y2}\tilde{W}_{yy}^{-1} \\ -C_1 & 0 & \tilde{W}_{zz}^{-1} & 0 \\ C_2 & 0 & -\tilde{W}_{yy}^{-1}\tilde{W}_{yz}\tilde{W}_{zz}^{-1} & -\tilde{W}_{yy}^{-1} \end{array}\right] \tag{9.81}$$

$\Rightarrow$

$$\begin{bmatrix} \Theta_{FC}(s) \\ \Pi_{FC}(s) \end{bmatrix} \stackrel{s}{=} \left[\begin{array}{c|ccc} A+H_zC_1+H_{y2}C_2 & I & \tilde{W}_{zz}^{-1}H_z & 0 \\ \hline 0 & 0 & I & 0 \\ 0 & 0 & 0 & I \\ \hline H_zC_1+H_{y2}C_2 & I & 0 & -H_{y2}\tilde{W}_{yy}^{-1} \\ -C_1 & 0 & \tilde{W}_{zz}^{-1} & 0 \\ C_2 & 0 & -\tilde{W}_{yy}^{-1}\tilde{W}_{yz}\tilde{W}_{zz}^{-1} & -\tilde{W}_{yy}^{-1} \end{array}\right]$$

where

$$\Theta_{FC}(s) = \begin{bmatrix} 0 & I & 0 \\ 0 & 0 & I \end{bmatrix} \stackrel{s}{=} \left[\begin{array}{c|ccc} A+H_zC_1+H_{y2}C_2 & I & \tilde{W}_{zz}^{-1}H_z & 0 \\ \hline 0 & 0 & I & 0 \\ 0 & 0 & 0 & I \end{array}\right] \tag{9.82}$$

$$\Pi_{FC}(s) = \begin{bmatrix} \Pi_{11}(s) & \begin{pmatrix} -H_{y2}\tilde{W}_{yy}^{-1} \\ 0 \end{pmatrix} \\ \Pi_{21}(s) & -\tilde{W}_{yy}^{-1} \end{bmatrix} \stackrel{s}{=} \left[\begin{array}{c|ccc} A+H_zC_1+H_{y2}C_2 & I & \tilde{W}_{zz}^{-1}H_z & 0 \\ \hline H_zC_1+H_{y2}C_2 & I & 0 & -H_{y2}\tilde{W}_{yy}^{-1} \\ -C_1 & 0 & \tilde{W}_{zz}^{-1} & 0 \\ C_2 & 0 & -\tilde{W}_{yy}^{-1}\tilde{W}_{yz}\tilde{W}_{zz}^{-1} & -\tilde{W}_{yy}^{-1} \end{array}\right]. \tag{9.83}$$

Since $\Pi_{\mathrm{FC}}(s)$ is a constant matrix, for $\Phi_{\mathrm{FC}} = 0$, one has

$$u_{\mathrm{FC}} = \mathrm{CSD_r}\left(\Pi_{\mathrm{FC}}, 0\right) y = \Pi_{12}\Pi_{22}^{-1} y = \begin{bmatrix} H_{y2} \\ 0 \end{bmatrix} y \tag{9.84}$$

and then $\mathrm{CSD_r}(\Theta_{FC},0) = \Theta_{12}\Theta_{22}^{-1} = 0$ (since $\Theta_{12} = 0$). Therefore, the closed-loop transfer function is given by

$$\begin{aligned} z &= \mathrm{LFT_l}\left(P_{FC}, K_{FC}\right) w \\ &= \mathrm{CSD_l}\left(\tilde{G}_{1,FC}, \mathrm{CSD_r}\left(\Theta_{FC}, 0\right)\right) w \\ &= \mathrm{CSD_l}\left(\tilde{G}_{1,FC}, 0\right) w. \end{aligned} \tag{9.85}$$

This concludes that the central H_∞ solution of this FC problem is in fact a static observer gain of (9.84) if $H = [H_z \; H_y]$ and $\tilde{W} = \begin{bmatrix} \tilde{W}_{zz} & 0 \\ \tilde{W}_{yz} & \tilde{W}_{yy} \end{bmatrix}$ are found such that $\tilde{G}_{1,FC}(s)$ is dual J-lossless.

It can also be verified that $G_{2,OE} = G_{2,FC} \begin{bmatrix} I & 0 & 0 \\ 0 & B_2 & 0 \\ 0 & I & 0 \\ 0 & 0 & I \end{bmatrix}$ and $\tilde{G}_{1,OE} = \tilde{G}_{1,FC}$. From

$$K_{FC} = \mathrm{CSD_r}\left(\Pi_{FC}, \Phi_{FC}\right) = \mathrm{CSD_r}\left(\begin{bmatrix} I & 0 & 0 \\ 0 & B_2 & 0 \\ 0 & I & 0 \\ 0 & 0 & I \end{bmatrix} \Pi_{OE}, \Phi_{OE}\right), \tag{9.86}$$

one then has

$$\begin{aligned} z &= \mathrm{LFT_l}\left(P_{FC}, K_{FC}\right) w \\ &= \mathrm{CSD_l}\left(\tilde{G}_{1,FC}, \mathrm{CSD_r}\left(G_{2,FC}, K_{FC}\right)\right) w \\ &= \mathrm{CSD_l}\left(\tilde{G}_{1,FC}, \mathrm{CSD_r}\left(G_{2,FC}, \mathrm{CSD_r}\left(\begin{bmatrix} I & 0 & 0 \\ 0 & B_2 & 0 \\ 0 & I & 0 \\ 0 & 0 & I \end{bmatrix} \Pi_{OE}, \Phi_{OE}\right)\right)\right) w \\ &= \mathrm{CSD_l}\left(\tilde{G}_{1,FC}, \mathrm{CSD_r}\left(G_{2,OE}, \mathrm{CSD_r}\left(\Pi_{OE}, \Phi_{OE}\right)\right)\right) w \\ &= \mathrm{CSD_l}\left(\tilde{G}_{1,FC}, \mathrm{CSD_r}\left(I, \Phi_{OE}\right)\right) w \\ &= \mathrm{CSD_l}\left(G_{1,OE}, \Phi_{OE}\right) w \end{aligned} \tag{9.87}$$

where $u_{FC} = \begin{bmatrix} B_2 \\ I \end{bmatrix} u_{OE}$. This shows that the closed-loop transfer function of the OE problem is in fact equivalent to that of the FC problem.

9.3.6 Output Injection (OI) Problem

Finally, one considers the output injection (OI) problems. The state-space formulae of the stacked matrix and the left coprime factorization can be calculated from (9.33) and (9.34) with $B_2 = I$ and $D_{12} = 0$, the *rcf* of $G_{2,OI} = \Theta_{OI}\Pi_{OI}^{-1}$ is realized by

$$
\left[\begin{array}{c|cc}
A+H_zC_1+H_{y2}C_2 & I & -H_{y2} \\
\hline
\tilde{W}_{zz}C_1 & 0 & 0 \\
\tilde{W}_{yz}C_1+\tilde{W}_{yy}C_2 & 0 & -\tilde{W}_{yy} \\
\hline
0 & I & 0 \\
0 & 0 & I
\end{array}\right]
\times
\left[\begin{array}{c|cc}
I & 0 & 0 \\
\hline
H_{y2}C_2+H_{y2}\tilde{W}_{yy}^{-1}\tilde{W}_{yz}C_1 & \beta & -H_y\tilde{W}_{yy}^{-1} \\
C_2-\tilde{W}_{yy}^{-1}\tilde{W}_{yz}C_1 & 0 & -\tilde{W}_{yy}^{-1}
\end{array}\right]
$$

$$
\Rightarrow \tag{9.88}
$$

$$
\begin{bmatrix} \Theta_{OI}(s) \\ \hdashline \Pi_{OI}(s) \end{bmatrix}
\overset{s}{=}
\left[\begin{array}{c|cc}
A+H_zC_1+H_{y2}C_2 & \beta & 0 \\
\hline
\tilde{W}_{zz}C_1 & 0 & 0 \\
0 & 0 & I \\
\hdashline
H_{y2}C_2+H_{y2}\tilde{W}_{yy}^{-1}\tilde{W}_{yz}C_1 & \beta & -H_{y2}\tilde{W}_{yy}^{-1} \\
C_2-\tilde{W}_{yy}^{-1}\tilde{W}_{yz}C_1 & 0 & -\tilde{W}_{yy}^{-1}
\end{array}\right]
$$

where

$$
\begin{bmatrix} \Theta_{OI}(s) \\ \hdashline \Pi_{OI}(s) \end{bmatrix}
=
\begin{bmatrix}
\Theta_{11}(s) & 0 \\
0 & \beta \\
\hdashline
\Pi_{11}(s) & -H_{y2}\tilde{W}_{yy}^{-1} \\
\Pi_{21}(s) & -\tilde{W}_{yy}^{-1}
\end{bmatrix}. \tag{9.89}
$$

The central solutions are given by an output injection gain $\mathrm{CSD_r}(\Pi_{\mathrm{OI}},0)=H_{y2}$ and therefore $z=-\tilde{G}_{12}w$. Similarly, for $\|\Theta_{11}\|_\infty=\beta$, the closed loop of $\mathrm{CSD_r}(\Theta,\Phi)$ belongs to BH_∞ if and only if $\beta\|\Phi\|_\infty<1$ since $\|\mathrm{CSD_r}(\Theta,\Phi)\|_\infty=\|\Theta_{11}\Phi\|_\infty$.

9.4 H_∞ Controller Synthesis with Coprime Factor Perturbations

In this section, one uses the H_∞ control theory to solve the robust stabilization problem of a perturbed plant. The perturbation considered here includes the discrepancy between the dynamics of real plant and the mathematical model of it, i.e., the nominal model, such as unmodeled dynamics (high-frequency dynamics) and neglected nonlinearities. Such perturbations are usually called "lumped" uncertainties or "unstructured" uncertainties in the literature. There are various ways to describe perturbations, including additive perturbation (absolute error between the actual dynamics and nominal model), input- and output-multiplication perturbations (relative errors), and their inverse forms [7]. Theoretically speaking, most of these perturbation expressions are "interchangeable," though a successful design would depend on, to certain extent, an appropriate choice of the perturbation (uncertainty) model. This section introduces a design technique which incorporates the so-called loop shaping design procedure (LSDP) [12] to obtain performance/robust trade-offs and a particular H_∞ optimization problem to guarantee the closed-loop stability and a level of robust stability based on the coprime factorization perturbation models.

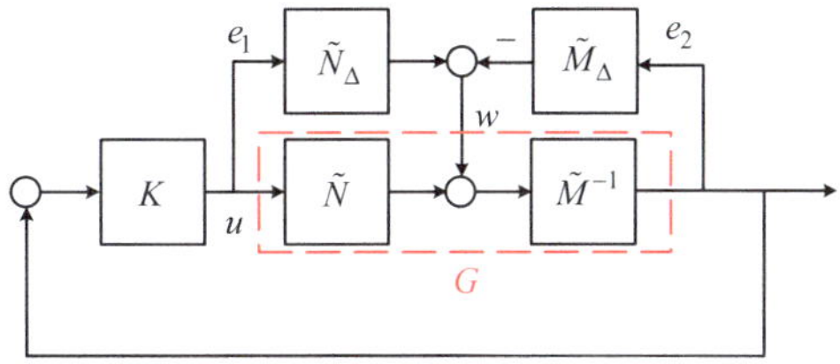

Fig. 9.7 Left coprime factorization perturbed feedback system

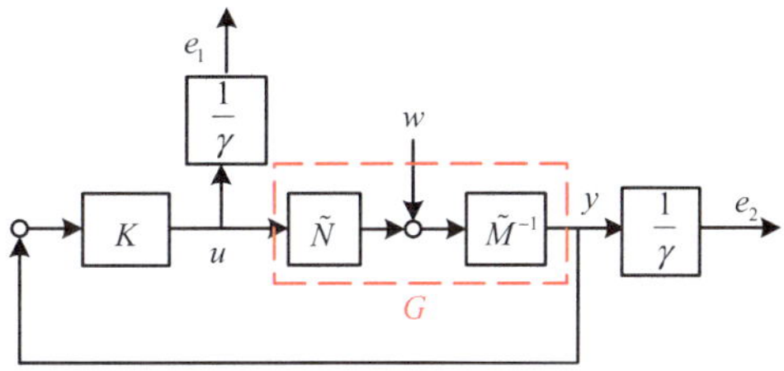

Fig. 9.8 H_∞ control problem of *lcf* plant description

9.4.1 *Robust Stabilization Problem of Left Coprime Factorization Case*

This subsection will show the robust stabilization of a plant, which is formulated by a left coprime factorization with perturbations on each factors. Let $G = \tilde{M}^{-1}\tilde{N}$ be an *lcf*. Figure 9.7 gives an H_∞ stabilization problem with an *lcf*-perturbed plant model, where $\tilde{M}_\Delta$ and $\tilde{N}_\Delta \in RH_\infty$ are stable transfer functions to represent the uncertainties on the nominal plant.

The objective of this problem is to stabilize not only the nominal plant but also a family of perturbed plants defined as

$$G_\Delta = \left\{ \left(\tilde{M} + \tilde{M}_\Delta\right)^{-1} \left(\tilde{N} + \tilde{N}_\Delta\right) : \left\| \left[\tilde{N}_\Delta \quad \tilde{M}_\Delta\right] \right\|_\infty < \varepsilon, \ \varepsilon > 0 \right\}. \tag{9.90}$$

The robust stability of the perturbed feedback system is guaranteed by

$$\|\mathrm{LFT_l}\,(P, K)\|_\infty = \left\| \begin{bmatrix} K(I - GK)^{-1}\tilde{M}^{-1} \\ (I - GK)^{-1}\tilde{M}^{-1} \end{bmatrix} \right\|_\infty \le \gamma = \frac{1}{\varepsilon}, \tag{9.91}$$

where $P = \left[\begin{array}{c|c} 0 & I \\ \tilde{M}^{-1} & \tilde{M}^{-1}\tilde{N} \\ \hline \tilde{M}^{-1} & \tilde{M}^{-1}\tilde{N} \end{array}\right]$. The control problem is equivalent to find a stabilizing controller K such that the H_∞ norm from w to $\begin{bmatrix} u \\ y \end{bmatrix}$ (or to $\begin{bmatrix} e_1 \\ e_2 \end{bmatrix} = \frac{1}{\gamma}\begin{bmatrix} u \\ y \end{bmatrix}$) is less than a specified value as shown in Fig. 9.8.

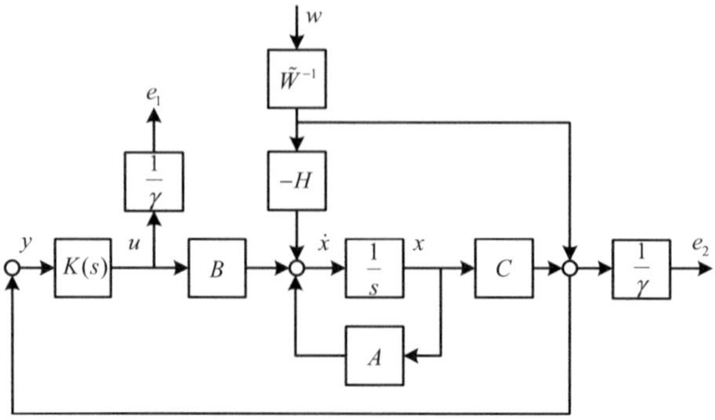

Fig. 9.9 State-space realization of the *lcf* plant description

Herein, without loss of generality, one considers $G(s) \overset{s}{=} \left[\begin{array}{c|c} A & B \\ \hline C & D \end{array}\right]$, with $D = 0$ for simplicity. A space-state realization of an *lcf* of this SCC plant is represented as Fig. 9.9.

Note that $P_{21}(s) = \tilde{M}^{-1}(s) \overset{s}{=} \left[\begin{array}{c|c} A & -H\tilde{W}^{-1} \\ \hline C & \tilde{W}^{-1} \end{array}\right]$ is square and its inverse is stable by the choice of H. The state-space realization of the SCC plant from w to $z = \begin{bmatrix} e_1 \\ e_2 \end{bmatrix}$ can be obtained from Fig. 9.9 by

$$P_\gamma(s) \overset{s}{=} \left[\begin{array}{c|c:c} A & -H\tilde{W}^{-1} & B \\ \hline \begin{pmatrix} 0 \\ \frac{1}{\gamma}C \end{pmatrix} & \begin{pmatrix} 0 \\ \frac{1}{\gamma}\tilde{W}^{-1} \end{pmatrix} & \begin{pmatrix} \frac{1}{\gamma}I \\ 0 \end{pmatrix} \\ \hdashline C & \tilde{W}^{-1} & 0 \end{array}\right]. \tag{9.92}$$

To apply the proposed approach, the first step is to find the stacked matrix $P_{\gamma*}(s) = \begin{bmatrix} P_{1*}(s) \\ \hdashline P_{2*}(s) \end{bmatrix}$ defined in Step 1 of Method I above and to perform the two successive J-lossless coprime factorizations, where from (9.3) and (8.17) one has

$$P_{\gamma*}(s) = \begin{bmatrix} P_{1*}(s) \\ \hdashline P_{2*}(s) \end{bmatrix} \overset{s}{=} \left[\begin{array}{c|cc} A & B & -H\tilde{W}^{-1} \\ \hline \begin{pmatrix} 0 \\ \frac{1}{\gamma}C \end{pmatrix} & \begin{pmatrix} \frac{1}{\gamma}I \\ 0 \end{pmatrix} & \begin{pmatrix} 0 \\ \frac{1}{\gamma}\tilde{W}^{-1} \end{pmatrix} \\ 0 & 0 & I \\ \hdashline 0 & I & 0 \\ C & 0 & \tilde{W}^{-1} \end{array}\right]. \tag{9.93}$$

A right coprime factorization is found by (9.11) with $F_{u_1} = F_u$

$$
\left[\begin{array}{c|cc}
A & B & -H\tilde{W}^{-1} \\
\hline
\begin{pmatrix} 0 \\ \frac{1}{\gamma}C \end{pmatrix} & \begin{pmatrix} \frac{1}{\gamma}I \\ 0 \end{pmatrix} & \begin{pmatrix} 0 \\ \frac{1}{\gamma}\tilde{W}^{-1} \end{pmatrix} \\
0 & 0 & I \\
\hline
0 & I & 0 \\
C & 0 & \tilde{W}^{-1}
\end{array}\right]
\times
\left[\begin{array}{c|cc}
I & 0 & 0 \\
\hline
F_u & W_{uu} & 0 \\
F_w & 0 & W_{ww}
\end{array}\right]
$$

$$
\Rightarrow \tag{9.94}
$$

$$
\left[\begin{array}{c} G_1(s) \\ \hline \tilde{G}_2(s) \end{array}\right] \overset{s}{=}
\left[\begin{array}{c|cc}
A + BF_u - H\tilde{W}^{-1}F_w & BW_{uu} & -H\tilde{W}^{-1}W_{ww} \\
\hline
\begin{pmatrix} \frac{1}{\gamma}F_u \\ \frac{1}{\gamma}C + \frac{1}{\gamma}\tilde{W}^{-1}F_w \end{pmatrix} & \begin{pmatrix} \frac{1}{\gamma}W_{ww} \\ 0 \end{pmatrix} & \begin{pmatrix} 0 \\ \frac{1}{\gamma}\tilde{W}^{-1}W_{ww} \end{pmatrix} \\
F_w & 0 & W_{ww} \\
\hline
F_u & W_{uu} & 0 \\
C + \tilde{W}^{-1}F_w & 0 & \tilde{W}^{-1}W_{ww}
\end{array}\right] \in RH_\infty
$$

Now, from (9.94),

$$
G_1(s) = \left[\begin{array}{c|c} A_{G_1} & B_{G_1} \\ \hline C_{G_1} & D_{G_1} \end{array}\right] \overset{s}{=}
\left[\begin{array}{c|c|c}
A + BF_u - H\tilde{W}^{-1}F_w & BW_{uu} & -H\tilde{W}^{-1}W_{ww} \\
\hline
\begin{pmatrix} \frac{1}{\gamma}F_u \\ \frac{1}{\gamma}C + \frac{1}{\gamma}\tilde{W}^{-1}F_w \end{pmatrix} & \begin{pmatrix} \frac{1}{\gamma}W_{uu} \\ 0 \end{pmatrix} & \begin{pmatrix} 0 \\ \frac{1}{\gamma}\tilde{W}^{-1}W_{ww} \end{pmatrix} \\
\hline
F_w & 0 & W_{ww}
\end{array}\right] \in RH_\infty \tag{9.95}
$$

where

$$
G_{11}(s) \overset{s}{=}
\left[\begin{array}{c|c}
A + BF_u - H\tilde{W}^{-1}F_w & BW_{uu} \\
\hline
\begin{pmatrix} \frac{1}{\gamma}F_u \\ \frac{1}{\gamma}C + \frac{1}{\gamma}\tilde{W}^{-1}F_w \end{pmatrix} & \begin{pmatrix} \frac{1}{\gamma}W_{uu} \\ 0 \end{pmatrix}
\end{array}\right] \in RH_\infty, \tag{9.96}
$$

$$G_{12}(s) \overset{s}{=} \left[\begin{array}{c|c} A+BF_u-H\tilde{W}^{-1}F_w & -H\tilde{W}^{-1}W_{ww} \\ \hline \begin{pmatrix} \frac{1}{\gamma}F_u \\ \frac{1}{\gamma}C+\frac{1}{\gamma}\tilde{W}^{-1}F_w \end{pmatrix} & \begin{pmatrix} 0 \\ \frac{1}{\gamma}\tilde{W}^{-1}W_{ww} \end{pmatrix} \end{array}\right] \in RH_\infty. \tag{9.97}$$

It can be found from (9.94) that

$$\tilde{G}_2(s) = \begin{bmatrix} \gamma G_{11}(s) & \gamma G_{12}(s) \end{bmatrix} \overset{s}{=} \left[\begin{array}{c|cc} A+BF_u-H\tilde{W}^{-1}F_w & BW_{uu} & -H\tilde{W}^{-1}W_{ww} \\ \hline F_u & W_{uu} & 0 \\ C+\tilde{W}^{-1}F_w & 0 & \tilde{W}^{-1}W_{ww} \end{array}\right] \in RH_\infty. \tag{9.98}$$

The second step is to find a J-lossless G_1. With the given H and $\tilde{W}$, the suboptimal H_∞ control is to obtain F_u, F_w, W_{ww}, and W_{uu} such that G_1 is J-lossless, i.e., to establish the following equations:

$$D_{G_1}^{\mathrm{T}} J_{z,w} D_{G_1} = J_{u,w}, \tag{9.99}$$

$$A_{G_1}^{\mathrm{T}} X_\gamma + X_\gamma A_{G_1} + C_{G_1}^{\mathrm{T}} J_{z,w} C_{G_1} = 0, \tag{9.100}$$

$$B_{G_1}^{\mathrm{T}} X_\gamma + D_{G_1}^{\mathrm{T}} J_{z,w} C_{G_1} = 0. \tag{9.101}$$

By solving (9.98), the constant matrix is given by (9.15) as

$$\begin{bmatrix} W_{uu} & 0 \\ 0 & W_{ww} \end{bmatrix} = \begin{bmatrix} \gamma I_u & 0 \\ 0 & \left(I_w - \frac{1}{\gamma^2}\tilde{W}^{-\mathrm{T}}\tilde{W}^{-1}\right)^{-\frac{1}{2}} \end{bmatrix}. \tag{9.102}$$

Solving (9.100) and (9.101) will give the following results, as described in (9.16):

$$\begin{bmatrix} F_u \\ F_w \end{bmatrix} = \begin{bmatrix} -\gamma^2 B^{\mathrm{T}} X_\gamma \\ \left(I_w - \frac{1}{\gamma^2}\tilde{W}^{-\mathrm{T}}\tilde{W}^{-1}\right)^{-1} \tilde{W}^{-\mathrm{T}} \left(H^{\mathrm{T}} X_\gamma - \frac{1}{\gamma^2}C\right) \end{bmatrix}; \tag{9.103}$$

and, from (9.17),

$$X_\gamma = \mathrm{Ric}\left(H_{X_\gamma}\right) \geq 0; \tag{9.104}$$

$$H_{X_\gamma} = \begin{bmatrix} A - \frac{1}{\gamma^2}H\left(\tilde{W}^{\mathrm{T}}\tilde{W} - \frac{1}{\gamma^2}I\right)^{-1}C & \left(H\left(\tilde{W}^{\mathrm{T}}\tilde{W} - \frac{1}{\gamma^2}I\right)^{-1}H^{\mathrm{T}} - \gamma^2 BB^{\mathrm{T}}\right) \\ -\frac{1}{\gamma^2}C^{\mathrm{T}}\left(I - \frac{1}{\gamma^2}\left(\tilde{W}^{\mathrm{T}}\tilde{W} - \frac{1}{\gamma^2}I\right)^{-1}\right)C & -\left(A - \frac{1}{\gamma^2}H\left(\tilde{W}^{\mathrm{T}}\tilde{W} - \frac{1}{\gamma^2}I\right)^{-1}C\right)^{\mathrm{T}} \end{bmatrix}. \tag{9.105}$$

Therefore, G_1 can be made J-lossless by the above.

The third step is to construct a dual J-lossless factorization of $\tilde{G}_2$. Notice that $\tilde{G}_2 \in RH_\infty$ and its inverse can be calculated via (9.13) as

$$\tilde{G}_2^{-1}(s) \stackrel{s}{=} \left[\begin{array}{c|cc} A+HC & -B & H \\ \hline W_{uu}^{-1}F_u & W_{uu}^{-1} & 0 \\ W_{ww}^{-1}(\tilde{W}C+F_w) & 0 & W_{ww}^{-1}\tilde{W} \end{array}\right] \tag{9.106}$$

is stable since H is any observer gain matrix, in that $\lambda(A+HC)$ are all in the left half plane. Hence, let $\tilde{\Theta}(s) = \begin{bmatrix} I & 0 \\ 0 & I \end{bmatrix}$ (dual J-lossless) and $\tilde{\Pi} = \tilde{G}_2^{-1}$, such that $\tilde{G}_2 = \tilde{\Pi}^{-1}\tilde{\Theta}$ is a J-lossless coprime factorization. Then the suboptimal solutions can be found by (9.28)

$$K = \mathrm{CSD_l}\left(\tilde{\Pi}, \Phi\right) = \mathrm{CSD_r}\left(\tilde{G}_2, \Phi\right), \quad \forall \Phi \in BH_\infty, \tag{9.107}$$

where

$$\tilde{\Pi}(s) = \tilde{G}_2^{-1}(s) \stackrel{s}{=} \left[\begin{array}{c|cc} A+HC & -B & H \\ \hline W_{uu}^{-1}F_u & W_{uu}^{-1} & 0 \\ W_{ww}^{-1}(\tilde{W}C+F_w) & 0 & W_{ww}^{-1}\tilde{W} \end{array}\right]. \tag{9.108}$$

Equivalently, by (5.123), $\tilde{\Pi}$ can be converted to an LFT matrix, $\tilde{\Pi}_P$, such that $\mathrm{CSD_l}\left(\tilde{\Pi}, \Phi\right) = \mathrm{LFT_l}\left(\tilde{\Pi}_p, \Phi\right)$ where

$$\tilde{\Pi}_P(s) \stackrel{s}{=} \left[\begin{array}{c|cc} A+HC+BF_u & -H & BW_{uu} \\ \hline F_u & 0 & W_{uu} \\ -W_{ww}^{-1}(\tilde{W}C+F_w) & W_{ww}^{-1}\tilde{W} & 0 \end{array}\right]. \tag{9.109}$$

One has the central solution (central controller) by inspection

$$K_0(s) = LFT_l(\tilde{\Pi}_P(s), 0) = \tilde{\Pi}_{P_{11}}(s) \stackrel{s}{=} \left[\begin{array}{c|c} A+HC+BF_u & -H \\ \hline F_u & 0 \end{array}\right]. \tag{9.110}$$

Note that (9.109) is essentially an observer-based structure. In fact, the central solution of the robust stabilization design problem exhibits an observer-based structure in nature. Also notice that the central controller is independent to F_w. For the case of $\Phi = 0$ (the central solution), the poles of the overall closed-loop system are determined by the A-matrix of $\mathrm{LFT_l}\,(P, K_0)$, which can be shown by a similarity transformation:

$$\begin{bmatrix} I & 0 \\ -I & I \end{bmatrix}\begin{bmatrix} A & BF_u \\ -HC & A+HC+BF_u \end{bmatrix}\begin{bmatrix} I & 0 \\ I & I \end{bmatrix} = \begin{bmatrix} A+BF_u & BF_u \\ 0 & A+HC \end{bmatrix}. \tag{9.111}$$

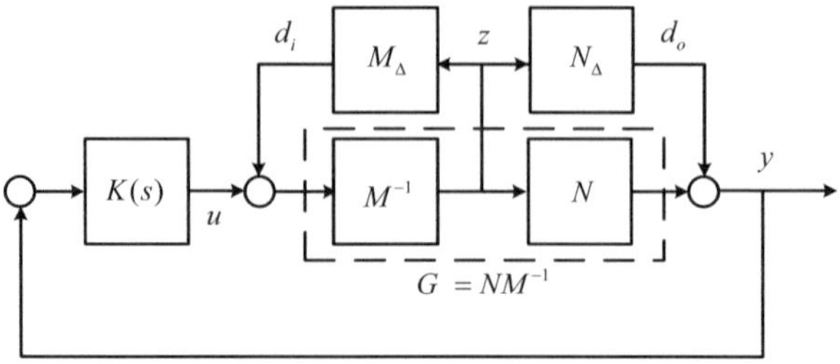

Fig. 9.10 Right coprime factorization perturbed feedback system

It shows that the eigenvalues of $A+HC$ and $A+BF_u$ are the closed-loop poles (which is similar to the separation principle) resulting from the central solution. Here $\lambda(A+HC)$ can be preassigned in the design problem, and $\lambda(A+BF_u)$ is solved by (9.102), (9.103), and (9.104). Apparently, the choice of $\lambda(A+HC)$ is a design freedom for which there exist several ways to determine H and $\tilde{W}$.

McFarlane and Glover [12] proposed a normalized coprime factorization design such that $G=\tilde{M}^{-1}\tilde{N}$ is a normalized coprime factorization (*ncf*), which gives a guideline to find H and $\tilde{W}$ to make $\begin{bmatrix}\tilde{N} & \tilde{M}\end{bmatrix}$ co-inner. By the definition of *ncf*, $H=-YC^{\mathrm{T}}$ and $\tilde{W}=I$ yield $\tilde{M}\tilde{M}^{\sim}+\tilde{N}\tilde{N}^{\sim}=I$, where $Y=\operatorname{Ric}\begin{bmatrix} A^{\mathrm{T}} & -C^{\mathrm{T}}C \\ -BB^{\mathrm{T}} & -A\end{bmatrix}$. In this case, the central controller is expressed as

$$K_0(s)\overset{s}{=}\left[\begin{array}{c|c} A-YC^TC-\gamma^2BB^TX_\gamma & YC^T \\ \hline -\gamma^2B^TX_\gamma & 0\end{array}\right] \tag{9.112}$$

where

$$X_\gamma=\operatorname{Ric}\left(\begin{bmatrix} A+\frac{1}{\gamma^2-1}YC^{\mathrm{T}}C & \gamma^2\left(\frac{1}{\gamma^2-1}YC^{\mathrm{T}}CY-BB^{\mathrm{T}}\right) \\ -\frac{1}{\gamma^2-1}C^{\mathrm{T}}C & -\left(A+\frac{1}{\gamma^2-1}YC^TC\right)^{\mathrm{T}}\end{bmatrix}\right). \tag{9.113}$$

9.4.2 *Robust Stabilization Problem of Right Coprime Factor Case*

Similarly, a dual problem is formulated by introducing the *rcf* plant description that attempts to ensure the robust stability with respect to the bounded uncertainty $\left\|\begin{bmatrix} N_\Delta \\ M_\Delta\end{bmatrix}\right\|_\infty<\varepsilon$ as shown in Fig. 9.10. Then a stabilizing K can be found such that

$$\|\mathrm{LFT_l}\,(P,K)\|_\infty=\left\|\ M^{-1}(I-GK)^{-1}K\ \ M^{-1}(I-GK)^{-1}\ \right\|_\infty\le\gamma \tag{9.114}$$

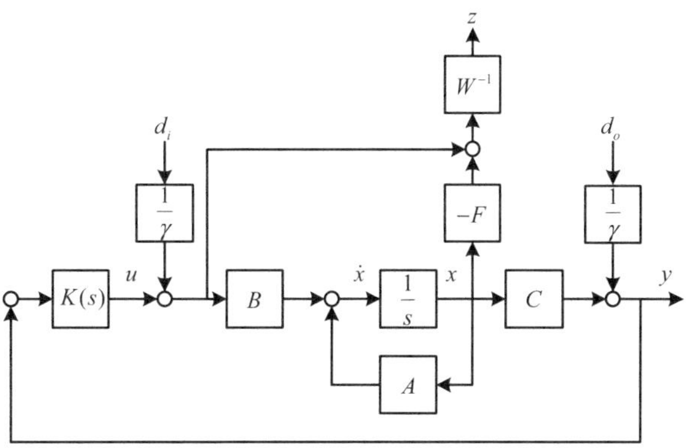

Fig. 9.11 State-space realization of the *rcf* plant description

with $P = \left[\begin{array}{cc|c} 0 & M^{-1} & M^{-1} \\ \hline I & NM^{-1} & NM^{-1} \end{array}\right]$.

In Fig. 9.11, the control problem is equivalent to find a stabilizing controller K such that the H_∞-norm from w to z is less than a specified value, where $\begin{bmatrix} d_o & d_i \end{bmatrix} = w$. Note that $P_{12}(s) = M^{-1}(s) \overset{s}{=} \left[\begin{array}{c|c} A & B \\ \hline -W^{-1}F & W^{-1} \end{array}\right]$ is square and its inverse is stable. Figure 9.11 illustrates a space-state realization of this SCC plant in *rcf*.

The state-space realization of the SCC matrix for Fig. 9.11 can easily be found as

$$P_\gamma(s) \overset{s}{=} \left[\begin{array}{c|c|c} A & \begin{pmatrix} 0 & \frac{1}{\gamma}B \end{pmatrix} & B \\ \hline -W^{-1}F & \begin{pmatrix} 0 & \frac{1}{\gamma}W^{-1} \end{pmatrix} & W^{-1} \\ \hline C & \begin{pmatrix} \frac{1}{\gamma}I & 0 \end{pmatrix} & 0 \end{array}\right]. \tag{9.115}$$

Different from the problem of *lcf* plant description, the solution process for the *rcf* problem will be facilitated by Method II. The first step of Method II is to give a coupled CSD representation as below

$$\tilde{P}_* = \left[\tilde{P}_{*1} \;\vdots\; \tilde{P}_{*2}\right] = \left[\begin{array}{cc|cc} -I & \frac{1}{\gamma}P_{11} & -P_{12} & 0 \\ 0 & \frac{1}{\gamma}P_{21} & -P_{22} & I \end{array}\right]. \tag{9.116}$$

The *lcf* of the matrix $\tilde{P}_* = \left[\tilde{P}_{*1} \; \vdots \; \tilde{P}_{*2}\right] = \tilde{M}_\infty^{-1}\left[\tilde{G}_1 \; \vdots \; G_2\right]$ can be obtained from (9.32) with $H_{y2} = H_y$ as

$$\left[\begin{array}{c|cc} I & H_z & H_y \\ \hline 0 & \tilde{W}_{zz} & 0 \\ 0 & 0 & \tilde{W}_{yy} \end{array}\right] \times \left[\begin{array}{c|c|c|cc} A & 0 & \begin{pmatrix} 0 & \frac{1}{\gamma}B \end{pmatrix} & -B & 0 \\ \hline -W^{-1}F & -I & \begin{pmatrix} 0 & \frac{1}{\gamma}W^{-1} \end{pmatrix} & -W^{-1} & 0 \\ C & 0 & \begin{pmatrix} \frac{1}{\gamma}I & 0 \end{pmatrix} & 0 & I \end{array}\right] \tag{9.117}$$

with that, from (9.33) and (9.34)

$$\begin{aligned} \tilde{G}_1(s) &\stackrel{s}{=} \left[\begin{array}{c|c|c} A - H_zW^{-1}F + H_yC & -H_z & \begin{pmatrix} \frac{1}{\gamma}H_y & \frac{1}{\gamma}B + \frac{1}{\gamma}H_zW^{-1} \end{pmatrix} \\ \hline -\tilde{W}_{zz}W^{-1}F & -\tilde{W}_{zz} & \begin{pmatrix} 0 & \frac{1}{\gamma}\tilde{W}_{zz}W^{-1} \end{pmatrix} \\ \tilde{W}_{yy}C & 0 & \begin{pmatrix} \frac{1}{\gamma}\tilde{W}_{yy} & 0 \end{pmatrix} \end{array}\right] \\ &= \left[\begin{array}{c|c} A_{\tilde{G}1} & B_{\tilde{G}1} \\ \hline C_{\tilde{G}1} & D_{\tilde{G}1} \end{array}\right] \in RH_\infty, \end{aligned} \tag{9.118}$$

$$G_2(s) \stackrel{s}{=} \left[\begin{array}{c|cc} A - H_zW^{-1}F + H_yC & -B - H_zW^{-1} & H_y \\ \hline -\tilde{W}_{zz}W^{-1}F & -\tilde{W}_{zz}W^{-1} & 0 \\ \tilde{W}_{yy}C & 0 & \tilde{W}_{yy} \end{array}\right] \in RH_\infty. \tag{9.119}$$

After multiplying $\tilde{M}_\infty$ at left-hand-side terminal, the closed-loop transfer function is obtained by $\mathrm{CSD_l}\left(\tilde{G}_1, \mathrm{CSD_r}\left(G_2, K\right)\right)$. The second step is to solve that $\tilde{G}_1(s)$ is dual J-lossless. By definition, $\tilde{G}_1(s)$ is dual J-lossless if there exists a matrix such that $Y_\gamma = Y_\gamma{}^{\mathrm{T}} \geq 0$ satisfies

$$D_{\tilde{G}_1} J_{z,w} D_{\tilde{G}_1}^{\mathrm{T}} = J_{z,y}, \tag{9.120}$$

$$A_{\tilde{G}_1} Y_\gamma + Y_\gamma A_{\tilde{G}_1}^{\mathrm{T}} - B_{\tilde{G}_1} J_{z,w} B_{\tilde{G}_1}^{\mathrm{T}} = 0, \tag{9.121}$$

$$C_{\tilde{G}_1} Y_\gamma - D_{\tilde{G}_1} J_{z,w} B_{\tilde{G}_1}^{\mathrm{T}} = 0. \tag{9.122}$$

Here let $W = I$ for simplicity. It is easy to work out via (9.35) that

$$\begin{bmatrix} \tilde{W}_{zz} & 0 \\ 0 & \tilde{W}_{yy} \end{bmatrix} = \begin{bmatrix} \left(1 - \frac{1}{\gamma^2}\right)^{-\frac{1}{2}} I & 0 \\ 0 & \gamma I \end{bmatrix} \tag{9.123}$$

and, with (9.36),

$$\begin{bmatrix} H_z & H_y \end{bmatrix} = \begin{bmatrix} \left(\frac{1}{\gamma^2} B - Y_\gamma F^{\mathrm{T}}\right)\left(1 - \frac{1}{\gamma^2}\right)^{-1} & -\gamma^2 Y_\gamma C^{\mathrm{T}} \end{bmatrix} \tag{9.124}$$

where $Y_\gamma = \mathrm{Ric}\left(H_{Y_\gamma}\right) \geq 0$ and the Hamiltonian is defined by (9.38) as

$$H_{Y_\gamma} = \begin{bmatrix} A^{\mathrm{T}} - \frac{1}{\gamma^2 - 1} F^{\mathrm{T}} B^{\mathrm{T}} & \gamma^2\left(\frac{1}{\gamma^2 - 1} F^{\mathrm{T}} F - C^{\mathrm{T}} C\right) \\ -\left(\frac{1}{\gamma^2 - 1}\right) B B^{\mathrm{T}} & -\left(A - \frac{1}{\gamma^2 - 1} B F\right) \end{bmatrix}. \tag{9.125}$$

Furthermore, with $\tilde{W} = I$, the corresponding Hamiltonian matrix can be reduced to

$$H_{X_\gamma} = \begin{bmatrix} A - \frac{1}{\gamma^2 - 1} H C & \gamma^2\left(\frac{1}{\gamma^2 - 1} H H^{\mathrm{T}} - B B^{\mathrm{T}}\right) \\ -\frac{1}{\gamma^2 - 1} C^{\mathrm{T}} C & -\left(A - \frac{1}{\gamma^2 - 1} H C\right)^{\mathrm{T}} \end{bmatrix}. \tag{9.126}$$

It is interesting, though logical, to notice that the solution of the *rcf* problem is a dual case of the *lcf* problem. Furthermore, since both G_2 and its inverse are stable (i.e., $G_2 \in GH_\infty$), one has $\Theta(s) = \begin{bmatrix} I & 0 \\ 0 & I \end{bmatrix}$ (J-lossless) and $\Pi = G_2^{-1}$ such that $G_2 = \Theta\Pi^{-1}$ where

$$\Pi(s) = G_2^{-1}(s) \overset{s}{=} \left[\begin{array}{c|cc} A + BF & (B + H_y)\tilde{W}_{11}^{-1} & H_y \tilde{W}_{yy}^{-1} \\ \hline -F & -\tilde{W}_{zz}^{-1} & 0 \\ -C & 0 & \tilde{W}_{yy}^{-1} \end{array}\right]. \tag{9.127}$$

Then the suboptimal solutions can be found as

$$K = \mathrm{CSD_r}\,(\Pi, \Phi)\ \forall \Phi \in BH_\infty. \tag{9.128}$$

The CSD matrix of Π can be expectably converted to an equivalent LFT matrix Π_P such that $\mathrm{CSD_r}(\Pi,\Phi) = \mathrm{LFT_l}(\Pi_P,\Phi)$ where

$$\Pi_P(s) \overset{s}{=} \left[\begin{array}{c|cc} A + BF + H_y C & H_y & (B + H_z)\tilde{W}_{zz}^{-1} \\ \hline -F & 0 & -\tilde{W}_{zz}^{-1} \\ \tilde{W}_{yy} C & \tilde{W}_{yy} & 0 \end{array}\right] \tag{9.129}$$

Table 9.2 Comparison between the *rcf* and *lcf* plant description problem

	The *lcf* plant description	The *rcf* plant description
SCC plant	$P=\left[\begin{array}{c:c} 0 & I \\ \tilde{M}^{-1} & \tilde{M}^{-1}\tilde{N} \\ \hdashline \tilde{M}^{-1} & \tilde{M}^{-1}\tilde{N} \end{array}\right]$	$P=\left[\begin{array}{cc:c} 0 & M^{-1} & M^{-1} \\ \hdashline I & NM^{-1} & NM^{-1} \end{array}\right]$
2-block condition	$P_{21}^{-1}=\tilde{M}(s)$	$P_{12}^{-1}=M(s)$
Solution method	A right CSD coupled a left CSD (Method I)	A left CSD coupled a right CSD (Method II)
Central controller form	$\left[\begin{array}{c\|c} A+HC+BF_u & -H \\ \hline F_u & 0 \end{array}\right]$	$\left[\begin{array}{c\|c} A+BF+H_yC & H_y \\ \hline -F & 0 \end{array}\right]$

Similar to (9.110), the central controller can therefore be obtained with $\Phi=0$ as

$$K_0(s)=LFT_l(\Pi_P(s),0)=\Pi_{P_{11}}(s)\overset{s}{=}\left[\begin{array}{c|c} A+BF+H_yC & H_y \\ \hline -F & 0 \end{array}\right]. \tag{9.130}$$

For the central solution, the poles of the overall closed-loop system are determined by the A-matrix of $\mathrm{LFT}_l\ (P, K_0)$, which can be shown by a similarity transformation,

$$\begin{bmatrix} I & 0 \\ I & I \end{bmatrix}\begin{bmatrix} A & -BF \\ H_yC & A+BF+H_yC \end{bmatrix}\begin{bmatrix} I & 0 \\ -I & I \end{bmatrix}=\begin{bmatrix} A+BF & -BF \\ 0 & A+H_yC \end{bmatrix}. \tag{9.131}$$

This illustrates that the eigenvalues of $A+H_yC$ and $A+BF$ are the closed-loop poles of the central solution where $\lambda(A+BF)$ is preassigned in the design problem, and $\lambda(A+H_yC)$ is solved from (9.120) and (9.121). Different from the *lcf* plant description problem, $\lambda(A+BF)$ is a design freedom indicated in terms of F and W. The partial pole assignment property of the coprime factor plant description problem is summarized as the following theorem.

The solutions shown in (9.109) and (9.129) have the same observer-based form, and the corresponding AREs present a similar structure. Due to the duality of the problems, the solution process for each problem is also in dual form. It is noted that the (2,1)-block of the *lcf* problem (i.e., P_{21}) is invertible, and hence Method I is adopted to solve J-lossless matrix in the first step and to deal with the inversion at the second step ($\tilde{\Pi}=\tilde{G}_2^{-1}$). Dually, since the (1,2)-block of the *rcf* problem (i.e., P_{12}) is invertible, Method II is employed. Therefore, the second steps of both problems do not have additional calculations to construct a (dual) J-lossless coprime factorization. Using Method I (or Method II) to solve the *lcf* (or *rcf*), problem is still achievable; however, excessive calculation is needed in the first step, which could lead to inaccurate solutions in real-world design cases.

Table 9.2 gives a comparison between the *rcf* and *lcf* plant description problems.

Exercises

1. Consider a unity feedback system with a proportional gain controller K, where $K = 3$ and the plant under control $G(s) = \frac{1}{s-2}$. Compute a normalized coprime factorization of $G(s)$. Considering perturbations Δ_N and Δ_M of the normalized coprime factors of $G(s)$, compute the stability radius ε with regard to the perturbations on coprime factors.
2. For a given SCC plant $P(s) \stackrel{s}{=} \left[\begin{array}{c|cc} A & B_1 & B_2 \\ \hline C_1 & D_{11} & D_{12} \\ C_2 & D_{21} & 0 \end{array}\right] = \left[\begin{array}{cc|c|c} 1 & 0 & 1 & 1 \\ 2 & -3 & 0 & 0 \\ \hline 2 & 0 & 0 & 0 \\ 0 & 0 & 0 & 1 \\ \hline 1 & 2 & 3 & 0 \end{array}\right]$ and $\gamma = 5$, compute a suboptimal H-infinity controller using the CSD approach.
3. For a given SCC plant $P(s) \stackrel{s}{=} \left[\begin{array}{c|cc} A & B_1 & B_2 \\ \hline C_1 & D_{11} & D_{12} \\ C_2 & D_{21} & 0 \end{array}\right] = \left[\begin{array}{cc|cc|c} 1 & 0 & 1 & 0 & 1 \\ 2 & -3 & 0 & 0 & 0 \\ \hline 2 & 0 & 0 & 0 & 0 \\ 0 & 0 & 0 & 0 & 1 \\ \hline 1 & 0 & 0 & 0 & 0 \\ 0 & 1 & 0 & 0 & 0 \end{array}\right]$, select an appropriate method (formulation) shown in Table 8.1 for the H-infinity suboptimal control problem.
4. Consider the model matching (or reference) control problem shown in the following figure. Formulate an H_∞ control problem that minimizes control energy u and output error e.

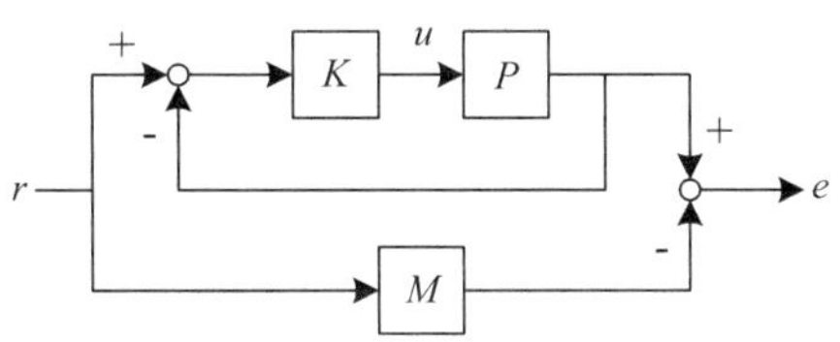

References

1. Bombois X, Anderson BDO (2002) On the influence of weight modification in Hinf control design. IEEE conference on decision and control, Nevada, USA
2. Doyle JC, Glover K, Khargonekar PP, Francis BA (1989) State-space solutions to standard H_2 and H_∞ control problems. IEEE Trans Autom Control 34:831–847
3. Glover K, Doyle JC (1988) State-space formulae for all stabilizing controllers that satisfy an H_∞-norm bounded and relations to risk sensitivity. Syst Control Lett 11:167–172
4. Glover K, Limebeer DJN, Doyle JC, Kasenally EM, Safonov MG (1991) A characterization of all the solutions to the four block general distance problem. SIAM J Control Optim 29:283–324
5. Green M (1992) H_∞ controller synthesis by J-lossless coprime factorization. SIAM J Control Optim 30:522–547

6. Green M, Glover K, Limebeer DJN, Doyle JC (1990) A J-spectral factorization approach to H_∞ control. SIAM J Control Optim 28:1350–1371
7. Gu DW, Petkow PH, Konstantinov MM (2005) Robust control design with Matlab. Springer, London
8. Hong JL (2004) An output feedback control for discrete-time state-delay systems. J Circuits Syst Signal Process 23:255–272
9. Hong JL, Teng CC (2001) Control for nonlinear affine systems: a chain-scattering matrix description approach. Int J Robust Nonlinear Control 11:315–333
10. Kimura H (1995) Chain-scattering representation, J-lossless factorization and H_∞ control. J Math Syst Estim Control 5:203–255
11. Lee PH, Soh YC (2005) Synthesis of stable H_∞ controller via the chain scattering framework. Syst Control Lett 46:1968–1972
12. McFarlane D, Glover K (1990) Robust controller design using normalized coprime factor plant descriptions. Springer, London
13. Tsai MC, Tsai CS (1993) A chain scattering matrix description approach to H_∞ control. IEEE Trans Autom Control 38:1416–1421

Chapter 10
Design Examples

In this chapter, several design examples are illustrated to demonstrate the validity of the CSD two-port framework. Two different design methodologies with respect to speed control of DC servomotors are presented. These examples will show how industrial controllers, such as pseudo derivative feedback (PDF) controllers and pseudo derivative feedback with feedforward (PDFF) controllers, can be formulated into the standard control design framework and then solved by the state-space solution procedures presented in previous chapters. By defining the transfer function from the load torque disturbance to the controlled output, the dynamic stiffness of a servo control system is characterized, and a scalar index value is defined by the inverse of the maximum magnitude of the transfer function with respect to frequency, i.e., the worst case in the frequency response. Thus, for performance measurement of robust design, maximizing the dynamic stiffness measurement implies minimizing the H_∞-norm in controller design. This chapter will also show how the dynamic stiffness of a servo system can be achieved by H_∞ design.

10.1 Mathematical Models of DC Servomotor

Mathematical models for describing the control system under investigation typically contain some inaccuracies when compared with the real plants. This is mostly caused by simplifications of the model, exclusion of some dynamics that are either too complicated or unknown, and/or is due to uncertain dynamics. These inaccuracies induce a significant problem in control system design. A possible, and proven useful, approach to dealing with this problem is based on modeling the real system dynamics as a set of linear time-invariant models built around a nominal one, i.e., the model is considered as uncertain but within known boundaries. The benefit of such a representation of the model is the possibility of designing a robust controller that stabilizes a closed-loop system with uncertainties under consideration. The ideal goal would be to design a controller capable of stabilizing even the "the worst-case scenario" representing the most degraded model. This section investigates, as

M.-C. Tsai and D.-W. Gu, *Robust and Optimal Control: A Two-port Framework Approach*, Advances in Industrial Control, DOI 10.1007/978-1-4471-6257-5_10,
© Springer-Verlag London 2014

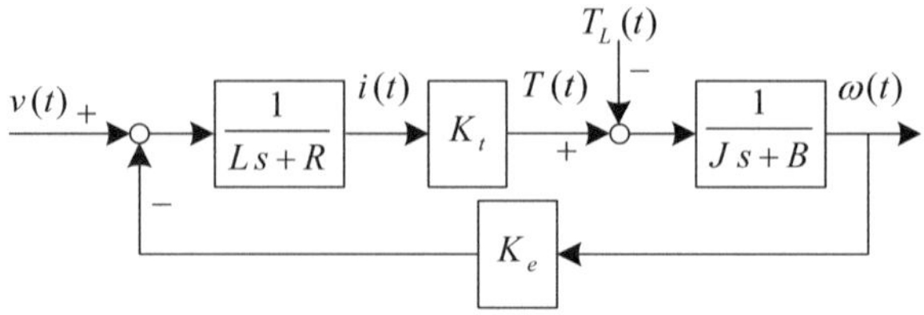

Fig. 10.1 Block diagram of standard DC servomotor

a real-world design example, the robust design of servo control systems, which are widely used in industries.

Consider a DC permanent magnetic (PM) servomotor. Let the rotor be characterized by the motor winding inductance L (unit: H) and the armature resistance R (unit: Ω). Then, the equation associated with such an electrical circuit is given by

$$v(t) = L\frac{\mathrm{d}i}{\mathrm{d}t} + Ri + e, \tag{10.1}$$

where the back EMF, e, of the motor has been taken into account. The torque generated at the motor shaft is proportional to the armature current, where the ratio is defined as the torque constant, K_t (unit: Nm/amp), as

$$T = K_t i. \tag{10.2}$$

Moreover, the proportional ratio between angular velocity of motor and the back EMF is defined as electromotive force constant, K_e (unit: V/s/rad), as

$$e = K_e \omega. \tag{10.3}$$

One can now deal with the mechanical representation of the motor. The motor exerts a torque while supplied by voltages. This torque acts on the mechanical structure, which is characterized by the rotor inertia J (kgm^2) and the viscous friction coefficient B (Nms/rad) as

$$T - T_L = J\frac{\mathrm{d}\omega}{\mathrm{d}t} + B\omega \tag{10.4}$$

where T_L denotes the load torque.

Based on the electrical and mechanical equations above, the system block diagram of a DC servomotor can be depicted as shown in Fig. 10.1 below.

10.2 Two-Port Chain Description Approach to Estimation of Mechanical Loading

In this section, the technique of two-port CSD formulations is employed for detecting the impedance of the mechanical loading in a DC motor. Let the block diagram of a DC motor be formulated in terms of a two-port LFT framework.

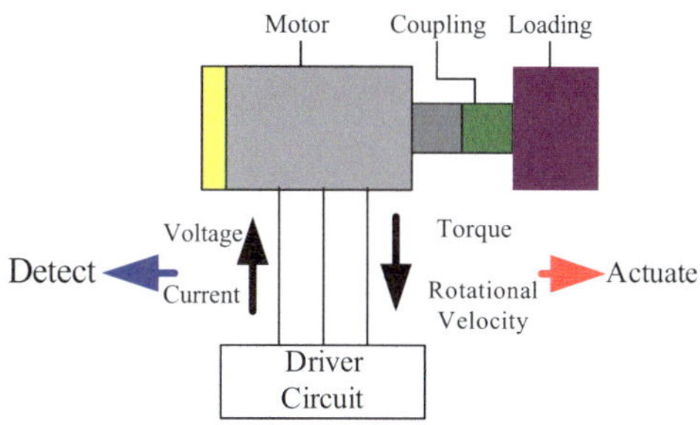

Fig. 10.2 Functions of actuator and sensor in DC motor

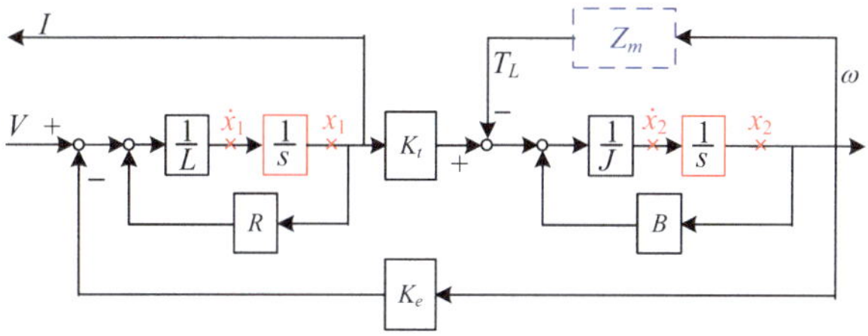

Fig. 10.3 Block diagram of DC motor

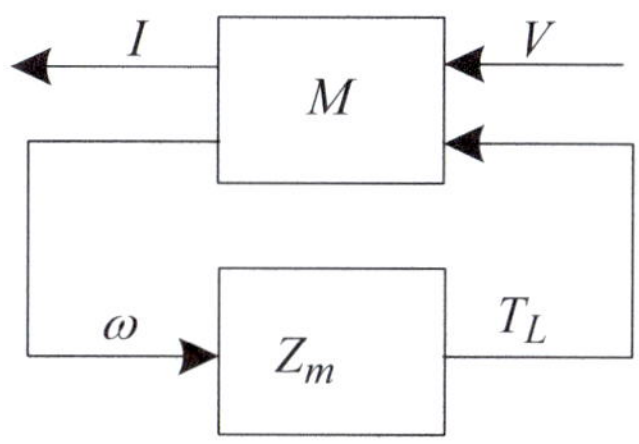

Fig. 10.4 LFT form of DC motor

Then the chain scattering description (CSD) as discussed earlier in this book can be adopted to characterize the relationship between the electrical impedance and the mechanical impedance for further analysis. This also implies that the motor not only can be employed to actuate the mechanical loading but can also monitor the operating condition, in which the mechanical loading can be found from the measured electrical impedance as depicted in Fig. 10.2.

Consider the block diagram of a DC motor shown in Fig. 10.3, where Z_m denotes the mechanical loading. Let V (voltage reference) and T_L (load) be input variables and I (motor current) and ω (motor angular velocity) be the outputs. A (2×2) LFT representation of Fig. 10.3 is depicted in Fig. 10.4, where

$$M(s) = \begin{bmatrix} \dfrac{Js+B}{LJs^2+(LB+JR)s+(RB+K_eK_t)} & \dfrac{K_e}{LJs^2+(LB+JR)s+(RB+K_eK_t)} \\ \dfrac{K_t}{LJs^2+(LB+JR)s+(RB+K_eK_t)} & \dfrac{-(R+Ls)}{LJs^2+(LB+JR)s+(RB+K_eK_t)} \end{bmatrix} \tag{10.5}$$

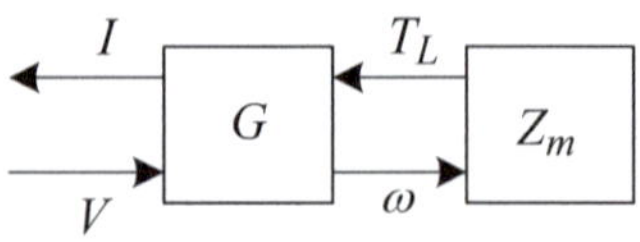

Fig. 10.5 Two-port chain description matrix of a DC motor

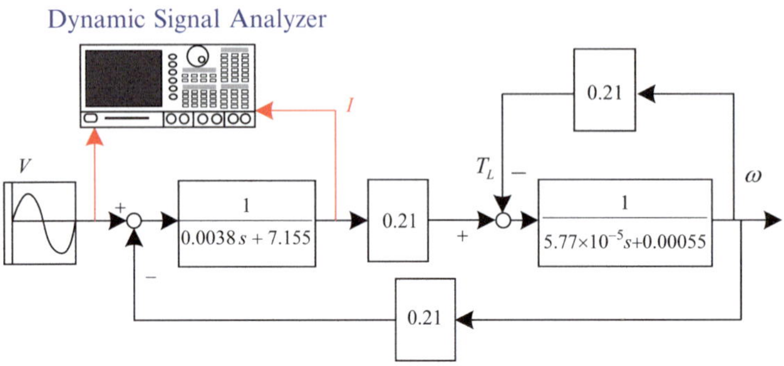

Fig. 10.6 Simulating current response by Simulink

As addressed in Chap. 5, the chain description of Fig. 10.5 with the "input" variables T_L, ω and the "outputs" I, V can be derived from (10.5), by (5.1), as

$$\begin{bmatrix} I \\ V \end{bmatrix} = \begin{bmatrix} \dfrac{1}{K_t} & \dfrac{Js+B}{K_t} \\ \dfrac{R+Ls}{K_t} & \dfrac{LJs^2+(LB+JR)s+(RB+K_eK_t)}{K_t} \end{bmatrix} \begin{bmatrix} T_L \\ \omega \end{bmatrix}$$

$$= \begin{bmatrix} G_{11} & G_{12} \\ G_{21} & G_{22} \end{bmatrix} \begin{bmatrix} T_L \\ \omega \end{bmatrix}. \tag{10.6}$$

Thus, for any mechanical impedance Z_m, the equivalent electrical impedance of the motor, denoted by Z_e, is given by

$$Z_e = \frac{V}{I} = \frac{G_{21}T_L + G_{22}\omega}{G_{11}T_L + G_{12}\omega} = \frac{G_{21}Z_m + G_{22}}{G_{11}Z_m + G_{12}}. \tag{10.7}$$

Reversely, if the electrical impedance Z_e is measured via V and I at the input port of Fig. 10.6, then the impedance of mechanical loading at the output port can be found as

$$Z_m = \frac{-G_{12}Z_e + G_{22}}{G_{11}Z_e - G_{21}}. \tag{10.8}$$

In the following, the parameter values of a DC motor are listed in Table 10.1, which will be used for computer simulations. Let the mechanical loading shown in Fig. 10.3 be a spring-damper.

Table 10.1 Parameters of the DC motor

Parameters	Values	Parameters	Values
J	5.77×10^{-5} (kgm^2)	R	7.155 (Ω)
B	0.00055 (Nms/rad)	L	0.0038 (H)
K_e	0.21(Nm/A)	K_t	0.21(Nm/A)

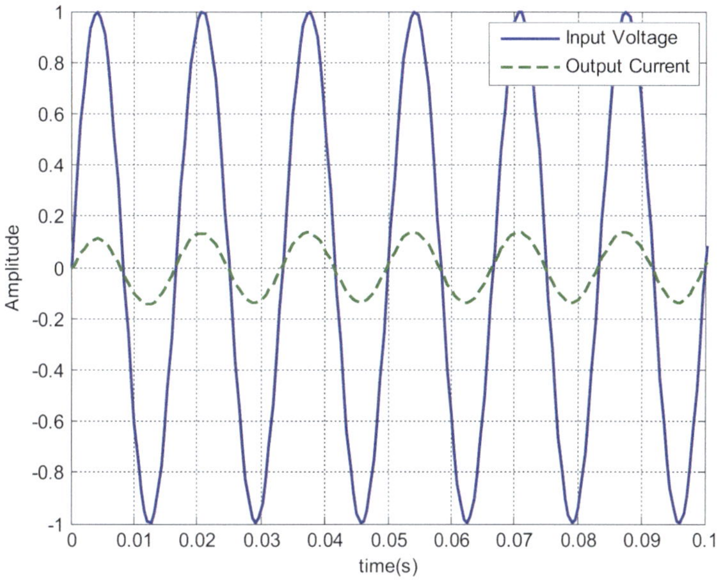

Fig. 10.7 60 Hz sinusoidal input voltage (*solid*) and output current (*dash*)

Consider the simple case of $Z_m = B$ (i.e., the damping loading) and $B = 0.00055$ (Nms/rad). For a 60 Hz sinusoidal input voltage with a magnitude of 1 injected into the DC motor, the response current can be investigated by using Simulink, as shown in Fig. 10.6 and Table 10.1.

With 60 Hz, (10.6) can be rewritten by

$$\begin{bmatrix} I \\ V \end{bmatrix} = \begin{bmatrix} 4.7619 & 0.0026 + 0.1036i \\ 34.0714 + 6.8217i & 0.0804 + 0.7449i \end{bmatrix} \begin{bmatrix} T_L \\ \omega \end{bmatrix}$$
$$= \begin{bmatrix} G_{11} & G_{12} \\ G_{21} & G_{22} \end{bmatrix} \begin{bmatrix} T_L \\ \omega \end{bmatrix}. \tag{10.9}$$

By comparing the amplitude and the phase between the input voltage and the output current in Fig. 10.7, the equivalent electrical impedance Z_e can be found as

$$Z_e = 7.2573 - 0.5896i. \tag{10.10}$$

Then the mechanical impedance Z_m can be computed, according to (10.8), as

$$Z_m = 0.00055. \tag{10.11}$$

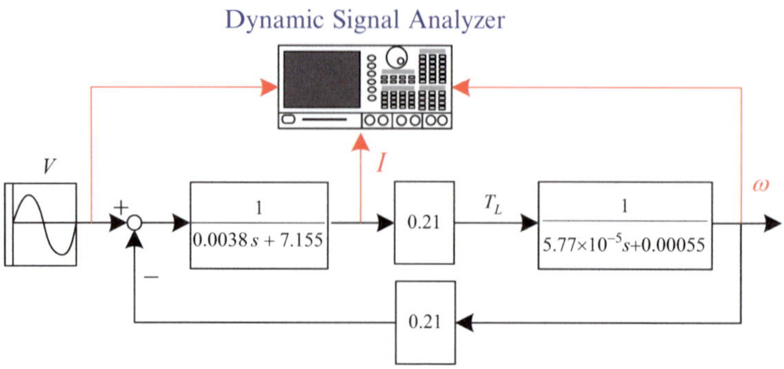

Fig. 10.8 Simulation of current and angular velocity with $T_L = 0$

Obviously, the computed mechanical impedance is identical to $B = 0.00055$ (Nms/rad). This implies that once the electrical impedance Z_e has been measured in practice, the mechanical loading can be computed by (10.8). For the special case that there is no loading (i.e., $Z_m = 0$), the electrical impedance will become $Z_e = 7.2062 - 0.5935i$.

Note that the CSD matrix G in (10.6) is called the transduction matrix [7], which describes the relationship between the inputs and outputs of an electromechanical transducer. In general, the transduction matrix can be obtained via mathematical analysis in theory or experimental data in practice. The theoretical analysis approach is given in (10.8). In the following, the procedure for obtaining the transduction matrix via a practical approach will be introduced.

Based on the definition of the transduction matrix G, one has

$$\begin{aligned} G_{11} &= \left.\frac{I}{T_L}\right|_{\omega=0} & G_{12} &= \left.\frac{I}{\omega}\right|_{T_L=0} \\ G_{21} &= \left.\frac{V}{T_L}\right|_{\omega=0} & G_{22} &= \left.\frac{V}{\omega}\right|_{T_L=0} \end{aligned}. \tag{10.12}$$

It should be noticed that it is difficult to maintain a DC motor with zero angular velocity ($\omega = 0$) in practice from an experimental setup. Therefore, an alternative method is to find the LFT matrix M of Fig. 10.4 as given by

$$\begin{aligned} M_{11} &= \left.\frac{I}{V}\right|_{T_L=0} & M_{12} &= \left.\frac{I}{T_L}\right|_{V=0} \\ M_{21} &= \left.\frac{\omega}{V}\right|_{T_L=0} & M_{22} &= \left.\frac{\omega}{T_L}\right|_{V=0} \end{aligned}. \tag{10.13}$$

Here, $V = 0$ and $T_L = 0$ can easily be implemented in the experimental setup. Then, the transduction matrix G can be calculated by transforming the LFT matrix M to a CSD matrix G. The method will be demonstrated with Simulink below.

At the condition of $T_L = 0$, for a 60 Hz sinusoidal input voltage with the magnitude of 1 injected to the DC motor, M_{11} and M_{21} can be simulated as shown in Fig. 10.8. By comparing the amplitude and the phase between the input voltage

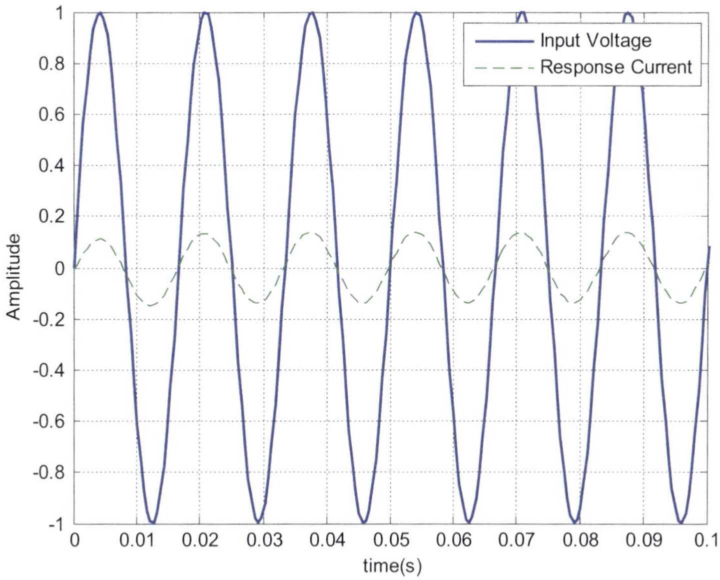

Fig. 10.9 Input voltage (*solid*) and response current (*dash*)

and the response current (and the angular velocity) as depicted in Fig. 10.9 (and Fig. 10.10), M_{11} (and M_{21}) can be calculated as

$$M_{11} = 0.1378 + 0.0114i \quad (\text{and} \;\; M_{21} = 0.1431 - 1.327i)\,. \tag{10.14}$$

On the other hand, M_{12} and M_{22} can be found by Fig. 10.11 under the condition of $V = 0$, where another motor is adopted to generate the input torque. From the results shown in Figs. 10.12 and 10.13, M_{12} and M_{22} can be computed as

$$\begin{aligned} M_{12} &= 0.1431 - 1.327i \\ M_{22} &= -13.93 + 44.2378i\,. \end{aligned} \tag{10.15}$$

The transduction matrix G can be obtained by the LFT matrix M, by (5.28), as

$$\begin{bmatrix} G_{11} & G_{12} \\ G_{21} & G_{22} \end{bmatrix} = \begin{bmatrix} 4.7619 & 0.0026 + 0.1036i \\ 34.0714 + 6.8217i & 0.0804 + 0.7449i \end{bmatrix}, \tag{10.16}$$

which is identical to (10.9) derived earlier.

This section has shown that the equivalent electrical impedance Z_e is a function of mechanical loading Z_m while the load is connected. In other words, if the dynamic properties of the motor and transmission are known in advance, then the operating condition of the mechanical loading can be monitored by the measurement of input

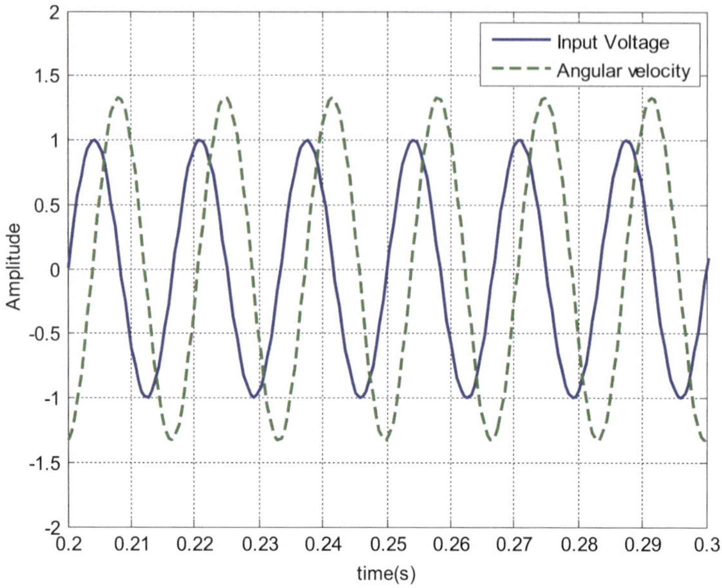

Fig. 10.10 Input voltage (*solid*) and response angular velocity (*dash*)

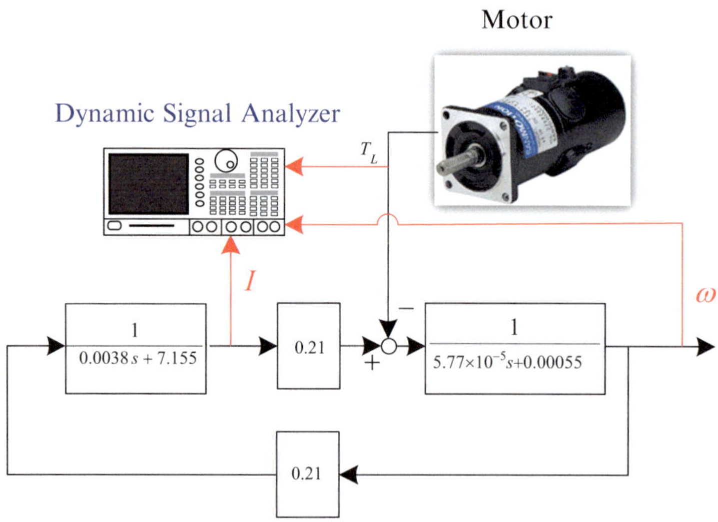

Fig. 10.11 Simulation of current and angular velocity with $V = 0$

electrical impedance of the DC motor. An application of this method for monitoring drill breakages in micro-drilling can be found in [3] where the drill breakage identified from variation of mechanical loading can be detected by the measurement of its equivalent electrical impedance.

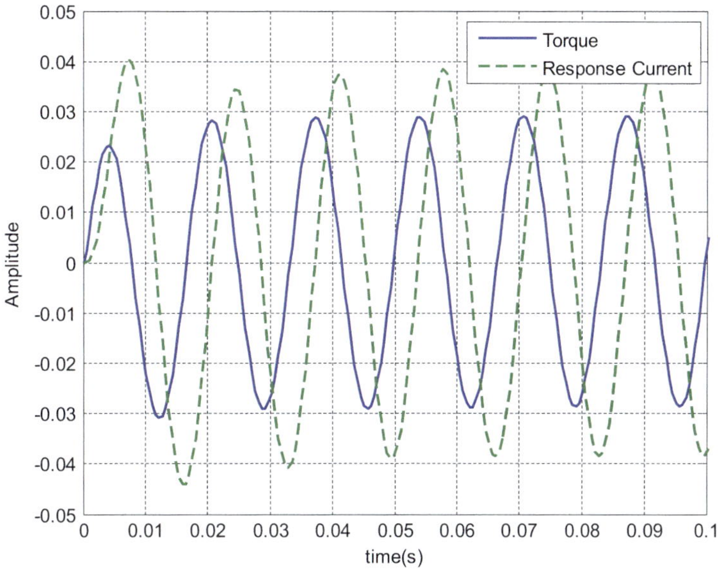

Fig. 10.12 Input torque (*solid*) and response current (*dash*)

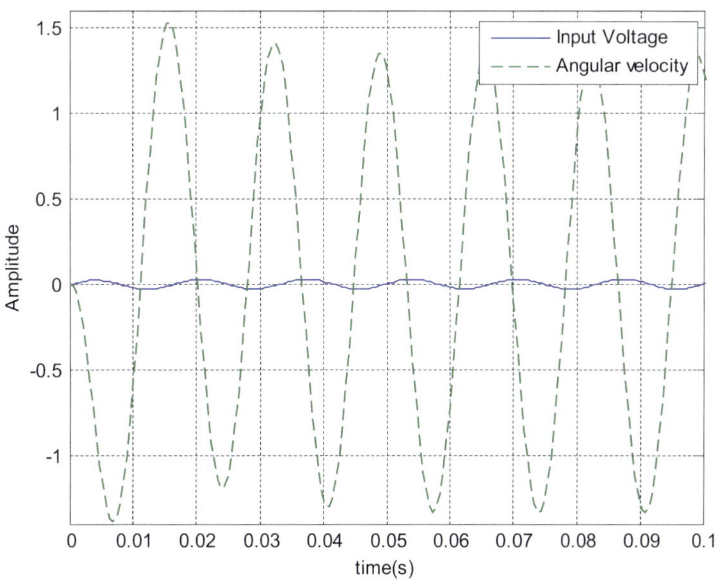

Fig. 10.13 Input torque (*solid*) and response angular velocity (*dash*)

10.3 Coprime Factorization Approach to System Identification

The inner most layer of a servo drive system is the current control loop. For the convenience of controller design in the velocity loop, the bandwidth of the current control loop must be made much higher than that of the velocity loop, e.g., ten times. When high-gain, closed-loop current control is implemented as a minor-loop control, the transfer function from the current command to the current output in the current control loop can be simplified as 1 [10]. Then, the motor model of Fig. 10.1 for speed control design can be simplified as a first-order model of $G_m = \frac{K_t}{Js+B}$, i.e., a simple motor torque constant, which converts current to torque, a single rotational inertia J, and a damping factor B. Obviously, the assumptions made here imply that the control of current in this servo loop equivalently generates the desired torque in that the magnitude of driving current is approximately proportional to the torque. Notice that the current controller of industrial servo drives, in practice, could not be tuned by the user. It is necessary to identify the controlled system for speed control design. The approach outlined in this section employs a plant representation in terms of a coprime factorization, which can estimate plant dynamics under feedback control. Identification methods for dealing with closed-loop experimental data have been developed; see [5] for an overview.

Consider the feedback configuration as depicted in Fig. 10.14, where G and K are the controlled plant to be identified and a stabilizing controller, respectively. Assume that the input signal r is available from measurements and a controller K is given such that the feedback system is stable. Then, the transfer functions from r to y and u, respectively, can be found as

$$y = H_{yr}r = \frac{GK}{1+GK}r \tag{10.17}$$

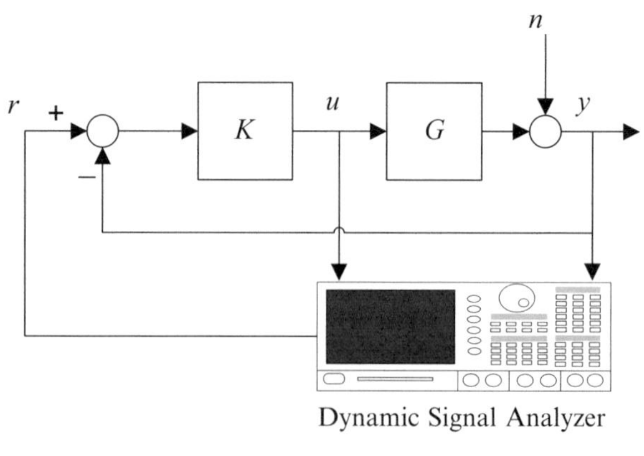

Fig. 10.14 Closed-loop system identification

Fig. 10.15 Experimental setup of servomotor

and

$$u = H_{ur} r = \frac{K}{1 + GK} r. \tag{10.18}$$

Thus, by measuring the black box transfer functions $\widehat{H}_{yr}$ and $\widehat{H}_{ur}$, an estimated G can be obtained as $\widehat{G} = \widehat{H}_{yr} \widehat{H}_{ur}^{-1}$. This shows that, in fact, the identification method based on the closed-loop data is derived from the concept of coprime factorization of the plant model, provided that the controller K has no unstable zeros. In practice, only measurements of frequency responses of signals u and y are required to obtain an estimated G within a certain bandwidth by a dynamic spectrum analyzer. Notice that the measured frequency responses of the coprime factors of a possibly unstable plant are from the closed-loop experimental data.

For example, consider the experimental setup with a servomotor and a current control power amplifier as shown in Fig. 10.15, where the transfer function from the current input to the velocity, $G_m(s)$, will be identified. Let $K(s) = 2 + \frac{4}{s}$ be chosen as a stabilizing controller and the excitation signal r be the swept sine to the closed-loop control system. The dynamic signal analyzer that measures the frequency responses of the motor velocity $y = \omega(t)$ and the motor current $u = i(t)$ simultaneously can calculate the Bode plot of G_m with experimental data. Figure 10.16 shows the measured Bode diagram; subsequently, the curve fitting gives the identified model of

$$\widehat{G}_m(s) = \frac{K_t}{Js + B} = \frac{0.2}{0.000058s + 0.00056}, \tag{10.19}$$

where $J = 0.000058$ (kgm^2), $B = 0.00056$ (Nms/rad), and $K_t = 0.2$ (Nm/amp). This model is close to basic specifications of the servomotor given in Table 10.1 and will be used for control design in the following sections.

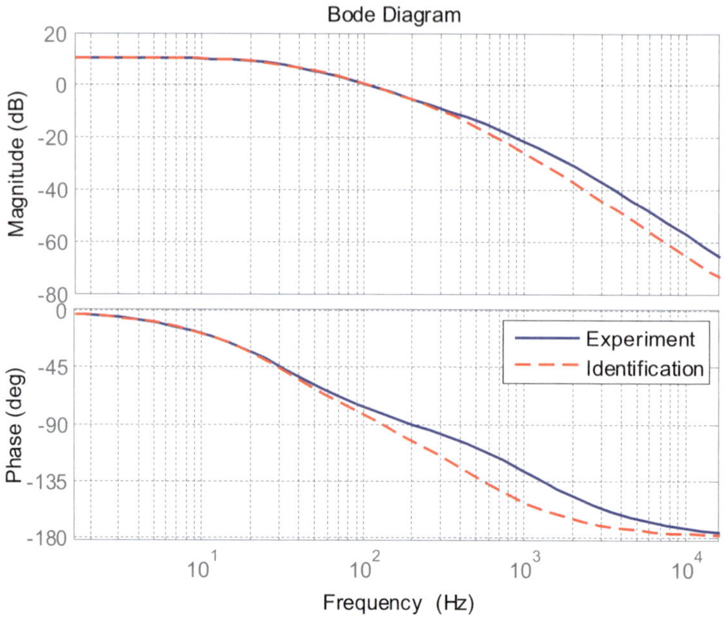

Fig. 10.16 Experimental data and identified model of DC motor

10.4 H_∞ Robust Controller Design for Speed Control

10.4.1 PDF Controller

10.4.1.1 Classical Controller Design

Consider a control law of the velocity loop, namely, a pseudo derivative feedback (PDF) controller [2] hereafter, which will be designed to generate the torque command. Figure 10.17 shows a PDF controller for the speed control of the DC motor where K_p denotes the proportional gain and K_i the integral gain constant. The closed-loop transfer function from ω^* to ω is given by

$$T(s) = \frac{\omega(s)}{\omega^*(s)} = \frac{(K_i K_t/J)}{s^2 + \left[\left(B + K_p K_t\right)/J\right]s + (K_i K_t/J)}. \tag{10.20}$$

Let

$$\zeta = \frac{\left(B + K_p K_t\right)}{2(K_i K_t J)^{0.5}}, \tag{10.21}$$

$$\omega_n = \left(\frac{K_i K_t}{J}\right)^{0.5}. \tag{10.22}$$

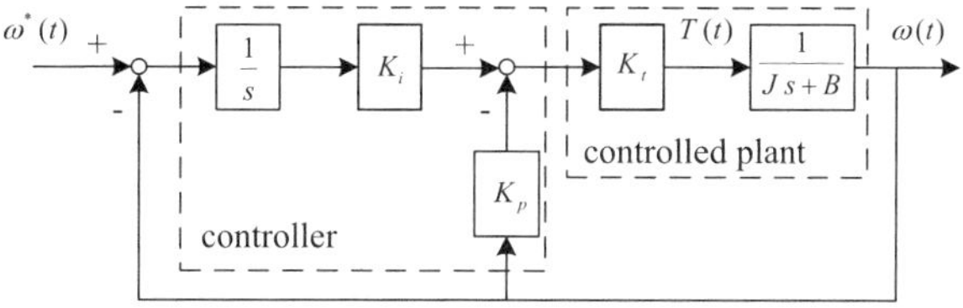

Fig. 10.17 Speed control with classical PDF controller

Then, $T(s)$ of (10.20) can be written into the standard second-order form as

$$T(s) = \frac{\omega_n^2}{s^2 + 2\zeta\omega_n s + \omega_n^2}, \tag{10.23}$$

where ζ is the damping ratio and ω_n the natural frequency. It is known [6] that the bandwidth of the standard second-order system can be found as

$$BW = \omega_n\left(1 - 2\zeta^2 + \sqrt{2 - 4\zeta^2 + 4\zeta^4}\right)^{0.5}. \tag{10.24}$$

Thus, by specifying damping ratio ζ and bandwidth BW based on design specifications and experiences, controller parameters (K_p, K_i) of the PDF controller can be determined directly, as shown in (10.25) below. The control performance resulting from this classical PDF design approach is considered here as a reference for being compared with the robust controller obtained from an H_∞ control design.

Consider the illustrated DC servomotor that has the nominal parameters given in Table 10.1. For the damping ratio of $\zeta^* = 0.9$ and the bandwidth of $BW^* = 100$ Hz, the PDF controller parameters are determined by

$$K_i = \frac{J\omega_n^2}{K_t} = 194.74, \quad K_p = \frac{2J\zeta\omega_n - B}{K_t} = 0.41. \tag{10.25}$$

However, due to the external load and/or the parameter inaccuracy, the moment of inertia of a servo drive system often varies in practice during operations. The variation of parameter J which is away from its nominal value $J_o = 5.77 \times 10^{-5}$ (kgm^2) will lead to the significant alteration in its output response of speed control. To illustrate the effect of model uncertainty, three cases of $J = 0.1J_o$, $J = J_o$, and $J = 10J_o$ are investigated, respectively, by computer simulations. Let the step speed command be 100 rad/s. Figure 10.18 shows the step responses based on the classical PDF design obtained by (10.25) above. Note that the integral term of the PDF controller is to assure a zero steady-state error due to the step input.

In Fig. 10.17, according to the closed-loop transfer function (10.20) from ω^* to ω, the characteristic equation of the speed control system can be obtained as $Js^2 + (B + K_pK_t)s + K_tK_i = 0$. To characterize the variations of the closed-loop poles with respect to J, let $1 + kL(s) = 0$, where $k = \frac{1}{J}$ and

$$L(s) = \frac{\left(B + K_pK_t\right)s + K_tK_i}{s^2}. \tag{10.26}$$

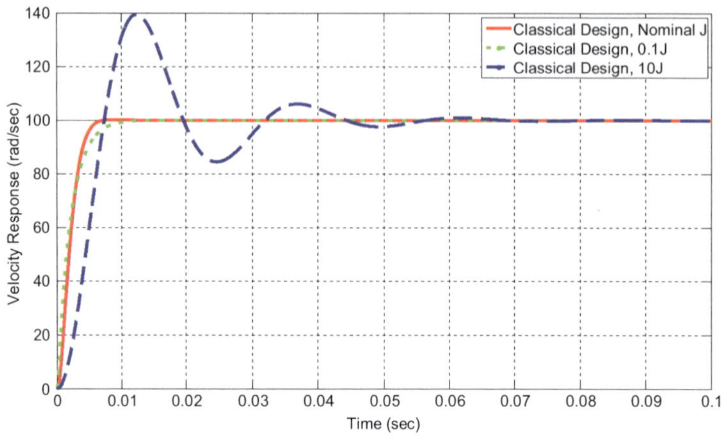

Fig. 10.18 Time responses subject to model uncertainty J

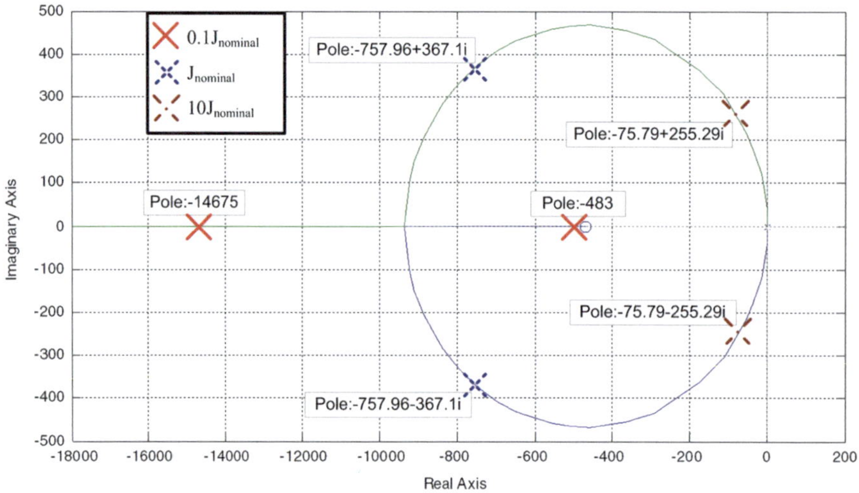

Fig. 10.19 Root locus plot according to variation of $1/J$

Figure 10.19 presents the root locus with a zero located at $-\frac{K_t K_i}{B+K_p K_i}$ and two open-loop poles at the origin. Then, for $J = J_o = 5.77 \times 10^{-5}$ (kgm^2), the closed-loop control system has the complex conjugate poles at $-757.96 \pm 367.1i$ (i.e., $\zeta^* = 0.9$ and $\omega_n^* = 842.18$ rad/s). As can be expected from the root locus of Figs. 10.19 and 10.18 shows that the speed response resulting from the conventional design with $J = 10J_o$ becomes slightly slower than that of the nominal case $J = J_o$, and its controlled output oscillates significantly with the overshoot around 40 % (i.e., 40 rad/s). However, the speed response resulting from $J = 0.1J_o$ has no overshoot,

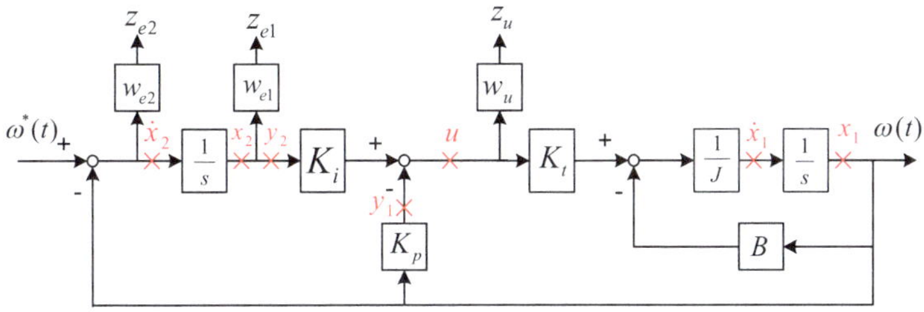

Fig. 10.20 Weighed PDF controller design

and it is almost the same as that of $J = J_o$, which can be expected from Fig. 10.19. This implies that if the controlled plant has certain parameter variations, how to ensure the robustness in control performance becomes an important design issue. In the following, H_∞ control design is adopted to find the PDF (or PDFF) controller in the velocity loop to enhance the dynamic stiffness and to reduce from the effect of system uncertainty.

10.4.1.2 Robust Controller Design with Constant Weighting Functions

Many control design approaches for improving the dynamic stiffness have been proposed. Undoubtedly, H_∞ control is one of the most appropriate techniques for dealing with robust stability with respect to parameter variations, which appear commonly in industrial drives [1]. In practice, high dynamic stiffness often results from large control efforts. As a matter of fact, robust H_∞ design often leads to high-order dynamic controllers. Hence, a trade-off between controller order and system performance should be considered in the formulation of design problems. In the following, H_∞ control design is adopted to find the PDF (or PDFF) controller in the velocity loop to enhance the dynamic stiffness.

Let $G_m(s) = \frac{K_t}{Js+B}$ be the transfer function of the DC servomotor from the torque to the velocity and $G_m(s) \overset{s}{=} \left[\begin{array}{c|c} A_m & B_m \\ \hline C_m & 0 \end{array}\right] = \left[\begin{array}{c|c} -\frac{B}{J} & \frac{K_t}{J} \\ \hline 1 & 0 \end{array}\right]$ denote a state-space realization.

Consider the H_∞ PDF design scheme of Fig. 10.20, where weighting functions w_{e1}, w_{e2}, and w_u should be chosen properly to satisfy the desired specifications. With the trade-off between the system performance and computational complexity, let the weighting functions all be positive constants under practical consideration.

Let P denote a SCC plant in the control framework shown in Fig. 10.21, where the closed-loop transfer function from ω^* to $z = [z_{e1}, z_{e2}, z_u]^{\mathrm{T}}$ is denoted by $\mathrm{LFT_l}(P,K_\infty)$. Then, the PDF control design of Fig. 10.20 is formulated into the

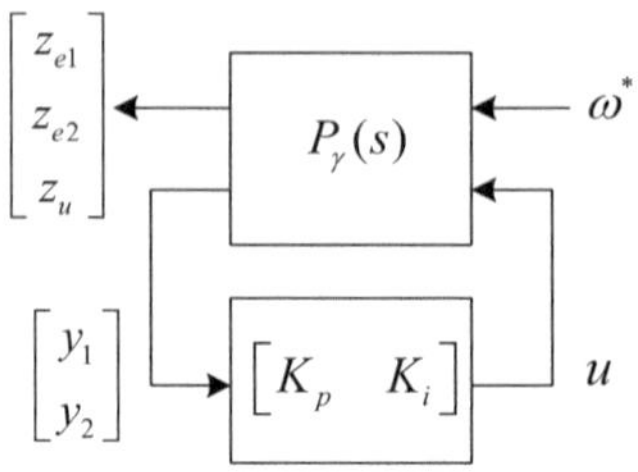

Fig. 10.21 Standard framework of PDF control problem

standard H_∞ control problem, and the stabilizing controller $K_\infty = [K_p, K_i]$ is found such that for a prespecified $\gamma > 0$,

$$\|\mathrm{LFT_l}(P, K_\infty)\|_\infty < \gamma \ \text{ or equivalently } \ \left\|\mathrm{LFT_l}\left(P_\gamma, K_\infty\right)\right\|_\infty < 1, \tag{10.27}$$

where

$$P_\gamma = \begin{bmatrix} \frac{1}{\gamma}P_{11} & \frac{1}{\gamma}P_{12} \\ P_{21} & P_{22} \end{bmatrix} \tag{10.28}$$

and

$$P(s) \overset{s}{=} \left[\begin{array}{c|cc} A & B_1 & B_2 \\ \hline C_1 & D_{11} & D_{12} \\ C_2 & D_{21} & 0 \end{array}\right] = \begin{array}{c} \\ \dot{x}_1 \\ \dot{x}_2 \\ \hline z_{e1} \\ z_{e2} \\ z_u \\ \hdashline y_1 \\ y_2 \end{array} \begin{array}{c} \begin{array}{cccc} x_1 & x_2 & \omega^* & u \end{array} \\ \left[\begin{array}{cc|c:c} -\frac{B}{J} & 0 & 0 & \frac{K_t}{J} \\ -1 & 0 & 1 & 0 \\ \hline 0 & w_{e1} & 0 & 0 \\ -w_{e2} & 0 & w_{e2} & 0 \\ 0 & 0 & 0 & w_u \\ \hdashline 1 & 0 & 0 & 0 \\ 0 & 1 & 0 & 0 \end{array}\right] \end{array}. \tag{10.29}$$

The SCC plant P above is a $(3+2)\times(1+1)$ transfer function matrix with $\dim(A) = 2$. Since $C_2 = I$ and $D_{21} = 0$, this is a special state feedback (SF) problem as mentioned in previous chapters. It can also be verified that

(a) (A,B_2) is stabilizable and (C_2, A) is detectable.

(b) $\operatorname{rank}\begin{bmatrix} A - j\omega I & B_2 \\ C_1 & D_{12} \end{bmatrix} = 1 + 2$ and $\operatorname{rank} D_{12} = 1$.

Based on the algorithm presented in Chap. 9, it can be expected that the static controller parameters (i.e., K_p and K_i) can be solved directly by the proposed SF H_∞ solution process.

Naturally, selecting different weighting functions will lead to different controller parameters. For illustrating the relationship between weighting functions and the resulting controllers, only one weighting function is adjusted at a time in the following case.

Case 1 Fixed $w_{e1} = 1$, $w_u = 1$, and $\gamma = 1$.

Controller	$w_{e2} = 0.2$	$w_{e2} = 0.4$	$w_{e2} = 0.8$
K_p	0.20	0.43	1.33
K_i	1	0.91	0.78

In this case, the controller parameter K_p is affected significantly by weighting function w_{e2}.

Case 2 Fixed $w_{e2} = 0.4$, $w_u = 1$, and $\gamma = 1$.

Controller	$w_{e1} = 0.1$	$w_{e1} = 1$	$w_{e1} = 10$
K_p	0.43	0.43	0.44
K_i	0.093	0.93	9.26

Obviously, the controller parameter K_i is proportional to the weighting function w_{e1}, while K_p remains almost the same for different w_{e1}.

Case 3 Fixed $w_{e2} = 0.4$, $w_{e1} = 1$, and $\gamma = 1$

Controller	$w_u = 0.1$	$w_u = 1$	$w_u = 10$
K_p	4.36	0.43	0.04
K_i	9.28	0.93	0.09

In this case, the ratio between K_p and K_i (i.e., K_i/K_p) remains almost the same, but both of them are inversely proportional to w_u. Recall that, in the root locus of Fig. 10.19 with varying $\frac{1}{J}$, the poles and zero of the loop transfer function of $L(s)$ are $p_1 = 0$, $p_2 = 0$, and $z = -\frac{K_t K_i}{B + K_p K_i} \approx -\frac{K_i}{K_p}$, respectively. Therefore, K_i/K_p, determined by the selections of w_{e1} and w_{e2} as discussed above, will naturally affect the tendency of the closed-loop poles. Thus, the weighting function w_u can be chosen properly to achieve the desired closed-loop poles, which characterize the system response.

Consequently, for $w_{e1} = 19.51$, $w_{e2} = 0.015$, and $w_u = 0.05$, solving the SF design problem of Fig. 10.21 with $\gamma = 1$, the controller parameters are found as

$$K_i = 272.9 \quad \text{and} \quad K_p = 0.55. \tag{10.30}$$

Analog to the classical PDF design, the step speed responses resulting from the H_∞ SF design are shown in Fig. 10.22. Note that the weighting functions are chosen purposely such that the step response with the nominal case $J = J_o$ is close to that of the classical PDF design as shown in Fig. 10.23. Compared with that of the classical PDF design, Fig. 10.24 shows the step responses with $J = 10J_o$ in which the maximum overshoot and settling time have been improved in the illustrated design example by using the H_∞ design approach in the case of the actual rotational inertia being ten times of its nominal value. Of course, in addition to better performance

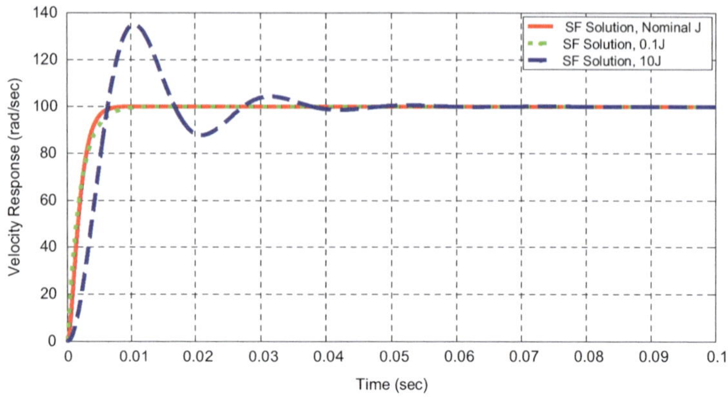

Fig. 10.22 Step responses with robust PDF design

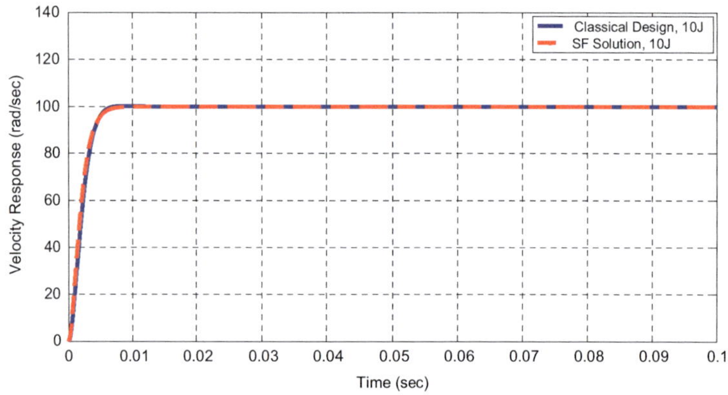

Fig. 10.23 Comparison of step responses with nominal $J = J_o$

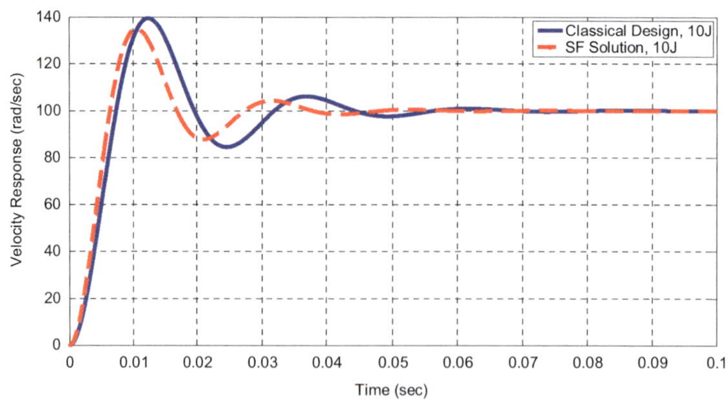

Fig. 10.24 Comparison of step responses with $J = 10J_o$

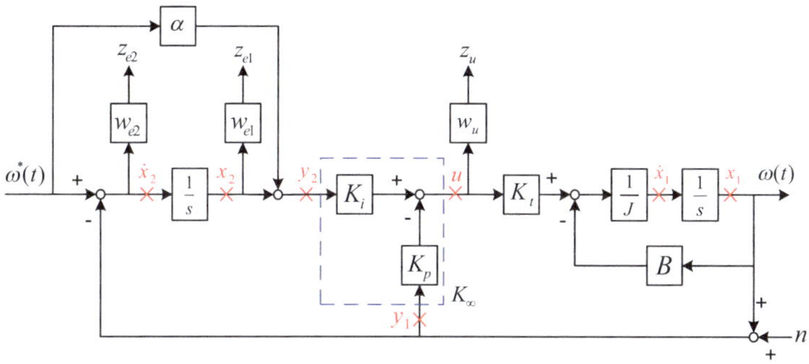

Fig. 10.25 H_∞ design of PDFF control scheme

achieved, the other prominent feature of the H_∞ approach is the systematic solution procedure which guarantees closed-loop stability and applicability for multivariable systems. No comparison is presented here for the case of $J = 0.1\,J_o$, because the simulations are very similar to those with the nominal inertia.

10.4.2 PDFF Controller

The pseudo derivative feedback with feedforward (PDFF) control scheme is an extended version of the PDF controller via augmenting a command feedforward path that provides extra freedom in the tuning procedure. As a controller scheme widely used in industrial practice, the classical PDFF control scheme has three constant gains (α, K_p, and K_i) for the use of performance tuning in that the feedforward term α makes the system more responsive to the command. The convenience and simplicity of tuning parameters in the PDFF scheme makes it appealing in many industrial drives. Nevertheless, the classical PDFF scheme may not easily yield satisfactory stability and performance simultaneously when the system has high-order dynamics and disturbances. To obtain higher dynamic stiffness and better stability robustness, this section investigates an extension of the classical PDFF control scheme by allowing the control parameters, K_p and K_i, to be dynamic controllers, denoted by $K(s) = \left[\, K_p(s)\ K_i(s) \,\right]$. Such a scheme is then called *advanced PDFF* in the following design examples and is recast as a particular two-degree-of-freedom configuration.

Analog to the PDF design of Fig. 10.20 above, consider the PDFF design problem of Fig. 10.25 involving an extra input n. This, in fact, becomes a weighted mixed-sensitivity design but has a particular controller scheme. Let the weighting constants w_{e1}, w_{e2}, and w_u all be positive. Then, the closed-loop transfer function from $[\omega^*,n]^T$ to $[z_{e1},z_{e2},z_u]^T$ is denoted by $\mathrm{LFT_l}(P,K_\infty)$, where the SCC plant P is a $(3+2)\times(2+1)$ transfer matrix and a state-space realization can be readily found in Fig. 10.25 as

$$P(s) \overset{s}{=} \left[\begin{array}{c|cc} A & B_1 & B_2 \\ \hline C_1 & D_{11} & D_{12} \\ C_2 & D_{21} & 0 \end{array}\right] = \begin{array}{c} \\ \dot{x}_1 \\ \dot{x}_2 \\ z_{e1} \\ z_{e2} \\ z_u \\ y_1 \\ y_2 \end{array} \begin{array}{c} \begin{array}{ccccc} x_1 & x_2 & \omega^* & n & u \end{array} \\ \left[\begin{array}{cc|cc|c} -\frac{B}{J_0} & 0 & 0 & 0 & \frac{K_t}{J_0} \\ -1 & 0 & 1 & -1 & 0 \\ \hline 0 & w_{e1} & 0 & 0 & 0 \\ -w_{e2} & 0 & w_{e2} & -w_{e2} & 0 \\ 0 & 0 & 0 & 0 & w_u \\ \hline 1 & 0 & 0 & 1 & 0 \\ 0 & 1 & \alpha & 0 & 0 \end{array}\right] \end{array} \tag{10.31}$$

Since D_{21} is nonsingular with the feedforward path gain $\alpha > 0$, this formulation is no longer a special SF design problem but rather a general case. It can be verified that

(a) (A,B_2) is stabilizable and (C_2,A) is detectable.

(b) $\text{rank}\begin{bmatrix} A - j\omega I & B_2 \\ C_1 & D_{12} \end{bmatrix} = 2 + 1$ and $\text{rank}\, D_{12} = 1$.

(c) $\text{rank}\begin{bmatrix} A - j\omega I & B_1 \\ C_2 & D_{21} \end{bmatrix} = 2 + 2$ and $\text{rank}\, D_{21} = 2$.

Note that the extra input n has been introduced in the PDFF design problem of Fig. 10.25 such that the above assumptions all are satisfied. Thus, the explicit solution of this PDFF control scheme problem can be solved by utilizing the CSD method presented in Chap. 9.

For $w_{e1} = 32.75$, $w_{e2} = 0.05$, $w_u = 1$, and $\alpha = 0.04$, solving the H_∞ suboptimal control problem of $\|\text{LFT}_l(P_\gamma, K_\infty)\|_\infty < 1$ with $\gamma = 1.75$ will yield

$$K_\infty(s) = [K_p(s), K_i(s)] \overset{s}{=} \left[\begin{array}{cc|cc} -1.5190\times10^{5} & 9.7567\times10^{7} & 0 & 0 \\ -4.9005\times10^{-10} & -25 & 1 & -25 \\ \hline 41.7021 & -2.6788\times10^{4} & 0 & 0 \end{array}\right]. \tag{10.32}$$

The system step responses resulting from the PDFF scheme controller in the H_∞ design are shown in Fig. 10.26, where the closed-loop poles with $J = J_o$ are given by -9.5392, -25, -53.8201, -151833.9 and the closed-loop zeros are -9.5392, -25. As can be seen, for the case of $J = 10J_o$, the step response becomes slower than that of the previous PDF design although there is a feedforward gain $\alpha = 0.04$ in this PDFF controller. This is due to the H_∞ design formulation that appears with pole-zero cancellations at $p_1 = -\frac{1}{\alpha} = -25$ and $p_2 = -\frac{B}{J_0} = -9.539$, which would affect the control performance significantly while uncertainties occur. This inherent pole-zero cancellation property in the closed-loop system resulting from the weighted mixed-sensitivity design was investigated in [12]. In the following, the concept of the H_∞ loop-shaping design addressed in [4, 8, 11] is employed to overcome this problem.

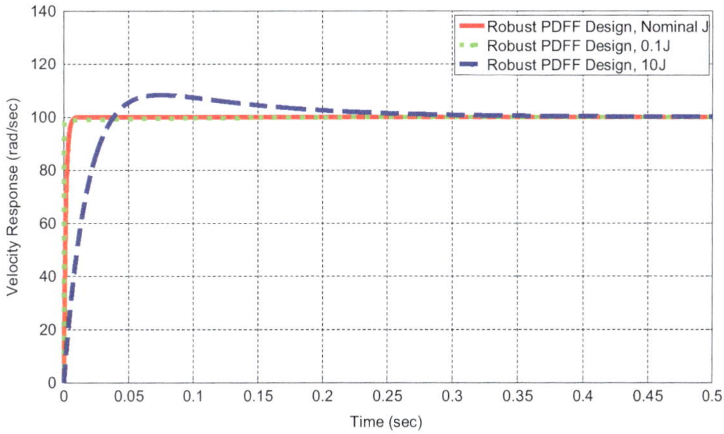

Fig. 10.26 Step responses resulting from robust PDFF design

10.4.3 Coprime Factorization Approach to Advanced PDFF Controller

The design example here employs the coprime factorization descriptions to formulate the advanced PDFF design into an H_∞ weighted mixed-sensitivity problem [13]. The advanced PDFF controller is found by a partial pole placement technique [9] and the loop-shaping design of the normalized coprime factorization [4]. The proposed design method provides a useful property in that the performance, such as bandwidth and stability, can be designed simultaneously.

Recall that $G_m(s) \stackrel{s}{=} \left[\begin{array}{c|c} A_m & B_m \\ \hline C_m & 0 \end{array}\right] = \left[\begin{array}{c|c} -\frac{B}{J} & \frac{K_t}{J} \\ \hline 1 & 0 \end{array}\right]$ denotes the transfer function of the servomotor from the current (torque) command to velocity. To retain the structure of the feedforward gain and the integrator in the PDFF controller scheme, consider the augmented one-input-two-output plant given by

$$G_s = \begin{bmatrix} -\frac{C_f}{s} \\ 1 \end{bmatrix} G_m \stackrel{s}{=} \left[\begin{array}{c|c} A_s & B_s \\ \hline C_s & 0 \end{array}\right]. \tag{10.33}$$

The problem formulation here employs the concept of the H_∞ loop-shaping design [8], where the original plant G_m is shaped by $\frac{C_f}{s} G_m$ to obtain an augmented plant G_s. To yield satisfactory stability and performance simultaneously, the coprime factorization (CF) description of the controlled plant discussed in Sect. 9.4 is employed to formulate the H_∞ weighted mixed-sensitivity problem.

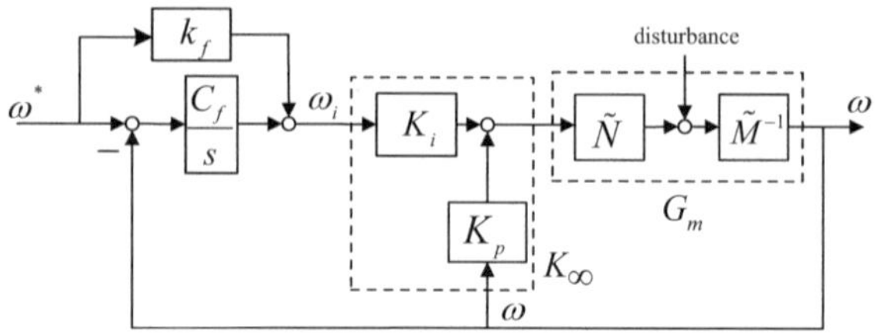

Fig. 10.27 Advanced PDFF scheme with CF description

Consider the control design problem shown in Fig. 10.27, where $G_m = \tilde{M}^{-1}\tilde{N}$ denotes a left coprime factorization. Notice that there are two outputs of the augmented plant G_s, i.e., ω, the output of G_m, and ω_i, the output of $-\frac{C_f}{s}G_m$, as shown in Fig. 10.27. From (6.30), one has

$$\begin{bmatrix} \tilde{N}(s) & \tilde{M}(s) \end{bmatrix} \overset{s}{=} \left[\begin{array}{c|cc} A_m + H_m C_m & B_m & H_m \\ \hline \tilde{w} C_m & 0 & \tilde{w} \end{array}\right], \tag{10.34}$$

where H_m is an observer gain matrix such that $A_m + H_m C_m$ is stable and $\tilde{w} > 0$. Furthermore, a coprime factorization of $G_s = \tilde{M}_s^{-1}\tilde{N}_s$ can be found as

$$\tilde{M}_s \overset{s}{=} \left[\begin{array}{cc|cc} A_m + H_m C_m & 0 & 0 & H_m \\ 0 & -k_f^{-1}C_f & -k_f^{-1} & 1 \\ \hline 0 & k_f^{-1}C_f & k_f^{-1} & 0 \\ \tilde{w} C_m & 0 & 0 & \tilde{w} \end{array}\right] = \left[\begin{array}{c|c} A_s + H_s C_s & H_s \\ \hline \tilde{W}_s C_s & \tilde{W}_s \end{array}\right] \tag{10.35}$$

$$\tilde{N}_s \overset{s}{=} \left[\begin{array}{cc|c} A_m + H_m C_m & 0 & B_m \\ 0 & -k_f^{-1}C_f & 0 \\ \hline 0 & k_f^{-1}C_f & 0 \\ \tilde{w} C_m & 0 & 0 \end{array}\right] = \left[\begin{array}{c|c} A_s + H_s C_s & B_s \\ \hline \tilde{W}_s C_s & 0 \end{array}\right], \tag{10.36}$$

where

$$H_s = \begin{bmatrix} 0 & H_m \\ -k_f^{-1} & 1 \end{bmatrix} \quad \text{and} \quad \tilde{W}_s = \begin{bmatrix} k_f^{-1} & 0 \\ 0 & \tilde{w} \end{bmatrix}. \tag{10.37}$$

Figure 10.28 shows the weighted mixed-sensitivity design problem of the coprime factorization approach. Note that similar to that of ω_d, the input command ω^* could be considered as one of the exogenous inputs in Fig. 10.28. This implies that both the command tracking and disturbance rejection can be taken into account simultaneously in this design method.

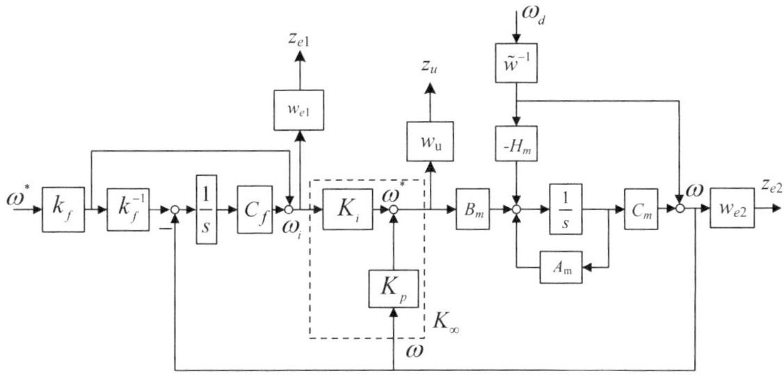

Fig. 10.28 PDFF design formulation of CF loop-shaping approach

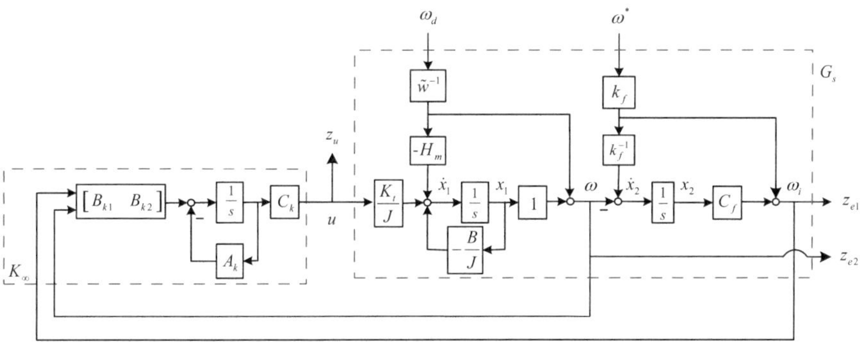

Fig. 10.29 State-space representation of CF design approach

To explore the loop-shaping approach with the internal weights, let the output weighting functions $w_u = 1.35$, $w_{e1} = 1$, $w_{e2} = 1$ for satisfying the design specification. By redrawing Fig. 10.28, the proposed design problem can be reconstructed as depicted in Fig. 10.29 which gives a clearer representation. An important feature of this formulation is that both of the feedforward gain k_f and the integrator of the PDFF controller structure with internal gain C_f have been taken into account as a part of the augmented plant. The existence of feedforward term k_f makes the system more responsive to commands and provides extra freedom in the tuning procedure. Note that the overall controller to be implemented should comprise an integrator with a gain of C_f, a feedforward gain k_f, and a computed dynamic H_∞ controller $K_\infty(s) = \begin{bmatrix} K_p(s) & K_i(s) \end{bmatrix} \overset{s}{=} \left[\begin{array}{c|cc} A_k & B_{k1} & B_{k2} \\ \hline C_k & D_{k1} & D_{k2} \end{array}\right]$ as

$$u = K_\mathrm{i}(s)\left[k_\mathrm{f}\omega^* + \frac{C_\mathrm{f}}{s}\left(\omega^* - \omega\right)\right] + K_\mathrm{p}(s)\omega. \tag{10.38}$$

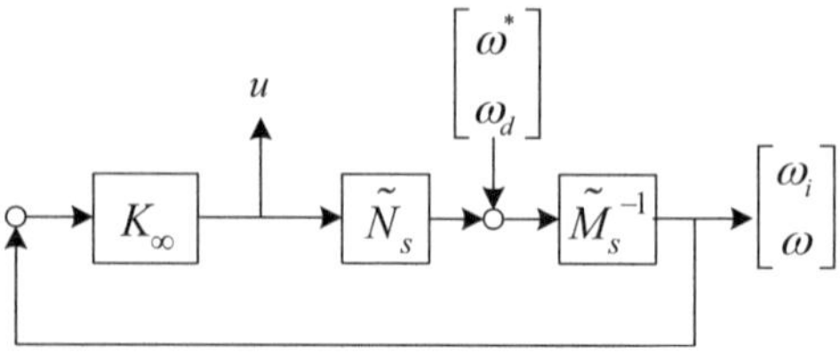

Fig. 10.30 Equivalent design problem

By the routine algebraic manipulations, the design problem can be transferred into Fig. 10.30. Then, the advanced PDFF control design problem becomes the standard H_∞ control problem of Fig. 10.30, where $K_\infty = [K_\mathrm{p}\ K_\mathrm{i}]$ and the closed-loop transfer function from $\begin{bmatrix} \omega^* \\ \omega_d \end{bmatrix}$ to $[w_i, w, u]^\mathrm{T}$ is denoted by $\mathrm{LFT}_\mathrm{l}(P, K_\infty)$. The H_∞ control problem is therefore to find a stabilizing controller $K_\infty = [K_\mathrm{p}\ K_\mathrm{i}]$ such that, for a prespecified $\gamma > 0$,

$$\left\| \mathrm{LFT}_\mathrm{l}\left(P_\gamma(s), K_\infty(s)\right) \right\|_\infty = \left\| \begin{bmatrix} S \\ K_\infty S \end{bmatrix} \tilde{M}_s^{-1} \right\|_\infty < 1, \tag{10.39}$$

where $S = (I - G_s K_\infty)^{-1}$ is the sensitivity function. A state-space realization of the SCC plant P can be readily found as

$$P(s) \overset{s}{=} \left[\begin{array}{c|cc} A & B_1 & B_2 \\ \hline C_1 & D_{11} & D_{12} \\ C_2 & D_{21} & 0 \end{array}\right] = \left[\begin{array}{cc|cc|c} -\frac{B}{J_0} & 0 & 0 & -H_m \tilde{w}^{-1} & \frac{K_t}{J_0} \\ -1 & 0 & 1 & -\tilde{w}^{-1} & 0 \\ \hline 0 & C_f & k_f & 0 & 0 \\ 1 & 0 & 0 & \tilde{w}^{-1} & 0 \\ 0 & 0 & 0 & 0 & 1 \\ \hline 0 & C_f & k_f & 0 & 0 \\ 1 & 0 & 0 & \tilde{w}^{-1} & 0 \end{array}\right]. \tag{10.40}$$

Now $k_f = 0.5$ and $C_f = 500$ are selected as an illustrated design example. Then, the normalized coprime factors of $G_m(s) \overset{s}{=} \left[\begin{array}{c|c} A_m & B_m \\ \hline C_m & 0 \end{array}\right] = \left[\begin{array}{c|c} -9.5392 & 3642.2291 \\ \hline 1 & 0 \end{array}\right]$ can be found by $\tilde{w} = 1$ and $H_m = -YC_m = -3{,}632.7024$, where the ARE of

$$YA_m^\mathrm{T} + A_m Y - YC_\mathrm{m}^\mathrm{T} C_m Y + B_m B_\mathrm{m}^\mathrm{T} = 0 \tag{10.41}$$

gives $Y = 3{,}632.7024$. Then, for the prespecified $\gamma = 1.75$, solving the H_∞ suboptimal control problem of $\|\mathrm{LFT}_\mathrm{l}(P_\gamma, K_\infty)\|_\infty < 1$ with P given by (10.40) will yield the center controller as

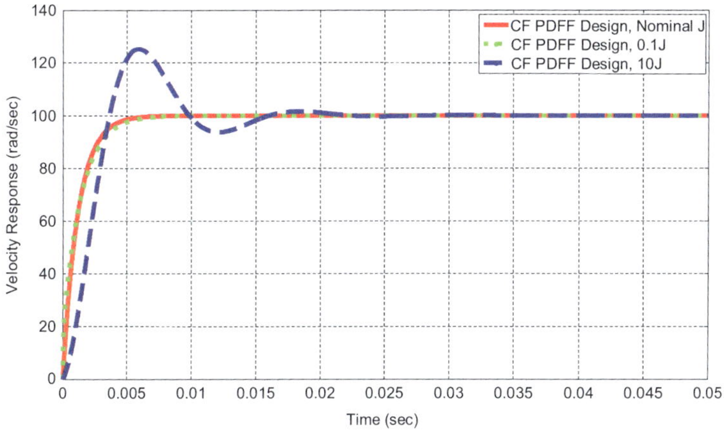

Fig. 10.31 Step responses with CF approach PDFF design

$$K_\infty(s) \overset{s}{=} \left[\begin{array}{c|c} A_k & B_{k1}\ B_{k2} \\ \hline C_k & 0\ \ 0 \end{array}\right] = \left[\begin{array}{cc|cc} -409900 & -628490 & 702.4671 & -286.3634 \\ -618640 & -951850 & 1032.8 & -373.3741 \\ \hline -758.59 & -1098.36 & 0 & 0 \end{array}\right] \tag{10.42}$$

It can also be found in Fig. 10.29 that the state-space realization of the closed-loop system from $\begin{bmatrix} T_L \\ \omega^* \end{bmatrix}$ to ω, denoted $T_{\mathrm{cl}}(s)$, is given by

$$T_{cl}(s) \overset{s}{=} \begin{array}{c} \\ \dot{x}_1 \\ \dot{x}_2 \\ \dot{x}_3 \\ \omega \end{array} \begin{array}{c} \begin{array}{ccc|c|c} x_1 & x_2 & x_3 & T_L & \omega^* \end{array} \\ \left[\begin{array}{ccc|c:c} -\dfrac{B}{J} & 0 & \dfrac{K_t}{J}C_k & \dfrac{1}{J} & 0 \\ -1 & 0 & 0 & 0 & 1 \\ B_{k1} & B_{k2}C_f & A_k & 0 & B_{ki}k_f \\ \hline 1 & 0 & 0 & 0 & 0 \end{array}\right] \end{array} \tag{10.43}$$

where the closed-loop poles are given by $-1{,}356{,}300$, -842.2846, -3642.2, and $-1{,}000$ and the closed-loop zeros are -3642.2 and $-1{,}000$, which are from $A_m + H_m C_m = -3{,}642.2$ and $-k_f^{-1} C_f = -1{,}000$. It was addressed in [9] that all of the eigenvalues of $A_s + H_s C_s = \begin{bmatrix} A_m + H_m C_m & 0 \\ 0 & -k_f^{-1} \end{bmatrix}$ are the closed-loop poles in the design problem of Fig. 10.28. This implies that the advanced PDFF design can provide an alternative for partial pole placement by assigning $\lambda(A_s + H_s C_s)$. Obviously, this CF design approach does not have pole-zero cancellation occurring at $-\frac{B}{J_0} = -9.539$, which would affect control performance. The system step responses resulting from the H_∞ design are shown in Fig. 10.31.

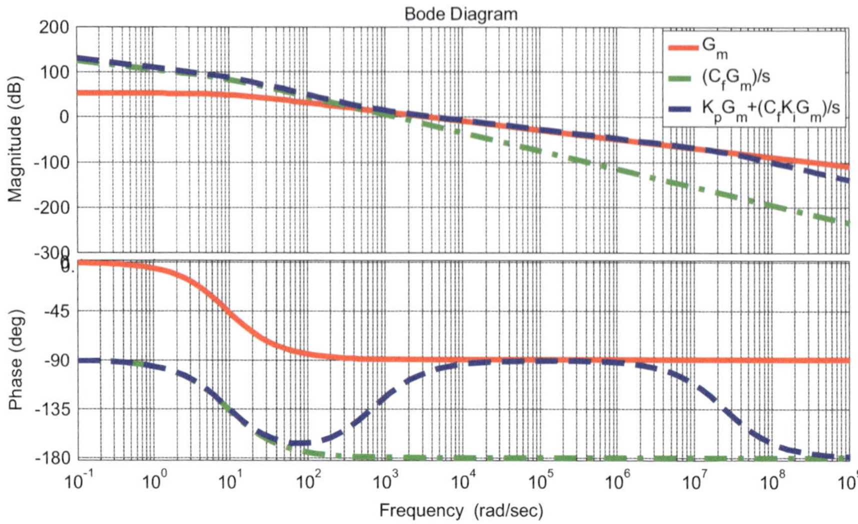

Fig. 10.32 Bode plots resulting from CF design approach

In the loop-shaping design, the structure and gains of the internal weighting function $\frac{C_\mathrm{f}}{s}$ can be selected to obtain the desired magnitude of the frequency response of a loop transfer function $\frac{C_\mathrm{f}}{s}G_m$, and then, the controller $K_\infty = [K_\mathrm{p}\ K_\mathrm{i}]$ is found with the guaranteed stability properties [8]. Figure 10.32 shows the Bode plots of the original plant G_m (the solid line), the shaped plant $\frac{C_\mathrm{f}}{s}G_m$ (the dashed line), and the loop transfer function $\left(K_\mathrm{p} + \frac{C_\mathrm{f}K_\mathrm{i}}{s}\right)G_m$ (the dotted line), respectively, where $\left|\left(K_p(s) + \frac{C_\mathrm{f}K_i(s)}{s}\right)G_m(s)\right| \approx \left|\frac{C_\mathrm{f}}{s}G_m(s)\right|$ for $s = j\omega$ within the frequencies of 1,000 rad/s. As can be seen in the dashed line of Fig. 10.32, the original plant G_m is shaped by $\frac{C_\mathrm{f}}{s}G_m$, such as the magnitude plot, to force the open-loop magnitude to approximate to infinity at very low frequencies and the stabilizing controller $K_\infty = [K_\mathrm{p}\ K_\mathrm{i}]$ to provide phase compensation, such as phase plots, to improve the transient response of the control system.

It should be noted that the weighting functions in the above design examples are chosen purposely such that the step response of the closed-loop transfer function from ω^* to ω resulting from the above three H_∞ designs, i.e., SF PDF, robust PDFF, and CF PDFF design, all are close to that of the classical PDF design at the nominal case $J = J_\mathrm{o}$, as shown in Fig. 10.33. The peak values of the control efforts in both robust PDFF and CF PDFF designs are bigger than the other two due to the added feedforward loop aiming at fast response. In the case of $0.1J_\mathrm{o}$ shown in Fig. 10.34, the robust PDFF design shows even faster response, and the control efforts in the two PDFF structures are naturally higher. Note that the peak value of control efforts due to the step input can be reduced in practice by the technique of soft starting such as the S-curve. Compared with that of the classical PDF design, Fig. 10.35 shows that the step response resulting with $J = 10J_\mathrm{o}$ has been improved in

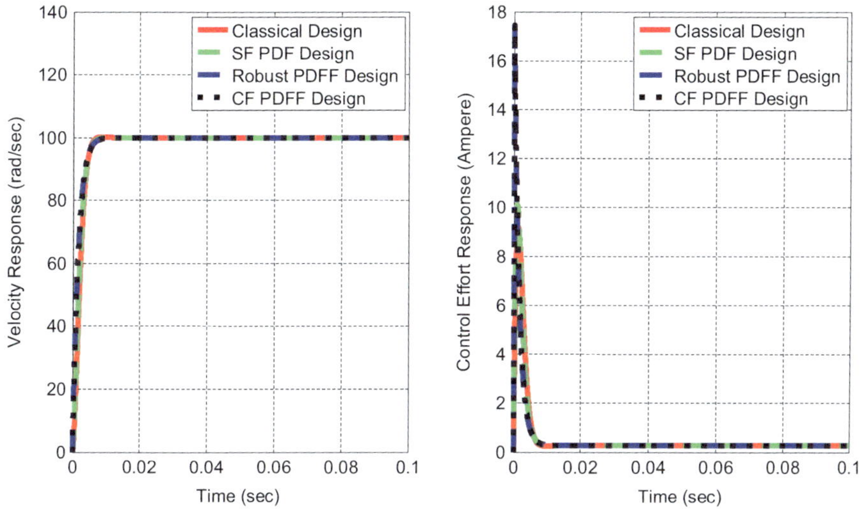

Fig. 10.33 Comparisons of step responses and control efforts with $J = J_o$ from four controllers

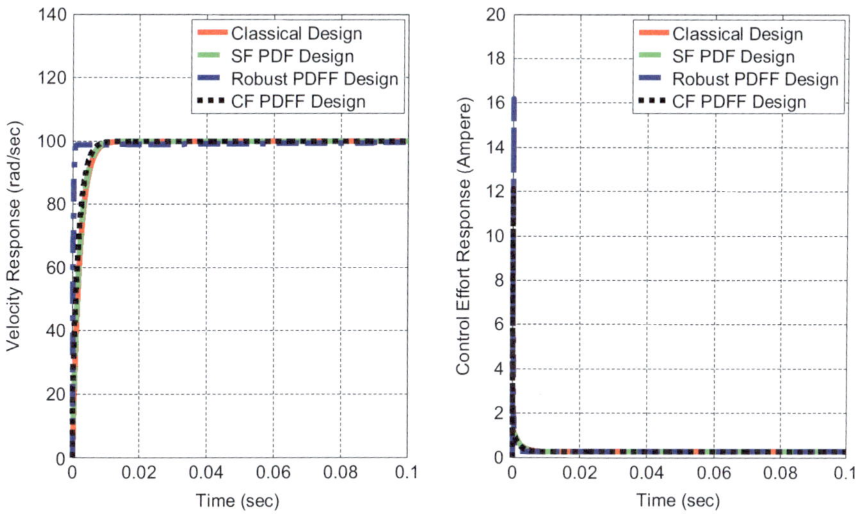

Fig. 10.34 Comparisons of step responses and control efforts with $J = 0.1J_o$ from four controllers

the maximum overshoot and settling time especially by the CF PDFF design. For the step response of the robust PDFF design in Fig. 10.35, the slower transient response with overshoot can be found. This is due to that the H_∞ design formulation appears with pole-zero cancellations, which affects the control performance significantly while uncertainties occur.

Disturbance rejection ability, also known as *dynamic stiffness*, is also an important index with which to evaluate the above servo controller design. Enhancing the

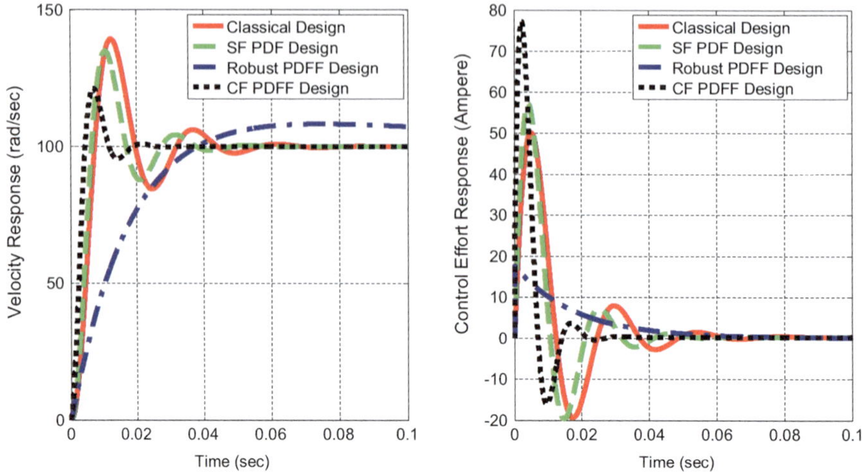

Fig. 10.35 Comparisons of step responses and control efforts with $J = 10J_o$ from four controllers

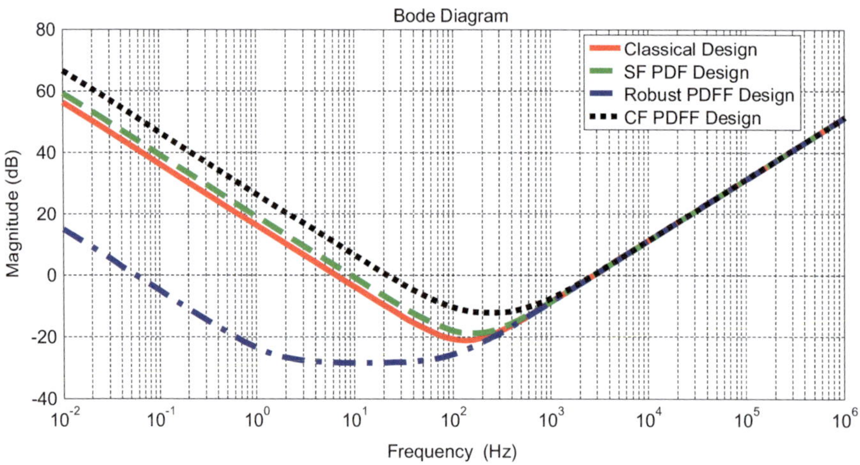

Fig. 10.36 Dynamic stiffness plots resulting from different controller designs

dynamic stiffness is equivalent to restoring torque to attenuate the output deviation such that the effect of process disturbance is reduced. Dynamic stiffness is characterized by the inverse of the magnitude of the closed-loop transfer function from the load torque disturbance T_L to the output speed ω. Thus, high dynamic stiffness is essential for the control system to obtain good disturbance rejection ability.

The inverse plots of the magnitude of the closed-loop transfer function from T_L to ω resulting from the above four designs are shown in Fig. 10.36, where the CF PDFF design shows the best *dynamic stiffness* among the four design examples. To result in higher dynamic stiffness and better stability robustness, the CF PDFF design extends the feedback loop control parameters to dynamic controllers. In practice, however, high dynamic stiffness often results in large control efforts; hence, a trade-off should be considered carefully. Due to the pole-zero cancellations, the dynamic stiffness of the robust PDFF design is less than the other three PDFF designs.

10.5 Summary

This chapter presents an application-oriented study that combines the frequently used PDFF control scheme in industry and the H_∞ control theory. Four different design methodologies, i.e., classical PDF, SF PDF, robust PDFF, and CF PDFF, for speed control of DC servomotors are presented. These examples showed how industrial controllers can be formulated into the standard control design framework and then solved by the state-space solution procedures presented in previous chapters. To achieve high dynamic stiffness design, the classical PDFF control design is formulated into the standard H_∞ control problem, i.e., a special state feedback (SF) problem. A weighted mixed-sensitivity design problem of the coprime factorization (CF) approach is provided. Both the command tracking and disturbance rejection can be taken into account simultaneously in this design method.

The concept of H_∞ loop-shaping design is also introduced to formulate the advanced PDFF control scheme into a weighted mixed-sensitivity design problem based on an augmented pseudo plant and the coprime factor description. This chapter investigates the peculiar pole-zero properties in the explicit form of the advanced PDFF design and characterizes the benefit and design procedure of special H_∞ pole placement property. The proposed design method provides a useful property that the bandwidth and the dynamic stiffness can be designed stage by stage, and the property has been successfully applied to design a servomotor system.

To sum up, the CF PDFF design can provide some freedoms of preventing/performing the pole-zero cancellation between the controller and plant. Although the pole placement is not a primary objective in the mixed-sensitivity design, such choice of weights for improving dynamic stiffness is arresting. Such design freedoms can be chosen to achieve higher dynamic stiffness with suitable tracking performance and also to maintain certain robust stability.

References

1. Alter DM, Tsao TC (1996) Control of linear motors for machine tool feed drives: design and implementation of H_∞ optimal feedback control. ASME J Dyn Syst Meas Control 118:649–656
2. Ellis G (2012) Control system design guide. Elsevier Science, Oxford
3. Fu L, Ling SF, Tseng CH (2007) On-line breakage monitoring of small drills with input impedance of driving motor. Mech Syst Signal Process 21(1):457–465

4. Glover K, McFarlane D (1989) Robust stabilization of normalized coprime factor plant description with H_∞-bounded uncertainty. IEEE Trans AC 34(8):821–830
5. Hof P, Schrama R, Callafon R, Bosgra O (1995) Identification of normalized coprime plant factors from closed loop experimental data. Eur J Control 1(1):62–74
6. Kuo BC (1986) Automatic control systems. Prentice Hall, Englewood Cliffs
7. Ling SF, Xie Y (2001) Detecting mechanical impedance of structure using the sensing capability of a piezoceramic inertial actuator. Sens Actuators 93:243–249
8. McFarlane D, Glover K (1992) A loop shaping design procedure using H_∞ synthesis. IEEE Trans AC 37(6):759–769
9. Shen BH, Tsai MC (2006) Robust dynamic stiffness design of linear servomotor drives. IFAC J Control Eng Prac 14(11):1325–1336
10. Tal J (1994) Step-by-step of motion control systems. Galil Motion Control, Inc., Rocklin
11. Tsai MC, Chang JY (1995) LQG/LTR loop shaping design with an application to position control. J Chin Inst Eng 18(2):281–292
12. Tsai MC, Geddes EJM, Postlethwaite I (1992) Pole-zero cancellations and closed-loop properties of an H_∞ mixed sensitivity design problem. Automatica 28(3):519–530
13. Whidborne J, Postlethwaite I, Gu DW (1993) Robust controller design using H_∞ loop-shaping and the method of inequalities. In: Proceedings of conference on decision and control, San Antonio, TX, USA, pp 2163–2168

Index

M.-C. Tsai and D.-W. Gu, *Robust and Optimal Control: A Two-port Framework Approach*, Advances in Industrial Control, DOI 10.1007/978-1-4471-6257-5,

© Springer-Verlag London 2014

MIX
Papier aus verantwortungsvollen Quellen
Paper from responsible sources
FSC® C105338

If you have any concerns about our products,
you can contact us on
ProductSafety@springernature.com

In case Publisher is established outside the EU,
the EU authorized representative is:
Springer Nature Customer Service Center GmbH
Europaplatz 3, 69115 Heidelberg, Germany

Printed by Libri Plureos GmbH
in Hamburg, Germany